APPLIED
FOOD
MICROBIOLOGY

G. Roland Vela
University of North Texas

PUBLISHING COMPANY

Star Publishing Company
P.O. Box 68
Belmont CA 94002-0068
USA

ISBN: 0-89863-185-8

Printed in the United States of America

Library of Congress Cataloging-in-Publication Data

Vela, G. Roland, date
   Applied food microbiology / G. Roland Vela.
     p.   cm.
   Includes bibliographical references and index.
   ISBN 0-89863-185-8 (alk. paper)
   1. Food—Microbiology.   I. Title.
QR115.V44   1996
576′.163—dc20

                95-43351
                CIP

Contents

Preface

Applied Food Microbiology was written to provide students knowledge of microbiology important in the preparation and provision of foods and beverages. Its scope is comprehensive and includes several disparate fields such as general characteristics of microorganisms, food, and water borne infections, intoxications, use of microorganisms as food sanitation, and the systems of laws and regulations that protect the public from unscrupulous practitioners. My intent is to provide a complete course in food microbiology at the introductory level accessible to everyone. Therefore it is suitable for students in hotel and restaurant management and also to students in allied health areas such as nutrition, pre-nursing, pre-chiropractic, kinesiology, and physical therapy.

Applied Food Microbiology provides knowledge regarding the transmission of communicable diseases by food and beverage to those who cannot spend several years studying biology, chemistry, and physics in preparation for specialized courses such as microbiology, virology, protozoology, and parasitology. On the other hand, it is not a "watered down" text for non-science students; it offers precise and rigorous scientific information on the material discussed. This approach gives the Instructor flexibility regarding the content and rigor of the course.

My debt of gratitude extends, but is not limited, to the following individuals: my wife, Emma Lamar, for help in every way and with everything; to my daughter, Yolanda Vela Welborn, for many of the drawings and illustrations and to Stuart Hoffman who can make anyone measure up to any task. I am also grateful to my graduate students who, much to my surprise, could do their work without my presence in the laboratory. I also want to acknowledge the contributions and help of many colleagues and I am especially grateful to the many hotel and restaurant management students at the University of North Texas for teaching me how to teach food microbiology to non-science students.

G. Roland Vela

Introduction

1

Without theory, practice is only routine governed by the force of habit.

Pasteur

The history of the microbiology of food and drink starts with the very beginning of human beings. Back in the unknown origins of time, that ancestral form of us all, early *Homo sapiens*, as they sat on the cave floor, first noticed that foods changed in taste, texture, and appearance, and eventually became inedible. These early human beings may have started the record keeping and data gathering that has led to the understanding that we are about to share in this text. They must have felt an urgency to acquire, preserve, and transmit knowledge of the condition and "keeping" quality of food and drink, as their lack of knowledge probably led them, as it does now, to stomach discomfort and even to serious disease and death. Whatever else human beings have had to learn while traveling through time from the beginning to now, it could wait and be put off for later—but not the knowledge of which plants and animals were suitable for eating and which were best left alone. The time for assessing new knowledge about food extended only from one meal to the next.

Early peoples could learn much by watching what the animals around them used for food and what they avoided, but this approach had severe limitations because of significant physiological differences. Mistakes often were final in that early individuals who made the wrong choices about food died from their errors and did not leave their genes in the developing population. This is a natural law still operative and in full force today.

FOOD PROCUREMENT

Human history can be divided into two eras according to the methods by which people procured the major part of their food. The first, and by far the longest, period, encompasses the time when humans captured and gathered their food. This first period extended from the origins of human beings, perhaps one to five million years ago, to the beginnings of recorded history. Most primitive societies still practice food gathering. Even the most sophisticated of us depend, albeit in a limited way, on the gathering of food from the sea and the forest.

The second period covers the time during which people learned to produce their own food. This period spans, at most, the last 10,000 years of our time on earth, to the present. During this time human beings have invented and developed all the techniques and processes of modern agriculture. This includes land and water management, genetic improvement of plant stocks, optimization of crop yields, domestication of wild animals, breed improvement, and the invention and production of countless machines and instruments for food production. The success of our efforts in producing food can be measured by the increase in the world's population.

FOOD MICROBIOLOGY

Both of these forms of food acquisition demanded that human beings acquire the knowledge and skill necessary to preserve food, for both activities are characterized by times of plenty and times of want. In the gathering society, food must be taken when available, rationed as a function of time, and consumed before it spoils. In the food-producing society, food must be harvested when it is ready, rationed as a function of time, and consumed before it spoils. Of these three activities, the last is the most important from the point of view of individuals concerned with preparing food and providing it to others.

As humans have evolved from more primitive, less sophisticated, animals to our present state of evolutionary development, the acquisition of food has become less and less the primary task of our work day, and our dependence on others to provide our daily bread has increased. Today, we look to specialists for the production, preservation, and preparation of almost all of our food. This dependence frees us from a major effort for survival, leaving more time for other, more intellectual activities. At the same time, the science and technology that support food production have increased in complexity to the point at which modern populations may not be able to survive without food production specialists.

The most wonderful thing that happened on planet Earth was the beginning, some three and one half billion years ago, of life as it is known and understood in modern science. And of this event a small facet was

Figure 1.1

Mood, music, and conversation are often the best part of the meal—going far beyond the need of early human beings to satisfy hunger and thirst.

doubly wonderful: the appearance of human beings and eventually of modern humans. Among the millions of living creatures great and small that have inhabited this planet, only humans have worried about their food. Only people plan ahead, invent new foods, improve old foods, move them in vast quantities over vast distances, and store them in prodigious amounts through a variety of strategies invented for that purpose.

Better than any other animal, human beings know and understand the logistics of food and time and have solved the most complex mathematical equations needed to get the proper kind of food at exactly the right time to precisely the right individual. At exactly twelve o'clock noon enough food of exactly the kind desired by each individual is available to feed nearly the entire population, whether these consumers are in New York City, Hong Kong, or Buenos Aires. The food consumed is of many different kinds; it is not produced in the place where it is consumed; and the shelf life of much of it is measured in mere hours.

Food is not the end of the story, though. It is said that the degree of sophistication of human beings can be measured by the time each takes to consume his or her food. The manner in which individuals take their food forms an extremely wide spectrum of habits, preferences, and customs, complicated by many extraneous factors. Although the modest table may be expressly for the purpose of taking sufficient food to dispel hunger, the better endowed setting may serve the purpose of mixing music, mood, conversation, and companionship with the act of satisfying hunger (Figures 1.1 and 1.2).

Figure 1.2

Town markets, often open-air, are very common in France and many other countries. In these countries, fresh foods are more common than prepared, packaged, or frozen foods. Photograph courtesy of French Government Tourist Office

THE PROBLEM WITH FOOD

All living things, including human beings, grow, live, and reproduce only if a constant stream of food in usable form is available to the organism. The food must provide all the substances needed to make new cells, as well as the energy to power all the necessary biochemical reactions. The combination of chemical substances and available energy is essential for growth and maintenance of life. When food is not available, growth and life cease.

In large animals with fastidious demands for special food chemicals, the need for a continual supply of a wide variety of substances is a matter of life and death. Human beings are among the most fastidious of animals in this regard. The problem is exacerbated in that the food human beings depend on often serves as excellent growth media for pathogenic microorganisms such as bacteria and fungi. As a result, humans must risk their life every time they eat. To satisfy the need for a continuing supply of food, people must eat one or more times a day. Yet, each time they eat, they take the risk that the food may contain organisms that produce intestinal discomfort, illness, or death. This is a paradox of problems because there are no alternatives. Death is a certainty without the food, and food may well be the bearer of death.

IN THE VERY BEGINNING

Early in their history human beings learned they were what they ate. They also learned that well-being and illness, to a large extent, depended on their diet.

The slow realization that people were of the soil and that their own body and substance were derived from the stuff of the soil and the fields—while not comforting to the theologian and the poet—was a major accomplishment. This understanding made easier the task of assessing the dietary needs of human beings and the methods by which food may be produced to satisfy these needs.

It seems fairly certain that life, in very simple form, originated on Earth from the chemical substances found on its surface. Strong evidence indicates that this occurred as the culmination of some one and a half billion years of **chemical evolution** during the time when the Earth's surface was cooling from a temperature of several hundred degrees of heat to its present-day temperature of 15°C. The first forms of life undoubtedly were quite similar to present-day **anaerobic bacteria**. Of the endless variety of forms that appeared, probably very few survived for even a brief time.

The first forms of life that can be recognized today existed some three and one half billion years ago, but by that time almost all of the critical processes of life were present in those early living things. Bacterial **photosynthesis**, **fermentation**, and primitive forms of **respiration**, as well as the phenomena of life and death can be imagined and reasonably understood with reference only to present-day knowledge. As the result of **mutations** during a time span of many millions of years, organisms similar to present-day **cyanobacteria** appeared on Earth. Their novel form of photosynthesis emitted oxygen as a waste product, and because living things were not capable of utilizing it, this oxygen accumulated in the environment. After many millions of years of this process, oxygen became one of the major constituents of the Earth's atmosphere, and all the chemicals on the surface of the Earth became associated with oxygen. They were **oxidized** rather than **reduced** as they had been previously.

Eventually, countless further mutations gave rise to more complex organisms capable of using oxygen to improve their metabolic efficiency (respiration). Later, large and more complex organisms, comparable to modern-day plants, appeared. Algae and plants together brought the concentration of oxygen in the air almost to its present-day level, enabling animals to evolve from the earlier, simpler life forms.

In this slow but inexorable process, new life forms—both plant and animal, larger and more complex than previous forms—appeared on the Earth. To experts in this area of research, the apex seems to have been reached several hundred million years ago with the appearance of the dinosaurs. Eventually, some one hundred million years ago, the primates, the group to which human beings belong, appeared on Earth. The forces that drive these processes of constant change by trial and error continue today and affect all living things in much the same way.

THE APPEARANCE OF HUMAN BEINGS

Although animals have inhabited the Earth for approximately one half billion years, human beings appeared on the scene relatively recently. Humans entered a scene in which every life form already had evolved, thrived, and filled almost every square centimeter of the planet's surface with its progeny. Of these, the most successful were the microorganisms. Each cubic centimeter of soil on most of the Earth's surface is thought to contain as many as one billion microorganisms of a large and bewildering variety. In a way, the planet was already old and replete with living things of all sizes and kinds when human beings appeared. This situation was not difficult for even the earliest human beings and, in the span of a few hundred thousand years, they rose from a weak, smallish, and insignificant form to be the undisputed lord of all. Primarily as the result of a superior brain and ability to accumulate knowledge, people conquered all of their natural enemies so that today, all beasts such as the lion and the wolf, which once preyed on humans and their offspring are essentially on the verge of extinction as a result of the ascent of human beings.

On the other hand, the major natural enemies of human beings, the microorganisms, are far from being vanquished even today. Human beings did not perceive the existence of these microscopic forms of life until approximately three hundred years ago and did not appreciate their importance as natural enemies until little more than one hundred years ago. A Dutch investigator named Antonie van Leeuwenhoek (1632–1723) was the first to see living things too small to be seen by the unaided human eye. He was able to make glass lenses capable of magnifying objects as much as 300 times their size. With these lenses he built primitive microscopes and discovered spermatozoa, red blood cells, pollen, and many other small things, including algae, protozoa, yeast, fungi, and the bacteria. There is no record to indicate that van Leeuwenhoek regarded the microorganisms he discovered as anything other than simple, small animals with no more power than that of small flies or ants. His intellectual satisfaction apparently rested on discovery rather than on further scrutiny. Not until the 1860s did the work of Louis Pasteur (1822–1895) and other European investigators give unambiguous proof that microorganisms were responsible for the fermentation of sugar into alcohol, the spoilage of food, and the putrefaction of almost all living matter. In the 1880s, Pasteur and Robert Koch (1843–1910), independently, performed experiments showing that bacteria and other microorganisms cause disease in animals and also in human beings.

Humans evolved and have developed to their present state in a world of microorganisms. Many experts in this field state categorically that if all the people, all the animals, and all the plants were to disappear from the Earth, life would continue undisturbed and unaltered as if nothing had happened. On the other hand, if the bacteria were to disappear by some comparable magic, experts say life on Earth would end within 40 years. With only micro-organisms such as fungi, algae, bacteria, and viruses, as it was in the beginning, however, the Earth would be a crushingly boring and uninteresting place, as it was before the coming of human beings.

Every aspect of human evolution and everyday existence has been affected by the microorganisms that surround us, live on us, and live in us. From the instant of birth, the body of each human being is colonized by uncountable billions of microorganisms that eventually establish such an intimate relationship with their host that together they form a human-microbe system, best explained in scientific terms as a **symbiotic relationship**. Human beings and microorganisms form a lifelong bond from which both benefit. The normal **microbial flora** of the human body probably serves to protect the host from invasion by other, disease-producing organisms by stimulating the immune state, crowding out invaders, and antagonizing organisms that are not members of the normal flora. The bacteria in the intestinal tract, for example, aid in the digestive process and produce important substances including vitamin K, essential for human nutrition.

HUMANS *VERSUS* MICROBES

From a different point of view, microscopic organisms that live in the soil and water compete with people for food and other valuable products. The development of an entire food preservation industry, including canning, packaging, bottling, refrigeration, and freezing, has but one objective: to protect our food from microorganisms that would consume it or spoil it by their metabolic activities. From time immemorial, people have looked for ways to store food in times of plenty for use when food is scarce. This desire may have originated by watching certain animals, but the superior human brain gave a completely new meaning to the idea of storing food. In their earliest attempts, people saved food by allowing it to freeze or to dry by natural processes. Food dried properly and stored carefully may last several years without noticeable change or decay. The scientific awakening of human beings can be measured accurately by the development of better and more sophisticated methods of preserving food. A clear trail can be seen from the inception of the art of freezing food to that of drying it, then to the preparation of preserves, pickling, salting, canning, sterilizing, deep-freezing, freeze-drying, and finally sterilizing by the use of radioactive isotopes.

The contention between human beings and microorganisms has another important aspect, focussing on the fact that microorganisms may cause illness in human beings. If humans consume food containing certain organisms, disease and even death may result. The view from the other side shows that, when human beings cook or heat food before eating it, they bring about the destruction of entire populations of microorganisms. As of now, it is a fair match with no advantage to either side.

THE SCOPE OF FOOD MICROBIOLOGY

Conflict   Protecting foods from the reach of microorganisms has two primary purposes: to prevent spoilage and to prevent the transmission of disease. Almost all food and drink can serve as an excellent vehicle for transporting many different kinds of microorganisms. Vast numbers and varieties of bacteria, fungi, protozoa, and viruses can survive in almost all the foods we consume, and many can grow and increase in numbers, even during brief storage of the food, if the temperature and other conditions permit. A few pathogenic bacteria in milk that, as such, cannot cause disease, may in the course of a few hours produce a population large enough to start an epidemic among those who drink the milk. In a different case, the growth of bacteria is secondary to the production of **toxins**. Even a small amount of growth of the organism *Clostridium botulinum* is enough to result in the production of sufficient toxin to cause the deaths of large numbers of human beings.

Among all the animals, and then only during the last one hundred years, only human beings have been able to understand and control the contamination of their food and water by microorganisms. Perhaps not merely coincidentally, during this same time the population of *Homo sapiens* on Earth increased dramatically and the average life span of individuals almost doubled.

The constant waves of epidemics caused by microorganisms carried in water, milk, and food, which have kept the population of human beings in check throughout their existence on Earth, are a tribute to the power of the microbe. Even though modern science and technology give the advantage to a few communities of human beings on Earth, it is precarious and can disappear under even slight stress. War, flood, or loss of electric power lead inevitably to epidemics of cholera, typhoid fever, and dysentery, and their appearance in the community requires only a few days.

Cooperation   Without knowledge of the presence of microorganisms, human beings learned early in their history that fruits, grains, and other foods could be fermented or "cured" and could be preserved in this condition for a long time. Preservation was not the only advantage. Often the food could be improved by fermentation. Products such as olives, palm petioles, and the juice of the maguey plant, which are not edible in their natural condition, can be made edible by microbial fermentation. The dining table has been enriched by a veritable cornucopia of fermentation products including cheese, buttermilk, beer, wine, sauerkraut, jalapeño peppers, sausages, and soy sauce. All these fermentations, it is now known, are brought about by the metabolic activities of bacteria and fungi. As human beings gain further knowledge of the microorganisms, new and ever better products almost certainly will be introduced.

Exploitation   As the third millennium of our measured time on Earth approaches, our burgeoning population may well deplete the traditional sources of food. In that unhappy situation, the very existence of human beings on Earth will be threatened by generalized and increasing famine. **Futurists** propose that human beings of the twenty-first or twenty-second century will be sustained by foods made from microbial cells. Contemporary scientists can easily produce microorganisms in the vast quantities necessary to serve as adequate and inexhaustible sources of food. Appetizing foods such as mushrooms, truffles, soy sauce, and tofu show clearly that human beings can use microorganisms as food. Moreover, no important impediments hinder the use of microorganisms as a sole source of food in lieu of the traditional vegetation and flesh of animals.

LEARN AND SURVIVE

Clearly, members of society who carry the responsibility for producing, storing, preparing, and serving food can better do their work when they have a thorough understanding of the microorganisms that are important in the various aspects of food microbiology. History presents ample evidence that lack of knowledge of bacteria, fungi, protozoa, and viruses can lead to dire and tragic consequences. For example, tuberculosis of the stomach killed large numbers of children before it was shown in the 1920s that milk cows infected with bovine tuberculosis shed the organisms into their milk and that drinking this milk caused tuberculosis in human beings. Once this was established, stomach tuberculosis could be eliminated from the population simply by avoiding milk from tuberculous animals.

Today, bacteriological diagnosis can reveal when cows in a diary herd have become infected and also when they have been cured. This permits the maintenance of tuberculosis-free dairy herds. In addition, the **pasteurization** of all milk and milk products assures that the causative agent of bovine and human tuberculosis, *Mycobacterium tuberculosis*, is killed before the product is consumed. Use of these two simple methods has made stomach tuberculosis a rare disease in the United States and other developed countries. Any country in the world that wishes to protect its children in like manner only has to adopt these well known practices to obtain the same result. Although a great problem was solved by a simple technique, the solution had to be supported by thousands of years of observation, trial and error, study, and intellectual effort before reaching the answer.

FURTHER READING

Bulloch, W. 1938. THE HISTORY OF BACTERIOLOGY. Dover Publications, New York.

de Kruif, P. 1940. MICROBE HUNTERS. Pocket Books, New York.

Lechevalier, H. A. and M. Solotorovsky. 1965. THREE CENTURIES OF MICROBIOLOGY. McGraw-Hill, New York.

Sigerist, H. E. 1943. CIVILIZATION AND DISEASE. University of Chicago Press, Chicago.

Tortora, G. J., B. R. Funke, and C. L. Case. 1982. MICROBIOLOGY AN INTRODUCTION. Benjamin/Cummings Publishing Company, Menlo Park, California.

Tuchman, B. W. 1978. A DISTANT MIRROR (Chapter 5: "This is the End of the World" The Black Death). Alfred A. Knopf, New York.

Zinsser, H. 1934. RATS, LICE, AND HISTORY. Little, Brown, Boston.

The World of Microorganisms

2

So, naturalists observe, a flea
Hath smaller fleas that on him prey;
And these have smaller still to bite 'em;
And so proceed ad infinitum.

Swift

From our earliest life experiences, human beings have been aware that we share the planet Earth with a great variety of other living things. The living things that have held our attention most firmly are those that we use for food and shelter, mainly the larger, more docile **herbivorous** mammals and a great variety of plants. As early as 2,000 years ago, humans had obtained sufficient knowledge about the fundamental characteristics of most living things to classify them and give each a specific name.

The science of **taxonomy** probably originated with Aristotle (384–322 B.C.), the great Greek philosopher, whom some consider the father of modern biology. Many others had similar ideas even before this time. In the 17th century, John Ray, an English scientist, conceived the idea of defining species. This novel concept enabled the Swedish physician Karl von Linne (1707–1778), Linnaeus, to create a new system of classification. His first formal presentation of this taxonomic system was published under the title *Systema Naturae*. Linnaeus's system of nomenclature, in which every living thing is given Latinized **genus** and **species** names, is still in use today and incorporates the vast majority of knowledge that accumulated from the time of Aristotle to the present.

Genus and species designations are made after the organism has been placed in a **kingdom**, **phylum**, **class**, **order**, and **family**. In this system human beings are named *Homo sapiens*. *Homo* is the genus, and *sapiens* the species. The names are derived from Greek or Latin and are designed to describe the organism. *Homo* = man and *sapiens* = knowledgeable. In like manner all microorganisms are designated by their binomial appellation. *Staphylo* = organisms that grow in grapelike clusters and *coccus* = spherical in shape. Combined, the term is *Staphylococcus*.

DISCOVERY OF THE MICROBIAL WORLD

Until 1675 A.D., people assumed that they shared the planet only with plants and animals. Skills in breeding, culturing, selecting, and exploiting all the known living things in the environment were developed to a high level, and *Homo sapiens* stood very much as the lord over all life on Earth. Just beyond their reach, however, barely beyond their field of perception, existed an entire world of living things of which they were totally unaware. The human eye can see only things that are larger than one tenth of a millimeter (0.1 mm) on the smallest axis. Consequently, the entire world of the microorganisms evaded people's view, as microorganisms are in the size range of 0.00003 to 0.1 mm. Table 2.1 lists common units of measurement used by scientists today.

In 1675, in the city of Delft in Holland, Antonie van Leeuwenhoek first glimpsed this microbial world. An amateur lens maker, he perfected lenses capable of magnifying objects to 300 times their size. With these he discovered the world of microscopic objects and small things such as spermatozoa, red blood cells, pollen, and the cellular nature of fungi. Van Leeuwenhoek also observed that his excrement contained numerous microorganisms, which he called **animalcules**. Through his observations he sought to understand how

Table 2.1   Metric Scale of Lengths Compared to English Scale

1 kilometer	=	1000 meters	=	M
1 meter	=	10 decimeters	=	dm
1 decimeter	=	10 centimeters	=	cm
1 centimeter	=	10 millimeters	=	mm
1 millimeter	=	1000 micrometers	=	μm
1 micrometer	=	1000 nanometers	=	nm
1 nanometer	=	10 Angstrom units	=	Å
1 mile	=	1760 yards	=	yds
1 yard	=	3 feet	=	ft
1 foot	=	12 inches	=	in
1 inch	=	1000 mils	=	mils

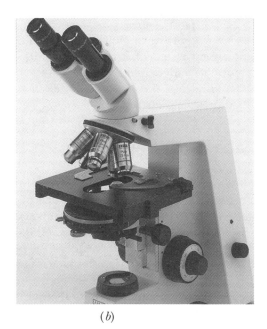

Figure 2.1

(a) *Microscope invented, built, and used by Antonie van Leeuwenhoek to discover bacteria and other microorganisms; (b) modern binocular, compound microscope used by microbiology students. Photograph courtesy of Carl Zeiss, Inc.*

(*a*)   (*b*)

these might be associated with the diarrhea from which he was suffering at the time. Leeuwenhoek was impressed with the size and the number of the animalcules he saw, stating in a letter that, "A thousand million of them would make the volume of a single grain of sand." Figure 2.1 illustrates one of van Leeuwenhoek's microscopes, as well as the modern microscope commonly used by microbiologists.

HUMAN–MICROBE INTERACTIONS

The most important aspect of our inability to perceive the vast populations of microorganisms lies in the fact that every action on every day of human life was affected severely by their presence. Microorganisms pervade all the areas, spaces, and crevices of the body and affect all of its activities and even the very nature of the human being. As a result of the activities of microorganisms in their environment, the food of early humans spoiled, their wine soured, their animals sickened, and their children died while they had to sit by helplessly, watching but not understanding.

Human beings, *Homo sapiens*, probably have existed for some 100,000 years, and the predominant strain at the present time, for approximately 35,000 years. During all this time, human beings have lived essentially at the mercy of these tiny, invisible living things and suffered plague and epidemic in defenseless and agonizing ignorance.

In desperate efforts to protect themselves from plague, tuberculosis, dysentery, leprosy, and a hundred other diseases caused by microorganisms, humans invented one useless cure after another, all to no avail. In their anguish to understand the nature of the illnesses that assailed them, they invented one false theory after another, but the resulting confusion only intensified their helplessness.

WHAT CAUSES DISEASE?

Most primitive societies of human beings assume that human disease is caused by supernatural forces such as the acts of gods or demons. According to this supernatural or **demonic theory** of the nature of disease, the living force in the body, the **anima**, may be replaced by an evil spirit or demon, and pain, suffering, and death are evidence of the presence of the evil in the body. The major harm associated with this idea is implicit in its very essence: People are powerless to intervene in what demons or gods will, and those who dare to intervene call the same evil unto their own body. Under these conditions curing of the sick entails appeasement of the spirits and not repair of the injured or infected body.

In the fourth century B.C., the **Hippocratic School of Medicine** departed from this notion, asserting that the origin of disease lay in imbalances of the body fluids or **humors**—blood, phlegm, yellow bile, and black bile. The most significant aspect of this **humoral theory** of the origin and nature of disease lay in the idea that people could intervene and indeed cure the sick simply by bringing the body humors back into balance. Although blood-letting (Figure 2.2) is no longer practiced, present-day **emetics** and **decongestants** are derived directly from those early attempts at healing the human body.

Figure 2.2

Red and blue stripes on barber pole symbolized blood vessels and advertised license for blood-letting in the United States well into the 1800s.

Countless other theories, each devised to meet either universal or local needs, were advanced. The compendia that describe these theories and notions provide one of the most fascinating outpourings of the product of the human imagination. Some of these ideas proposed the existence of small, invisible, living "germs," the **contagium vivum**, which, by their presence in the body, caused disease. Lucretius (96–55 B.C.) called these the seeds of disease and assumed they grew in the body of one individual and were passed to another in much the same way that lice or mites growing on the body of a person may pass to that of another.

In 1546, Fracastoro de Verona (Girolamo Fracastoro, 1483–1553) proposed that disease was transmitted from sick persons to healthy ones by **seminaria** (seeds) and described the mechanisms by which these were transmitted. The methods of transmission Fracastoro described are still an essential part of medical science. These and many other "germ" theories were based on conjecture only or at best, like Fracastoro's, on circumstantial evidence. Thus, they were of little use in understanding the principal causes of disease and less in curing it.

Respected physicians of Fracastoro's time still prescribed medicine according to the phases of the moon and diagnosed disease according to the positions of the planets in the zodiac. Even Fracastoro, a man well ahead of his time in understanding the nature of disease, practiced medicine according to the traditions of his profession. He was truly a man with one foot in the future and the other securely planted in the past.

WHAT CAUSES FOOD TO SPOIL?

The same confusing and chaotic situation that persisted in medicine even to the middle of the 19th century was found in other areas of knowledge important to human beings. In spite of the urgency of the problem, people made little or no progress in understanding how or why food spoiled or how this might be prevented. Food and other desirable materials acquired at great cost often rotted when they were stored even for brief periods. This represented great financial loss at its best and, at its worst, famine and starvation. The storage of food in time of plenty in no way guaranteed survival during times of need, because microorganisms often consumed food before the owners were ready to eat it.

Scientists who turned their attention to the study of **putrefaction**, or the spoilage of foods, generally became discouraged and took refuge in the conclusion of **Deus vult**, or God wills it. Many theories were developed over the ages, but none were applicable to the problem of food rotting and none could be used to protect valuable property from the ravages of microorganisms. One popular theory rested on the statement that food had come from the soil and therefore must return to whence it came, an explanation similar to saying that it rots because it rots.

Little progress was made in this area until Theodor Schwann (1810–1882), Louis Pasteur, and others began to report that food spoiled only when bacteria and fungi grew in it. Although the data presented by these investigators were irrefutable, they were ignored until the last part of the 19th century.

WHAT CAUSES FERMENTATION?

One of the most interesting and oldest of human activities is that of fermenting grapes and other fruits into a great variety of wines. Human beings most likely learned the art of **fermentation** by accident. Not long after that, natural processes were improved and refined, and the refinements developed into sophisticated, new techniques. This knowledge increased to the point of mastery even though people knew nothing of the role that fungi and bacteria play in the overall processes. After the acquisition of much knowledge on the nature of the process, the fermentation of fruits was followed by the fermentation of cereal grains for the production of beer. Beer and wine are major items in the diets of human beings in almost all parts of the world.

Wine, now as in the past, holds a special fascination, and reference to it can be found in all walks of life. The descriptions wine connoisseurs use are often closer to poetry than to analytical chemistry, employing descriptive words such as "mellow," "bright," "lilting," "enchanting," and "piquant."

In a different vein, the central place of wine in religious ceremonies goes back to prehistoric times. Many references in the Bible (the wedding in Cana, for example) give evidence of the central role of wine in religious and social rites. Many of these are observed in Western societies even today. Wine as the symbol for the blood of Jesus Christ in the communion of some Christian religions is but one example of this. Another example of the cult and mood of mysticism that often surrounds wine can be seen in its identification with some elusive, deeply felt, and probably indescribable passion. In the words of the American poetess Emily Dickinson:

> I had been hungry all the years;
> My noon had come to dine;
> I trembling, drew the table near;
> And touched the curious wine.

Even though people had learned how to produce vast quantities of an endless variety of fermented products many centuries or millennia before, they had not come to understand, even as late as the middle of the 19th century, the underlying principles that made fermentation possible. As a consequence, fermentations often went awry, causing severe loss and often financial

Figure 2.3

Transubstantiation—a rite in which wine is believed to be changed into the blood of Christ—is one of the major points of faith of the Catholic Church.

ruin. As in medicine, many theories abounded, but all were false, and success often rested in the hands of fate alone.

Up to and including most of the 19th century, the most honored scientists of Europe assumed that wine was produced by the rearrangement of sugar molecules and that the microorganisms seen in fermenting liquids were a byproduct of the reaction. When Charles Cagniard de Latour (1777–1859), Theodor Schwann (1810–1882), and Friedrich Kutzing (1807–1893), independently and after extensive study, proposed in 1836 and 1837 that microorganisms converted sugar into alcohol, their idea and their data were rejected by the leading scientists of Europe. To ridicule de Latour, Schwann, and Pasteur, the great chemists Justus von Liebig (1803–1873) and Friedrich Wohler (1800–1882), performed a skit in public in which Liebig recited: "The fermentation of sugar was accomplished . . . by animalcules that ate sugar and ejected alcohol through their anus and carbon dioxide through their genitals."

ORIGINS OF THE SCIENCE OF BACTERIOLOGY

Fermentation

The first significant step toward understanding the nature of or the principles involved in the three phenomena described above was based on Leeuwenhoek's discovery of microorganisms in the late 1600s. His contemporaries, however, did not completely understand the discoveries he made. The relationship between microorganisms and disease, spoilage of food, and fermentation was not appreciated completely until some 200 years later. In papers published in 1857, 1858, and 1860, Pasteur proved to the satisfaction of most European scientists that microorganisms produce lactic acid, butyric acid, alcohol, and carbon dioxide by the fermentation of carbohydrates. He also showed that different microorganisms produced different fermentation products from the same sugar. The impact of this discovery has not been exploited fully to this day, but a staggering variety of products has been realized nevertheless—commercial solvents, medicinals, cosmetics, hormones, and antibiotics, among many others.

The fermentation industry is one of our most important industrial activities and often is considered to be a measure of technological development. Although all countries have sizable fermentation industries, those that produce the larger amounts of the most sophisticated products are deemed the most highly developed. The products of fermentation range from beer and bread at the low end of the technological spectrum to the production of insulin by bacterial cells containing human genes at the higher end.

Putrefaction

Many scientists realized that the spoilage of food and other products could be avoided by excluding microorganisms from putrescible materials. Studies of putrefaction and of **abiogenesis** (the origin of living things from inanimate substances) became one and the same in that the theories and experiments pertinent to one were the same as those developed for the other. As early as 1715, Louis Joblot (1645–1723) had shown that fresh hay that was boiled and sealed in clean flasks did not putrefy until the flasks were opened again. The same results, but with further refinements, were noted by Lazzaro Spallanzani (1729–1799) in 1765, by Theodor Schwann, and, independently, by Franz Schulze (1815–1873) in 1837 and by H. G. F. Schroder (1810–1885) and Theodor von Dusch (1824–1890) in 1854. These experiments failed to convince the majority of European scientists that microorganisms caused spoilage.

Finally, in 1861 Pasteur proved that any substance that had been sterilized would remain free of living things and would not putrefy as long as it remained protected from microorganisms that were carried by the air. The significance of these studies is seen today in aseptic surgery, the canning of foods, and the preservation of blood and other tissues in organ banks.

Disease

The search for the cause of disease and the nature of communicable diseases, which started in prehistoric times and consumed the efforts and even the lives

Figure 2.4

Pasteur proved that bacteria exist in the air and that they were not the product of abiogenesis as then was believed.

of hundreds of dedicated scientists, finally culminated in the experiments performed by Pasteur (pictured in Figure 2.4) and Robert Koch (1843–1910) in the period from 1860 to 1880. Pasteur proved that bacteria caused the disease called anthrax in sheep, and Koch identified the specific bacterium. At the same time, Koch showed that other bacteria were not involved in the disease process and that the characteristics of anthrax were attributable to the properties of the causative organism alone (see Figure 2.5). The era from 1880 to 1927 is known as the Golden Age of Bacteriology, as many major discoveries were made during this time. Some of the most important are listed in Table 2.2.

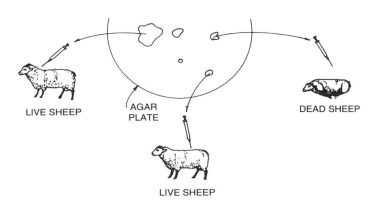

LIVE SHEEP

AGAR PLATE

DEAD SHEEP

LIVE SHEEP

Figure 2.5

Koch showed that sheep died from anthrax only when bacteria from a specific colony were injected into test animals.

Table 2.2   Major Discoveries in Bacteriology

1867	Joseph Lister English surgeon	Initiated practice of aseptic surgery using phenolic acid aerosol to kill bacteria in the air and bandages dipped in phenol to avoid infection
1875	Ferdinand Cohn German botanist	Published first work on the classification of bacteria; discovered the bacterial spore and how it affects sterility
1879	Albert Neisser German physician	Discovered the bacterium that causes gonorrhea
1880	Louis Pasteur French chemist	Discovered nature of spoilage, fermentation, disease, and immunity; accomplished attenuation of microorganisms and developed anthrax, rabies, and fowl cholera vaccines
1881	John Tyndall British physicist	Confirmed Pasteur's discovery of bacteria in the air and method of "fractional sterilization"
1881	Alexander Ogsten English physician	Discovered *Staphylococcus aureus* and nature of bacteremia
1881	Robert Koch German physician	Established study of pure cultures; proved that specific bacteria cause specific disease; discovered *Vibrio cholera*; pioneered use of agar medium
1882	Friedrich Fehleisen German physician	Discovered *Streptococcus pyogenes*
1884	Elie Metchnikoff Russian physiologist	Discovered *phagocytosis* or the killing of bacteria by white blood cells
1884	Hans C. J. Gram Danish physician	Studied differential stains for visualization of bacteria and invented Gram stain in 1884
1884	Georg Gaffky German physician	Discovered *Salmonella typhi* (*Bacillus typhosus*), the causative agent of typhoid fever
1885	Theodor Escherich German pediatrician	Discovered *Escherichia coli* in the stool of human beings

The science of bacteriology, now microbiology, was born from this work and rapidly became one of the most important of the modern sciences. Microbiology is significant in all the affairs of human beings, and the more that is learned of organisms, the greater is the importance to modern humans. Our view of nature and even of ourselves has changed entirely from 1850 to now as a result of the study of microorganisms.

Table 2.2 Major Discoveries in Bacteriology *(continued)*

1886	Daniel E. Salmon and Theobald Smith, American veterinary pathologists and bacteriologists	Discovered *Salmonella cholera-suis*; introduced veterinary vaccines; studied Texas fever in cattle
1887	Richard J. Petri German scientist	Published on the use of a special plate for agar media
1887	David Bruce English physician	Discovered *Brucella melitensis*, the etiologic agent of undulant fever or Malta fever
1890	Emil von Behring German physician and Shibasaburo Kitasato Japanese bacteriologist	Discovered antitoxins and first used them in the treatment of diphtheria infections
1892	Dmitri Ivanovski Russian botanist	Published first report on isolation of viruses that cause tobacco mosaic disease
1897	Emile van Ermengen Belgian bacteriologist	Discovered *Clostridium botulinum*, which produces botulism
1898	Kyoshi Shiga Japanese bacteriologist	Discovered *Shigella shiga* and made it possible to distinguish between amebic and bacillary dysentery
1906	August von Wasserman German physician	Formulated blood test for detecting syphilis in human beings
1911	Paul Ehrlich German chemist	Theoretical work led to discovery of antibiotics; the "magic bullet"
1915	Frederich W. Twort English bacteriologist	Discoverer of viruses that attack bacteria
1927	Alexander Fleming English bacteriologist	Discovered penicillin

THE NATURE OF MICROORGANISMS

As a result of the work of a small number of dedicated scientists during the last half of the 19th century, much of the old world was transformed into the world as it is today. The concept of abiogenesis was rendered false by the work of Pasteur, who proved definitively that bacteria reproduce in much the same manner as other living things. The study of life no longer could be divided into the kingdoms of plants, animals, and human beings as in previous times. For the first time humans were seen as animals among many other animals. Soon thereafter microorganisms were recognized as a major, and perhaps the most important, group of living things. In modern biology the microorganisms, not *Homo sapiens*, hold the central position in the over-all scheme of life on Earth. The properties common to all living things on Earth are:

- genetic information in DNA or RNA, or both
- genetic information interchangeable among all living things
- structure and function based on proteins
- made from the matter of the environment in which they exist
- react to changes in the environment
- evolved from previous form to present form
- evolution based on cumulative mutations

Although small in size and bulk, microorganisms far outnumber all other forms of life and encompass a bewildering variety of physiological types. A gram of garden soil may contain a larger number of bacteria than the total of all human beings on Earth. Not only are microorganisms present in extremely large numbers, but they also are found in an endless variety of kinds and types. The variety of microorganisms now known is so great that classification has not been completely established. To some microbiologists classification of the microorganisms seems to be an impossible task by today's standards and limits of knowledge.

Intense and continuous study during the last hundred years has yielded sufficient information to understand something of the nature of the micro-organisms and their relationships to plants and animals (**metazoans**). These studies have shown that microorganisms are not simply small plants or small animals as first was believed but, rather, unique living forms as different from plants and animals as the latter two are from each other. The scheme given in Table 2.3 is not intended to show all the differences between the micro-organisms and the metazoans but only those necessary to establish the fact that they are different forms of life. In spite of the great differences between these two life forms, all living things share many important and fundamental characteristics.

Because the vast majority of microorganisms are invisible to the unaided eye, these were not detected until recently. In contrast, mushrooms and algae are large aggregates or colonies of microorganisms and, as such, can be seen easily with the unaided eye. These were well known to the earliest sci-entists, but their nature and fundamental structure were not understood. Until well into the 19th century, they were assumed to be primitive plants. Even in later times the microorganisms were classified with the plants, but this

Table 2.3 Some Major Differences between Metazoans and Microorganisms

Metazoans	Microorganisms
Complex, large	Simple, microscopic
Cells interdependent	Cells not interdependent
Extensive differentiation	Limited differentiation
Complex life cycles	Limited life cycles
Limited metabolic capacities	Broad metabolic capacities

was done more to avoid taxonomic difficulties than to assert the fact that microorganisms were small, simple plants.

The first successful attempt to bring order to the classification of microorganisms, and to recognize them as neither very small animals nor very primitive plants, was made by Ernst H. von Haeckel in 1866. He proposed the name *protista* and suggested three kingdoms. These did not include the bacteria or the viruses, as they were not yet well understood.

Many other systems for the classification of living things have been proposed since von Haeckel's time, including some that place the fungi and the viruses in kingdoms of their own. With a certain amount of updating and enlarging, von Haeckel's system is still quite useful. The version shown in Table 2.4 satisfies not all, but many, of the problems associated with classification of the microorganisms. The system described schematically in that table and in Figure 2.6 delineates the interrelationships among the many different forms of life on Earth and gives an attractive evolutionary map suggesting the steps

Table 2.4 Division of All Living Things Into Three Kingdoms and Into Seven Kingdoms According to Criteria of 1993

I. Plantae	I. Plantae (plants and multicellular algae)
II. Animalia	II. Animalia
III. Protista	III. Protista (protozoa and microalgae)
A. Higher Protista	IV. Fungi (molds and yeasts)
1. Algae	V. Monera
2. Protozoa	1. Archaebacteria
3. Fungi	2. Eubacteria
B. Lower Protista	3. Cyanobacteria
1. Bacteria	VI. Viruses
2. Viruses	VII. Viroids, Plasmids, (Prions?)
3. Other?	

Figure 2.6

Common ancestry of all living things as it is understood today.

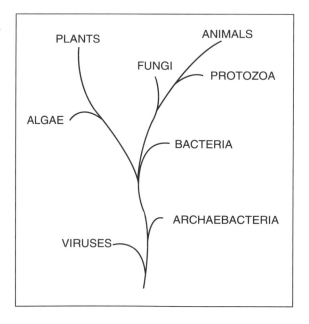

by which each kind of organism arose from a more simple entity. Plasmids and prions are not living things by today's definition of the term. They are molecules of deoxyribonucleic acid (DNA) and protein, respectively, which have the ability to increase in number in the same way as do some forms of life.

At the beginning of the 20th century, biologists began to recognize the relationships that exist among all forms of life and presented this idea in the statement, "There is only one kind of life on Earth." This statement was an overt rejection of the previous belief that Earth contained different kinds of life and that human beings were a special and superior form of living thing.

This modern idea was confirmed in 1936 with the finding that the chemical reactions required for life are similar in all kinds of organisms. The name given to it was the theory of **biochemical universality**. Finally, during the 1960s the genetic apparatus of different living things was shown to be chemically identical. The DNA of all organisms is able to function in other organisms. This gives rise to the notion that, "If you understand how *Escherichia coli* works, you'll know how an elephant works."

THE ORIGIN OF LIFE ON EARTH

It is impossible to know when, where, and in what form living things first appeared on Earth. It seems certain, however, the planet was formed some four and one half billion years ago, and no living thing existed at that time. The material of the planet was extremely hot at the time of formation, and water existed only as steam at extremely high temperatures. As the millennia passed, the planet cooled, allowing chemical compounds to form and accumulate in ever increasing quantities and varieties. When the temperature of the surface of the Earth fell below 100°C, water vapor in the atmosphere condensed and collected in newly formed seas and oceans of very hot water. As a result of the high temperature and the presence of ultraviolet radiation, chemical substances present in the water reacted with each other, forming an infinite variety of compounds that, in their turn, joined the unending, infinite series of chemical reactions.

As the temperature of the planet continued to decrease, the chemical compounds formed became more and more complex and also more and more stable. These events characterized the phenomenon of **chemical evolution** on Earth. Among the substances found in this "primordial soup of chemical substances" were compounds such as glucose, galactose, ribose, amino acids of many kinds, fatty acids, glycerol, purines, and pyrimidines. All the chemical substances that can be found in living things today were **synthesized** and existed in the waters of the Earth at that time. Because these complex substances were molecules that react readily with each other, they could interreact to form large molecular aggregates of ever increasing complexity.

Although it is impossible to know how, it is logical to assume that these early complex aggregates of carbohydrates, lipids, proteins, and nucleic acids eventually became stabilized and thereafter changed only by becoming more complex. Because a chemical reaction takes place in less than one billionth of a second, and because the primordial chemical soup lasted approximately one billion years, it is reasonable to assume that some of the complex aggregates acquired characteristics similar to those now found in living things. With the passage of time, these characteristics became the determining force and, as such, acquired selective properties. Eventually chemical structures evolved which modern scientists would classify as primordial life forms.

This was not the end of chemical evolution but, rather, the beginning of **biological evolution** of life on Earth. In the span of the next three and one half billion years, life as we see it today emerged by evolutionary processes. Modern scientists propose that the first living organisms were similar to the **archaebacteria** and that all other forms of life had their evolutionary origins in these early forms.

THE BACTERIA

Bacteria form the largest and most complex group of living organisms on Earth. Although quite small, they grow rapidly and form large populations in a short time. These organisms range from 0.6 to 100 μm in length but on

Figure 2.7

Scanning electron micro-scope pictures of Bacillus subtilis; *bar at lower right represents 1 μm. Photo courtesy of Munhyeong Choi, University of North Texas.*

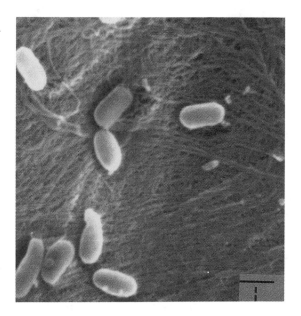

the average measure only 1 by 2 or 3 μm. The weight of a single live bacterial cell is approximately 0.000,000,000,001 gram (1×10^{-12} g); that is, 1×10^{12} bacterial cells weigh 1 gram. Figure 2.7 shows electron microscope pictures of a strain of bacteria.

Under laboratory conditions, certain bacteria divide four times in one hour and will continue to divide at that rate as long as the environment is suitable for growth. Bacteria divide by **binary fission**, producing two identical cells for each one that undergoes division. The result of such division can be expressed as a **geometric progression** with a base of 2. The equation

$$\frac{dN}{dt} = kt$$

describes the kinetics of bacterial division and shows that the number of new cells that can appear in a growing population depends on the number of cells present at a given, beginning, time, the length of time, t, in which growth proceeds, and the specific growth rate constant, k. Specific growth rate constant, a complex term, is affected strongly by the generation time of the organism. When the generation time is brief, say 20 minutes, vast populations of cells can be produced in a few hours.

Bacteria are **procaryotic**, simple organisms with vast physiologic and metabolic potential. They are ubiquitous in nature and accumulate in large numbers in the soil. They have the ability to survive for long periods in the dormant condition and can withstand greater environmental stress than any other form of life. The **bacterial spore** is a specialized structure that allows the organism to resist desiccation, freezing, chemical stress, and mechanical insult. Spores are known to survive for more than 140 years under laboratory conditions and may survive in nature for hundreds of years. Many spores withstand exposure to boiling water for several hours, and at least one kind will survive **autoclaving**, heating to 121°C for 15 minutes.

It is assumed that the original forms of life on Earth were bacteria. It is also assumed that when the planet Earth no longer is able to support life, the bacteria will be the last to perish.

Classification

The earliest attempts to classify bacteria were those proposed in 1872 by Ferdinand Cohn (1828–1898). This system was based on biological and physiological grounds but was abandoned in 1974. The classification of microorganisms has been recorded in *Bergey's Manual of Determinative Bacteriology*. This manual has been revised numerous times since its introduction in 1923.

A new system of classification was introduced in 1980. In it, the bacteria are arranged in groups according to artificial, or statistical, criteria. The taxonomic basis used in *Bergey's Manual* is quite complex, leading to the division of the experts into two opposite conclusions:

1. Classification is of extreme importance. "Nothing is more important in research than the identity of the organism. To work without knowing the precise identity of the subject is to throw the work away. No one who cannot do bacterial identification belongs in research."
2. Classification is of little importance. "At any rate, identity lasts only until the next mutation. You can't call bacteria by names like pets. Anything you give a name to, you can't incinerate."

All systems of classification of the bacteria are designed to resolve the compromise between these two extremes. The following lists give a brief view of some of the characteristics of bacteria used in their classification.

List 1. Some of the terms used to describe the morphology of bacterial cells:

Coccus:	spherical
Bacillus:	rod-shaped
Coccobacillus:	intermediate
Spirillum:	coiled like a corkscrew
Pleomorphic:	irregularly shaped

List 2. Terms used to describe the arrangement of cells:

single cells:	most cells appear singly in slide preparations
diplo-:	two cells attached to each other (e.g., diplococcus or diplobacillus)
tetra-:	four cells attached to each other

sarcina:	cubic packet of eight cells; only in coccus
strepto-:	many cells in a chain (e.g., streptococcus or streptobacillus)
palisade:	cells arranged like a fence; only in bacillus

List 3. Staining reactions of cells:

Gram positive:	stain purple with Gram's staining method
Gram negative:	stain red with Gram's staining method
Acid-fast:	not decolorized by exposure to a mixture of alcohol and acid after staining with a heated dye solution

List 4. Description of motility:

non-motile or	
motile by:	flagella
	gliding
	flexion

List 5. Arrangement of flagella on the cell:

monotrichous:	—one flagellum at end of cell
lophotrichous:	more than one flagellum at end of cell
amphitrichous:	one or more flagella at both ends of cell
peritrichous:	flagella uniformly distributed over surface of cell

List 6. Terms that describe the relation of bacteria to oxygen:

aerobic:	need oxygen to live
anaerobic:	cannot grow in the presence of oxygen
facultative:	can live with or without oxygen
microaerophilic:	must have oxygen, but at low concentrations

List 7. Relation of bacteria to temperature:

psychrophilic:	live at temperatures below 15°C
mesophilic:	live at room temperature; 15° to 45°C
thermophilic:	live at temperatures of 45° to 95°C

List 8. Terms that describe the growth of bacteria in relation to energy source:

photosynthetic:	obtain energy from sunlight
chemolithotrophic:	obtain energy from oxidation of inorganic substances
chemoorganotrophic:	obtain energy from oxidation of organic substances
paratrophic:	obtain energy from a host as parasites

Hundreds of other criteria are employed in the identification and classification of bacteria. In the last five years, identification by **DNA fingerprinting** has replaced older methodologies.

It is estimated that there are several million species of bacteria, although only a small fraction of this number has been identified. Finding previously unidentified organisms in almost any sample of soil or other natural material is not hard to do. On the other hand, organisms that are of some importance to human beings were isolated and identified long ago. Almost all the bacteria that cause major diseases in human beings are well known and well characterized. In like manner, all organisms of importance in the food and beverage industry have been identified and studied. Because these bacteria are well known, it is not difficult to find and identify them readily if they exist in mixed populations with many other, closely related organisms.

Ecology

Bacteria can be found in all the habitats the Earth provides. The vast majority of bacteria are **saprophytic** organisms that live in soil and water everywhere on Earth. Because the air carries bacterial cells in great numbers, these go wherever the air goes. It is as if there were a great pressure of bacteria and other microorganisms, which makes it almost impossible to resist their entry. To maintain materials in a sterile condition, microbiologists with specialized training are needed, and even their best work offers no guarantee, just a greater probability of success.

Importance of the Bacteria

The bacteria are of major concern to human beings for many reasons, including their role in human disease, in the spoilage of food and other valuable substances, in the growth of plants, in the production of valuable substances such as alcohol and antibiotics, and in the cycles of matter on Earth. Of the hundreds of thousands of species of bacteria, only about 200 are important as pathogens. Even so, these organisms are responsible for plagues and epidemics that in the past killed as many as half of the inhabitants of Western Europe within a few years. Some of the **pandemics** (epidemics that encompass large areas of the planet) that have changed the course of the history of human beings are noted in Table 2.5. The first three probably were caused by *Yersinia pestis*, the next by smallpox, and the last by the influenza virus.

The bacteria are involved in almost everything that happens in nature. They are responsible in large part for the decay of all matter everywhere on Earth; conversion of chemical substances from one form to another, as in the changing of proteins to nitrate, carbon dioxide, and water; and production

Table 2.5   Major Plagues That Have Assailed Human Beings

Years	Location	Number Dead
200–300 A.D.	Rome and vicinity	5,000 per day
530–600	Mediterranean	one third of population
1330–1400	China, Europe	one third of population
1895–1915	India, etc.	10,000,000 (Indian)
1914–1918	worldwide	20,000,000

of many substances not found otherwise in nature. Two examples of the latter are **dextrans** and **levans**. From ample observation it has been inferred that the bacteria may be an important selective force in the evolution of the human race. Childhood epidemics have the effect of removing many individuals from the species before they have contributed their genetic traits to the evolving population. By this mechanism only the individuals in a population who are resistant or who can survive microbial infection live long enough to reproduce and thereby contribute their genetic characteristics to the human race as it evolves through time.

Bacteria also function as geochemical agents in the deposition of sulfur, oil, and possibly peat in the soil. They are active in leaching insoluble chemicals from the soil and converting them into soluble form. Bacteria also move nitrogen from the air into the protein substance of living things and then back into the air in an all-important nitrogen cycle that makes life possible on Earth. Finally, the bacteria are **symbionts** of human beings, and without them human life might be quite different or perhaps not possible at all.

The most prominent features of the cell, shown in Figure 2.8, are the **pili**, which project from the surface of the cell. The nuclear matter is seen

Figure 2.8

Electron micrograph of Escherichia coli *K12; cell is magnified 100,000 times its actual size. Photograph courtesy of Munhyeong Choi, University of North Texas.*

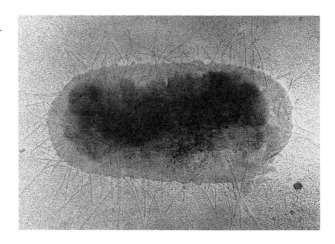

easily in the interior of the cell, as are some of the **ribosomes**. Careful examination will reveal a small part of the cell wall on the top edge of the cell.

A large part of the concern with the quality of food and drink is made necessary by the presence of the bacteria. As they are present everywhere in large populations, all food and drink always carry a large and varied bacterial load. Normally this does not matter because most bacteria are not harmful, but if even a small number of **pathogenic** bacteria is present, the situation changes rapidly. Ground crops such as lettuce and carrots that are consumed raw must be selected carefully lest the soil in which they grow contain human or animal fecal matter containing pathogens. Meat and other foods that have to be cut and trimmed must be protected from contamination by the very process used in their preparation.

Knowledge of the nature of microorganisms has made it possible to reduce the incidence of some diseases caused by these mechanisms. Historically, major epidemics of intestinal diseases responsible for the death of hundreds, thousands, and hundreds of thousands of human beings have resulted from transmission of microorganisms in food and beverage. Some major bacterial diseases transmitted to human beings through food and water are:

Typhoid fever	*Salmonella typhi*
Salmonellosis	*Salmonella spp.*
Dysentery	*Shigella dysenteriae*
Cholera	*Vibrio cholerae*
Campylobacteriosis	*Campylobacter jejuni*
Listeriosis	*Listeria monocytogenes*
Diarrheal disease	*Yersinia enterocolitica*

THE VIRUSES

Viruses are either living things at the very edge of life or inanimate things that closely resemble living organisms. They differ from all other living things, and special definitions have to be applied to justify placing them among the world of living organisms. Viruses are **obligate parasites** that exist inside the cells of a host as parts of the host's genes or in nature simply as dormant, nonliving, complex chemical molecules. Although they are all parasitic, they do not all produce disease in the host they parasitize. Many viruses, however, do cause human disease. The four major diseases transmitted by food and water are poliomyelitis, hepatitis, Coxsackie virus infection, and Echo virus infection.

Viruses were discovered by Dmitri Ivanowski in 1892, although Benjamin Jesty and Edward Jenner worked successfully with cowpox virus in the late 1700s (see Figure 2.9). In the 1860s, Pasteur found the appropriate techniques for immunizing chickens against fowl cholera and, in 1885, the method for immunizing dogs and human beings against rabies. Because

Figure 2.9

Testament to Benjamin Jesty's discovery of vaccination with cowpox virus before Jenner's work.

Benjamin Jesty (1737–1816), son of Robert Jesty of Yetminster and grandson of John Jesty of Leigh, Lived at Upberry farm for somewhile.

He took his son Robert to London in 1805 and there publicly vaccinated him before the doctors of the Vaccine Pock Institution, and so pleased were they with his pioneering spirit that they had his portrait painted in token of their respect.

On July 2, 1806, Lord Henry Petty, in the House of Commons upheld Jesty's claim to be the inventor of vaccination. His tombstone at Worth Matravers churchyard near Swanage reads:

An upright and honest man, particularly noted for having been the first person (known) that introduced the Cow Pox by inoculation, and who, from his great strength of mind, made the experiment from the cow on his wife and two sons in the year 1774.

these diseases are caused by viruses, those workers must have understood the nature of the disease agents well enough to work with them successfully. Most likely, however, all scientists who worked with viruses before Ivanowski, including Pasteur, assumed that the organisms they had under investigation were small bacteria and did not think of them as previously undescribed forms of life.

The outstanding characteristic of viruses—one that attracted the attention of many early investigators—is their filterability. The undescribed agents that could cause diseases such as rabies, smallpox, and measles would pass through Chamberland filters while even the smallest bacteria were uniformly retained. Because of their very small size and inability to grow on laboratory media, the viruses could not be observed or grown in the laboratory by conventional methods. Early workers developed techniques by which they could grow animal cells in the laboratory and use these as the host cells on which viruses then would grow.

As a class, most viruses are smaller than 0.1 μm, and even the best of light microscopes cannot resolve objects smaller than approximately 0.23 μm. The

Table 2.6 Differences Between Viruses and All Other Living Things

All Other Living Things	Viruses
Cellular structure	Not cells; like complex molecules
Have DNA and RNA	Have either DNA or RNA, not both
DNA double-stranded	DNA and RNA single- or double-stranded
Have more than 5,000 genes	Have fewer than 250 genes
Made of many different proteins	Made of one to 50 different proteins
React to environment	Do not react to environment
Metabolize and assimilate nutrients from the environment	Do not metabolize or assimilate nutrients from the environment
Grow	Do not grow
Reproduce	Cause themselves to be produced by a host cell

nature of viruses—their size, shape, and chemical properties—were all determined by circumstantial evidence. Not until 1930, when the electron microscope became available, could humans see real images of the viruses. Early electron microscopes had resolving power of approximately 0.01 m, compared to those in use today, which can render detailed images of objects as small as 0.5 nm (5 Angstrom units, Å) in diameter. Figure 2.10 shows the size of several bacteria and viruses in comparison to the human red blood cell.

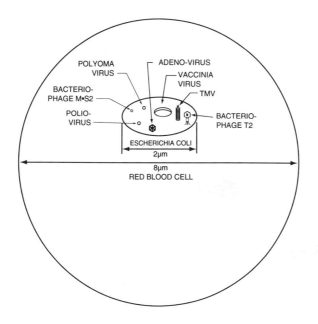

Figure 2.10

Comparison of size of human red blood cell and procaryotic organisms.

Physiological Characteristics

The chemical structure of viruses is more like that of highly complex molecules than that of living things. In the first place, viruses are not active physiologically. They are unable to react to their environment, to assimilate matter, to grow, or to reproduce. Second, because no chemical reactions take place in the viral particle, the virus is not able to sense the passage of time. Viruses are essentially an inanimate, dormant, complex of large molecules with the ability to cause a host cell to replicate the same arrangement of molecules. During this replicating process viruses often cause great damage to the host cell. The damage often is manifested as some disease process such as rabies, poliomyelitis, hepatitis, or the acquired immunodeficiency syndrome (AIDS).

Viral particles are composed of one nucleic acid (either DNA or RNA) and one to 50 different proteins. Some of the larger viruses may have 40 or more different proteins and even lipids and carbohydrates in small amounts. One of the characteristics of living things is their ability to repair parts of the organism that wear out with the passage of time. By contrast, viruses cannot repair changes caused spontaneously by outside forces and, once they leave the host cell where they were produced, exist as inactive or dormant forms.

Viruses consist of only a few components, in contrast to other living things, which generally are made of thousands or hundreds of thousands of different kinds and species of chemical substances. The viral particle consists of a molecule of nucleic acid surrounded by a **capsid**. There are DNA viruses and RNA viruses. Both have highly specialized mechanisms for entering the host's cells and interfering with normal cell functions to the point at which the cell turns all its efforts to producing viral progeny. The capsid is made of protein and can be simple or complex. The simple capsid is composed of several hundred or several thousand identical molecules of one kind of protein, whereas the complex capsid is made of several different kinds of protein molecules. The protein molecules that make up the capsid are called **capsomeres**. In both cases these are arranged in symmetrical order resembling the structure of crystalline substances, and the virus obtains its form or morphology from this arrangement. Figure 2.11 shows the structure of a viral particle.

The larger, more complex viruses may have an envelope surrounding the capsid. The envelope may be made from viral proteins and lipids, but in some cases the envelope is part of the cell membrane from the host cell that produced the virus. Some viral particles have islands of protein molecules attached to the outside of the envelope, called **spikes**. These function as sites of attachment of the virus to the surface of the host cell to facilitate infection.

The function of the capsid, and **envelope** when one is present, is to protect the nucleic acid during transit from the host cell that produced it to a new host cell that is susceptible to infection. A second function of the capsid is to attach to the proper site on the surface of a new cell in which the nucleic

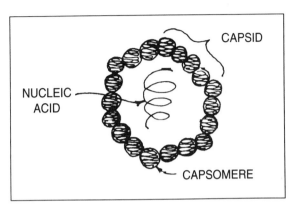

Figure 2.11

Structure of a typical viral particle such as poliovirus.

acid can function. Once attached to the surface, the capsid aids in inserting the nucleic acid into the **cytoplasm** of the host cell. The nucleic acid moves about in the cytoplasm until it finds the host cell's DNA and then attaches at a specific site on the DNA molecule. From this point on, the viral DNA becomes like a set of genes on the host's chromosome. These viral genes direct the cell in making new viral particles. The viral particles can be released a few at a time during the life of the infected host cell, or they all may escape at the same time when the cell dies.

All viruses are obligately parasitic, and many cause diseases in plants and animals, including human beings. They also infect and kill microorganisms such as algae, protozoa, fungi, and bacteria. Traditionally, viruses that affect bacteria are called **bacteriophages**, or bacteria eaters. Of course, they do not eat the bacteria, but they do destroy them. Some of the bacteriophages have a complex morphology involving several distinct parts, such as a tail tube, tail fibers, tail plate, and collar. Each part has its own function and some, like the tail fibers, even move, but they still have no similarity to living cells in complexity of function.

Classification

The classification of viruses was not well established until just a few years ago. Recently developed techniques in molecular biology have enabled the study of viruses from the molecular perspective. All viral agents are divided into two major groups on the basis of their nucleic acid:

1. DNA viruses
   a. single-stranded DNA   = ssDNA
   b. double-stranded DNA = dsDNA

2. RNA viruses
   a. single-stranded RNA   = ssRNA
   b. double-stranded RNA = dsRNA

A much older, but still valuable, classification of the viruses is made on the basis of the host organisms they infect:

    1. Animal viruses
        2. Plant viruses
            3. Bacterial viruses

Other criteria are employed to distinguish viruses from each other, and these form the taxonomic basis for a complete system of classification. The result is a nomenclature based on the Linnean binomial system, using genus and species as the final nominative. It can be argued that, if the viruses are not living things, they should not be classified as if they were. Even so, the system works as if they were living organisms. A small part of the criteria employed in classifying viruses is:

1. The infection or disease they cause
2. The shape of the virus particle
3. Presence or absence of an envelope
4. Effect on red blood cells of various animals
5. Particle size
6. Antibodies produced on injection into animals

    The basic description of one of the genera of viruses, *Picornaviridae,** is:

1. Enterovirus
    Poliovirus, Coxsackievirus, Echovirus
2. Simple, naked virus particles
3. Resistant to chemicals; pass through digestive tract, inactivated by heat, desiccation, and chlorine
4. Single-stranded RNA
5. Spherical; 25–30 nm dia.

Many other tests that depend on other characteristics of the viruses are available. Often the virus cannot be isolated from body fluids for identification, but the specific **antibodies** produced during infection can be used to identify the causative agent. In the same way, the production of antibodies can be a measure of the progress of an infection. Assessing the antibody titer of a representative portion of the population can serve as a measure of the progress of an epidemic as it moves across the land.

Ecology

Viruses survive outside of the host cell for variable times. Some are limited to a survival period of only a few hours, and others may remain alive for many days or even weeks. They often are resistant to desiccation, heat, and chemical attack. This permits them to survive treatment as harsh as passage through the

---

*pico = small; rna = RNA; viridae = the virus family

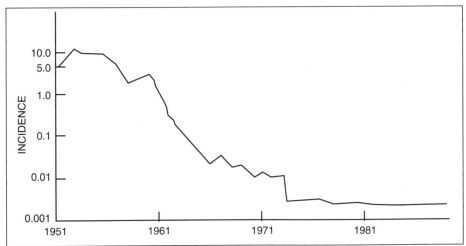

Figure 2.12

Effect of vaccination on infection rate of paralytic poliomyelitis in the United States from 1951 to 1990.

sewage treatment plant. The poliovirus, for example, is sufficiently resistant to survive for long periods in sewage effluence.

Viruses can be cultured in the laboratory as parasites of living cells. Embryonated chicken eggs have been used for many years to identify specific viruses and also to grow large quantities of the virus for vaccine production. Another method is to culture the cells that the virus normally parasitizes, such as human skin cells or monkey kidney cells in the laboratory. These cells can be grown easily in special nutrient solutions under controlled conditions in the laboratory. Once the cells are growing, the virus inoculum is added, and infection usually follows. Virus multiplication takes place in the infected cells, and a crop of viral particles can be harvested.

Viral particles can be separated from the host cells and cell debris by filtering the whole culture through filter membranes of 0.45 µm porosity, as the viral particles pass with the filtrate and host cells are retained. Further purification can be accomplished by **differential centrifugation** and other, more difficult, modern laboratory techniques. Through these methods, highly purified virus particle preparations can be obtained for use in immunization, identification, and chemical study. Figure 2.12 depicts the dramatic drop in the incidence of paralytic poliomyelitis as a result of vaccination.

Importance of the Viruses

Several viral diseases transmitted by food and water are listed in Table 2.7. In addition, the route of transmission is not established for some diseases, but the food and drink route is suspected. Lacking better knowledge, epidemiologists must assume that the suspected route of transmission is, in effect, the

Table 2.7   Viral Diseases Caused by Food- and Water-Borne Viral Agents

Medical Name	Common Name	Causative Agent
Poliomyelitis	polio, infantile paralysis	Poliovirus
Hepatitis type A	infectious hepatitis	HAV
Gastroenteritis	intestinal flu	ECHO virus
Gastroenteritis	Coxsackievirus	Coxsackievirus
Viral diarrhea	winter vomiting disease	Norwalk agent
Infant diarrhea	viral diarrhea	Rotavirus

correct one. In this case, then, the precautions taken have to be the same as those in which the exact mechanism of transmission is known.

Viruses cause many other diseases not associated with food and water and in this manner are of major importance to human beings. For example, some experts say that the epidemic of AIDS that is now affecting the human race may well exterminate it. This, of course, is a prospect of the greatest importance and one that requires immediate attention.

Plasmids

Plasmids are small molecules of double-stranded, circular DNA found in the cytoplasm of many bacteria. They also may be present in other organisms but so far have been found only in bacteria, fungi, and some protozoa. Plasmid DNA is produced in the cytoplasm of the host cell and then may become integrated into the host chromosome. There the plasmid genes manifest themselves as a set of the host's genes. Plasmids carry genes that endow the host with some property it lacks when no plasmids are present. Plasmid infection can be cured by growing the host in the presence of chemicals such as sodium dodecyl sulfate, acridine orange, or ethidium bromide in concentrations that inhibit plasmid DNA synthesis but not host cell DNA synthesis. Plasmids are infective in that, if a cell that carries the plasmids comes in close contact with one that does not, the new cell will acquire the plasmid and all the genes the plasmid carries.

Plasmids were discovered in 1962, when some bacteria were found to be resistant to several antibiotics at the same time even though they never had been in contact with most of them. The resistance was found to be infective, and whole populations of bacteria in an entire geographical area became resistant during a brief time. The phenomenon was discovered by the Japanese microbiologist Watanabe, in 1964, in clinical isolates of *Shigella dysenteriae*. These resistant organisms, with their plasmids, were brought to the United States by returning servicemen, and antibiotic resistance plasmids soon spread to native American shigellas.

THE ALGAE

The algae form a large and complex group of organisms with many fundamental similarities to the plants. Foremost among these is the presence of chlorophyll and the nature of the photosynthetic process. Plants and algae have the ability to use water in photosynthesis and to give off oxygen as a waste product. Although one kind of bacteria, the **cyanobacteria**, do this as well, all of the other photosynthetic bacteria are unable to live in the presence of oxygen and do not produce oxygen during photosynthesis. These **anaerobic** organisms utilize various chemicals such as **hydrogen sulfide** (H_2S) instead of water (H_2O). As a consequence, photosynthetic bacteria produce sulfur as the waste product of photosynthesis, whereas organisms such as the cyanobacteria, algae, and plants that use water produce oxygen as the waste product. The generalized reaction of photosynthesis found in all bacteria, cyanobacteria, algae, and plants is given below in its simplest form.

$$\text{LIGHT}$$
$$\downarrow$$
$$2H_2A + CO_2 \rightarrow (CH_2O) + A_2 + H_2O$$

In the reaction above, H_2A is any reduced substance that can donate H^+ to CO_2, i.e., reduce carbon dioxide. Through evolutionary processes the cyanobacteria, and then the algae and plants developed the ability to use water in this reaction. The use of water is, therefore, a special form in the reaction above, and the waste product is oxygen.

$$\text{LIGHT}$$
$$\downarrow$$
$$H_2O + CO_2 \rightarrow (CH_2O) + O_2$$

The evolutionary advantage of the cyanobacteria, algae, and plants rests on the availability of water on the planet's surface. The effect on the planet was drastic and irreversible. The liberated oxygen accumulated in the environment, converting the planet's surface from the reduced condition to the **oxydized** condition. This made the planet suitable for organisms that require oxygen.

Physiological Characteristics

The algae are **eucaryotic** organisms and, consequently, are regarded as higher protists. Like other eucaryotic organisms, algae have a nucleus that contains most of the cell's DNA in the form of **chromosomes**. The chromosomes exist in pairs and the **genes** in **allelic pairs**. Algal cells also contain mitochondria, vacuoles, and inclusions of various kinds, and the chlorophyll is in **chloro-plasts**. These organisms also contain other, nonphotosynthetic pigments including several **carotenoids**, **xanthophyls**, and **phycobilins**.

Many algae are motile by flagella at some stage of their life cycle, and some are indistinguishable from the protozoa during this stage because they can grow as **heterotrophic** organisms. Many algae will not produce chlorophyll when glucose or other utilizable carbohydrate is available. In general, most algae have cell walls made of cellulose like the plants, although some have no walls and others have walls made of different substances.

Classification

The algae are classified according to the colors that result from the interactions of light and the various pigments found in their cells. They form a heterogeneous group in terms of color, size, morphology, and structure. Many are unicellular, microscopic, and fairly simple. Others are large, complex colonies made of vast numbers of cells acting in unison. Some show a certain degree of differentiation of cell types and even some interdependence among the cells of the colony or "plant." The unicellular algae, by contrast, are more like the bacteria and fungi than the plants. The smallest algae are a few micrometers in cell diameter. The largest (for example, the sea kelp) can measure more than 50 meters in length. Because of their size, the latter formerly were thought to be aquatic plants but now are classified as microorganisms because the distinction is based more on physiological and biological properties than on size alone. Figure 2.13 shows the variance in algae.

Figure 2.13

Algae vary in shape and size; some are simple and others complex; some are microscopic in size, and others measure 50 meters in length.

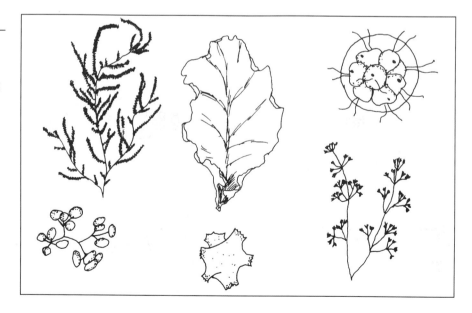

Table 2.8 Classification and Characteristics of Algae

Division Name	Common Name	Useful Material	Structure
Chlorophyta	Green algae	Starch	Unicellular to plant-like
Phaeophyta	Brown algae	Laminarin and fats	Plant-like
Rhodophyta	Red algae	Starch	Unicellular to plant-like
Euglenophyta	Euglenoids	Paramylum and fats	Unicellular
Pyrrophyta	Dinoflagellates	Starch and oils	Unicellular, filamentous
Chrysophyta	Diatoms, yellow-green golden algae	Leucosin and oil	Unicellular, filamentous

  In evolutionary terms the many similarities between plants and algae are difficult to overlook, and the rationale used by those who say that plants arose from the algae approximately one to two billion years ago is difficult to ignore. Microbiologists generally study the small, single-celled algae using techniques similar to those used in the study of bacteria and fungi, and botanists study the large, complex algae using methods developed for the study of aquatic plants. Not surprisingly, microbiologists refer to the algae as "plant-like microorganisms," whereas botanists call them "single-cell plants." It can easily be assumed that one is as wrong as the other or, perhaps, that both are wrong. Table 2.8 gives the classifications and characteristics of the algae.

Ecology

The algae are found everywhere on Earth, including all the permanent ice masses such as the Arctic and the Antarctic, as well as the driest of deserts from the Sahara to the Antofagasta of Chile. Although most algae are aquatic, many are terrestrial. Some algae are native to fresh waters and also native to the marine environment. Many others are found only in brackish waters. Some algae are extremely **halophilic** and are found only in places such as the Salton Sea and the Red Sea. Some algae are extremely **thermophilic**, growing at temperatures as high as 95°C, and others are extremely **xerophilic**, growing in desert soils with little or no water.

Importance of the Algae

The algae are of prime importance to the existence of life on Earth. They produce more oxygen in the environment than the plants do and are said to be the photosynthetic machine that makes life possible for the animals and other oxygen-dependent organisms. Together with the plants, they form one side of the complex and universal relationship that makes all life possible on the Earth's surface. The other side of this relationship contains the animals and all other oxygen-utilizing organisms. These oxygen consumers respire oxygen and produce carbon dioxide as a product of their metabolism. The "formula for life" can be written as:

LIGHT ENERGY

$\downarrow$

$$CO_2 + H_2O \rightleftharpoons (CH_2O) + O_2$$

$\downarrow$

WORK ENERGY

Plants and algae live by using light energy to produce carbohydrates and oxygen from carbon dioxide and water. Animals and other living things live by converting carbohydrates and oxygen into carbon dioxide and water. Algae form the basis of all food chains because they constitute the major part of the vast quantities of **phytoplankton** that inhabit all the oceans of the planet.

These organisms are of great importance to the food and drink industry. Many of the red algae, **Rhodophyta**, and the brown algae, **Phaeophyta**, are used as food in various parts of the world. They are eaten raw as salad vegetables and also are cooked in many different forms. Many species of algae are used as diet additives, as they are rich in vitamins B_1, C, and K. They also are rich in iodine and potassium and long have been used as nutritional sources of these minerals.

Some algae, both marine and fresh-water, are cultivated in large quantities as sources of food and other valuable products in Japan and other oriental countries. **Agar** and **carrageenan** are extracted from red algae and are used universally as thickening agents in foods, medicines, and cosmetics. These extracts are found in lesser amounts in many household products, and microbiological laboratories would be at a loss without agar for the isolation of bacterial cells. Agar is also used as a clarifying agent in the preparation of fruit juices and other liquid foods.

Extensive studies have been conducted in Japan with the objective of producing massive quantities of food from the microscopic, unicellular alga **chlorella**. Many scientists think that microscopic algae may be the most economical food crop for future generations. The simplest plan involves growing algae on decontaminated sewage water in large fields such as those used presently for the cultivation of rice.

Dinoflagellates are algae of the class *Pyrrophyta*. Those of the genus *Gonyaulax* produce toxins that have adverse effects on human beings and many animals. *Gonyaulax* normally are found in small numbers in sea water. Their presence usually is of no significance, but under certain environmental conditions the rate of reproduction is high and the number of cells becomes so great that the water takes on the red color of the cells. This "red tide" may be large enough to cover many square kilometers of ocean surface.

When shellfish consume the gonyaulax cells, the toxin in the cells accumulates in the shellfish and is passed on to animals, including human beings, that feed on the shellfish. In the past, this presented a major problem of food intoxication and resulted in many deaths, but today the problem is solved by official warning and closing of all fishing areas covered by the red tide.

Algae as Geochemical Agents

Diatoms are microscopic, free-swimming algae covered by two shell-plates made of **siliceous** material. The shells, or valves, fit like the two halves of a Petri plate, protecting the organism from hazards in its environment. The valves are made of an organic matrix with an outer layer of silica (SiO_2) and other minerals that are insoluble in water. When the organism dies, the organic material is dissolved by the action of bacteria and fungi and only the glass-like "skeletons" remain. The insoluble remains of diatoms accumulate in vast quantities on ocean bottoms.

The accumulations of many millennia formed structures such as the White Cliffs of Dover and the deposits at Lompoc, California, which are approximately 500 meters thick. This material, called **diatomite** or **diatomaceous earth**, has a porous but tightly packed consistency. From early times, cuttings quarried from these deposits were used as filters for clearing and purifying drinking water. These filters are still essential in many parts of the world where water is not made potable before distribution. Diatomaceous earth also is used as an insulating material in high-temperature boilers and as an abrasive substance in household and industrial cleansers. Many dentifrices and facial cleansing creams contain finely ground diatomaceous earth.

THE PROTOZOA

The protozoa form a large, complex group of microscopic organisms that exist as single cells or as small colonies of single cells attached to one another. They are ubiquitous in nature and are found in large numbers in all parts of the planet regardless of altitude, temperature, or humidity. Protozoa almost certainly evolved from the algae by an evolutionary bridge such as that provided by the **euglenoids**. Euglenoids are animal-like, motile algae (see Figure 2.14) that are photosynthetic but, under the proper conditions, can live by respiration like the animals do. Thus, they appear to be related closely to the algae but share many physiological characteristics with the animals.

Many zoologists think of protozoa as single-cell animals, and the name is derived from protos = first, and zoon = animal (i.e., the most primitive animals). Ample modern evidence suggests that animals evolved from protozoa, probably from a **flagellated** organism, as these bear a closer relationship to the

Figure 2.14

All protozoa are motile at some stage of their life.

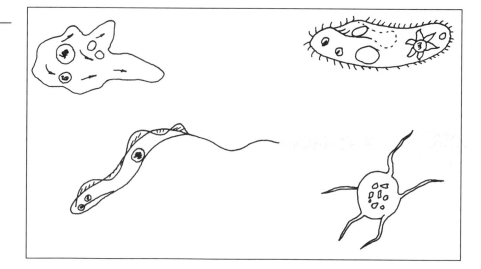

animals than the other protozoa do. Beginning students of biology surely must be confused to see that zoology textbooks consider the protozoans as animals and place them in the first phylum of the animal kingdom, whereas microbiology textbooks group them with the algae and the fungi and place them in the kingdom called the "higher protista." This is a traditional point of argument, albeit a moot one, but it does emphasize that the lines of demarcation of taxonomic divisions exist not so much in nature as in textbooks.

Physiological Characteristics

All the protozoa are aerobic heterotrophs associated with the degradation of organic matter in nature. Some obtain nutrients from the environment by diffusion in the same manner as the bacteria, fungi, and algae do. Others have specialized cell openings, **cytostomes**, which allow them to ingest particulate matter. The ingested particles enter the cell inside digestive **vacuoles** that contain **hydrolytic enzymes**. These enzymes degrade the particles systematically until their chemical constituents become water-soluble and pass into the cytoplasm, where they are used in the metabolic reactions of the cell.

Reproduction in the protozoa takes place by at least three well known methods. In asexual reproduction, a cell that has matured undergoes a building process in which all its parts are duplicated and divided into two equal parts in the cell. The two halves then separate by forming a new wall that divides the cell into two equal parts, which then separate, resulting in two new and identical cells. This is the simplest kind of reproduction and is essentially the same as that found in the bacteria.

Other protozoa, such as those of the genus *Paramecium*, undergo sexual reproduction in which two cells conjugate (fuse) and their nuclei fuse into a

heterologous zygote. When the cells separate, some of the genetic material from one of the mating partners appears in the other, and vice versa. Still other protozoa, such as the sporozoa, reproduce by a complex process called **schizogony**. This method is seen primarily in organisms similar to those that cause malaria.

Many protozoa can enter into a dormant condition in which the viable cell is encased in a hard, protective coat. The dormant form is called a **cyst**. Cysts are formed in response to inhospitable conditions such as dryness, dearth of nutrients, presence of toxic substances, and other environmental stresses. The organisms can survive for extended periods in the cyst form and will become active and **excyst** when conditions again are conducive to growth. In some protozoa the cyst may be part of a complex life cycle, and in others, such as *Entamoeba histolytica*, the cyst also may serve as a reproductive structure. In these, the nucleus divides as the cyst develops, resulting in **multi-nucleate** cysts that yield several new vegetative cells when the cyst germinates. On excystment *Entamoeba histolytica* cysts give rise to four **trophozoites**.

Classification

Approximately 20,000 species of protozoa live at the present time. Many more existed in the past but became extinct long ago. These are known only by their fossilized remains, which geologists employ to identify different strata of the Earth's surface. This practice is useful particularly in the oil exploration industry, as many oil-bearing soils can be identified by the number and kind of protozoan fossils found therein. The vast majority of protozoa are saprophytic in nature, and few are found living as parasites. Probably fewer than twenty have the ability to live as parasites in the bodies of human beings, but these cause three of the most devastating diseases of humankind. One of them, **amebic dysentery**, is transmitted by food and water, and the other two, malaria and sleeping sickness, are transmitted by the bites of the **anopheles mosquito** and **triatomid bugs**, respectively.

The protozoa can be divided into four classes according to their method of motility. These include ameboid motion, flagellar locomotion, and motility by the use of cilia. The sporozoans are not motile throughout most of their life cycle, but all of them pass through a motile phase at one point or another. The four classes of the protozoa are described briefly in Table 2.9.

Ecology of Protozoan Diseases

The protozoa are ubiquitous in nature. They are found in all fresh waters and in the marine environment as well. The ability of many protozoa to form cysts allows the aquatic organisms to survive the drying out of water habitats and most of them to survive lack of nutrients or the accumulation of toxic products in the environment. Most of these organisms, however, live in the soil

Table 2.9   Major Pathogens of Classes of Protozoa

Class Name	Locomotion	Etiology, Human Beings
Sarcodina		
Entamoeba histolytica	Ameboid	Amebic dysentery
Ciliata		
Balantidium coli	Cilia	Ciliary diarrhea
Mastigophora		
Giardia lamblia	Flagella	Diarrhea
Sporozoa		
Plasmodium vivax	(Non-motile)	Malaria

and water as saprophytes. Their role in degradation of particulate matter is extremely critical in the cycles of matter that make life on Earth possible for all living things. Some of the protozoa live as parasites in animals or plants and have adapted to the parasitic form of life. Still others exist as **commensals** with plants or animals.

Importance of the Protozoa

The protozoa are important to all the processes in the soil necessary for maintaining life on Earth. The main role of the protozoa is associated with degradation and solvation of particulate matter in recycling organic materials. All living things that die and fall to the earth must be processed and reconverted to the original chemical substances from which they were formed. Small animals, such as insects and beetles, along with microorganisms such as protozoa, fungi, and bacteria, are primarily responsible for converting all complex matter to the oxidized, mineral form.

Protozoa also are important to human beings because they cause severe diseases such as dysentery and malaria. Many protozoa cause diseases in animals that are valuable to human beings, such as cattle, poultry, horses, and household pets. **Toxoplasmosis** was largely unrecognized even as late as the 1950s, but diligent study utilizing newly developed techniques showed that it is contracted by household pets, cattle, and human beings. Toxoplasmosis and other protozoan diseases are of further importance to humans in that these diseases often are transmitted from animals to human hosts.

THE FUNGI

The third group of eucaryotic protists comprises a large and diverse family of microorganisms collectively called the fungi. Other names applied to the fungi—drawn from historical use, common use, and erroneous indications—

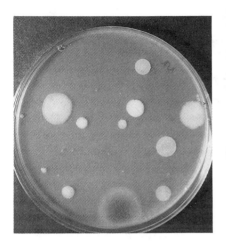
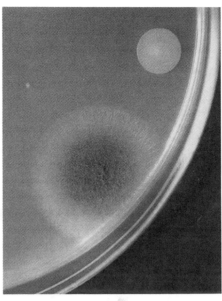

Figure 2.15

Laboratory cultures of common air and soil fungi. Photos courtesy of V. G. Nevarez, University of North Texas.

include mushrooms, yeasts, morels, molds, wilts, rusts, smuts, and mildews. One of the reasons for this variety of names is that these organisms differ from one another in so many ways that the unifying characteristics of the group have been recognized only recently. They vary in size from single-cell yeasts, which are 5 μm or smaller in diameter, to mushrooms that may be more than 1 meter long and weigh as much as 15 kilograms. Figures 2.15 and 2.16 show some laboratory cultures and scanning electron micrographs, respectively, of common air and soil fungi.

The fungi affect human beings in many ways. They cause serious diseases and also the spoilage of huge quantities of food and other valuable products. They also serve as a source of food and are the basic material for a vast industry of medicinal substances, cosmetics, foods, and solvents. In addition, some fungi produce powerful toxins that cause death when ingested, and others produce hallucinations and other psychic phenomena when consumed in measured quantities. Still others are responsible for the production of **neoplasia** and **teratogenesis** in human beings and animals.

Physiologic Characteristics

The great majority of fungi are aerobic, heterotrophic, saprophytic organisms that are ubiquitous in nature. Fungi of many classes have the ability to live without air and bring about the fermentation of sugar. Outstanding among these are the yeasts, which have been used from our earliest times on Earth for leavening bread and producing wine. Fungi grow on simple

Figure 2.16

Scanning electron micrographs of various air and soil fungi; bars indicate magnification. Photos courtesy of Munhyeong Choi, University of North Texas.

compounds such as glucose and inorganic nitrogen, but many of them also can utilize complex substances such as celluloses and proteins.

Asexual reproduction is by binary fission as in the bacteria, by budding, or by asexual spores. Sexual reproduction also is common. Figure 2.17 depicts the life cycles and reproduction of asexual and sexual fungi. In asexual reproduction, fungal cells do not disperse when they divide but, rather, form long chains of cells called **hyphae**. As these continue to grow, they spread and form a large mass of hyphae called a **mycelium**. Specialized cells in the mycelium give rise to **fruiting structures**, which produce large numbers of **reproductive spores**. As the spores mature, they are dispersed in the air or water, and each

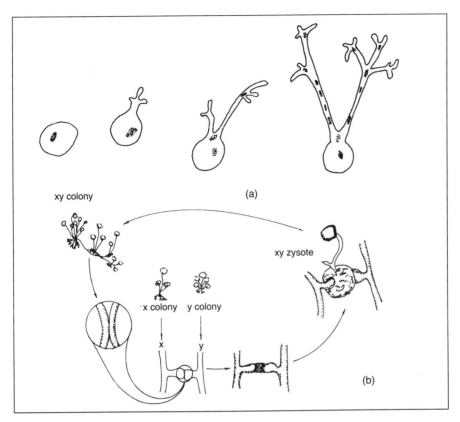

xy colony

(a)

xy zysote

x colony   y colony

x   y

(b)

Figure 2.17

Life cycles and reproduction of asexual (a) and sexual (b) fungi.

Zygomyces are found in soil and water. The spores are formed in a sac or **sporangium**.

Basidiomyces reproduce by sexual processes. All the mushrooms are in this group. Spores are produced on a **basidium** which is a fleshy structure analogous to the stem of mushrooms.

Deuteromyces (Fungi imperfecti) form a heterogeneous group with no known sexual stage of reproduction.

Ascomyces reproduce by sexual fusion forming an **ascus** which ruptures when mature to liberate the spores. The common bread mold, *Neurospora crassa* belongs to this group.

Figure 2.18

The four classes of fungi.

new spore may give rise to a new colony. In some fungi, if the mycelium of one colony comes in contact with that of another, hyphae will fuse, producing zygotes that give rise to a new organism with genetic traits from both colonies. This form of sexual reproduction accounts for the great variety of fungi found in almost any environment in nature and also for their ability to inhabit extreme and hostile environments.

Most fungi grow better in acid conditions than in alkaline or neutral conditions. This trait is useful to human beings, as fungi can be enlisted to produce acids in various foods, protecting them from spoilage by bacteria or even other fungi. One example of this is the conversion of sugar mill tailings into edible acids such as lactic and butyric acids. Sugar solutions spoil readily while the acids, valuable in the food industry, have long shelf lives.

The **aerial mycelium** of fungi is recognized commonly as mold growth on fruits, vegetables, bread, meat, and all other putrescible materials. The aerial mycelium grows from the surface of the **substrate** into the air for the purpose of disseminating reproductive spores. The **submerged mycelium** penetrates into the substance of the substrate. All fungi secrete a variety of enzymes, which degrade and digest the substrate until it loses its structure, texture, and form. As the substrate is degraded, simpler molecules are released, and these serve as nutrients for the fungal colony. They are taken up by the cells of the submerged mycelium and are distributed to all the cells of both the submerged and the aerial mycelia. Nutrition for the entire fungal colony is obtained by this means.

The aerial mycelium produces fruiting structures in which the spores are formed, and these are carried by the air for great distances. When spores fall on substrates that can support their growth, they **germinate** and form new colonies. The spores may have originated asexually or from the sexual fusion of closely related organisms. Spores are produced in vast quantities and are carried by the air over large distances. Fungal spores constitute a substantial part of what is reported as pollen by some weather forecasters on television news programs. In addition, any part of the mycelium that may be broken off and transported to an environment where growth can take place will give rise to a new colony, as the cells will simply continue to grow.

Classification

Approximately 100,000 species of fungi have been classified, but probably just as many more have not been catalogued. The classification of fungi is complex, but the first division is made on the basis of reproductive structures. The four classes of fungi, illustrated in Figure 2.18, are:

1. **Zygomyces**, found in soil and water; the spores are formed in a sac or **sporangium**.

2. **Basidiomyces**, which reproduce by sexual processes. All the mushrooms are in this group. Spores are produced on a **basidium**, a fleshy structure analogous to the stem of mushrooms.
3. **Ascomyces**, which reproduce by sexual fusion, forming an **ascus**, which ruptures when mature to liberate the spores. The common bread mold, *Neurospora crassa*, belongs to this group.
4. **Deuteromyces** (*Fungi imperfecti*), which form a heterogeneous group with no known sexual stage of reproduction.

Some of the fungi are similar to algae, and others closely resemble the protozoa, leading many experts to consider them as the evolutionary source of both algae and protozoa. Characteristics of the fungi, however, are sufficiently different from those of the other organisms and their diversity is so broad that some experts place them in a kingdom by themselves.

Ecology

Because the spores produced on fungal fruiting structures flow constantly into the air, they normally are present there in large numbers. They can be found at altitudes of 2 or 3 kilometers above the earth's surface and, in decreasing numbers, even higher. They have been found as high as 200 kilometers above the earth. Because air movements carry fungal spores over large distances, constant and thorough seeding of fungi occurs in all parts of the planet.

Fungi can adapt to extreme conditions and grow in many habitats that other organisms find inhospitable. They often are found in acid environments where nothing else will grow, some grow on minute particles of dust suspended in the air, and others grow on window panes obtaining nutrients from vapors in the air.

Importance of the Fungi

Of the 100,000 species of fungi that have been identified, only 70 or 80 are known to cause disease in human beings. Fungal diseases are found most commonly on the skin and are not considered of great importance, but some cause systemic infections that lead to serious illnesses and even death. In addition, some fungi that are not pathogenic produce toxins in foods and beverages, which have severe effects on many animals including human beings. The best known of these diseases, called **ergotism**, results from the ingestion of ergot, a toxin produced by *Claviceps purpurea*.

Aside from toxin production, the major significance of fungi in the food and beverage industry is economic and is twofold. One is concerned with spoilage of valuable products, and the other with the production of these products. The extent of financial damage done by the fungi yearly is probably too great to assess accurately. Losses are associated not only with spoilage

Figure 2.19

Modern 1000-liter industrial fermenter (courtesy of New Brunswick Scientific Co., Edison, New Jersey) and simplified, schematic diagram of an older machine.

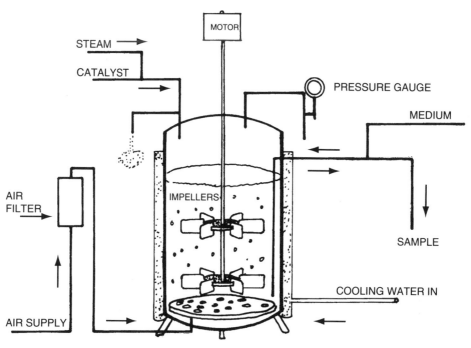

of stored materials but also with diminution of food harvests when crop plants are infected by pathogenic fungi.

Because fungi are ubiquitous in nature and because the air carries large numbers of spores, all food and drink contain fungi once they are exposed to the air. When conditions are suitable, fungi grow rapidly and produce extensive populations in a matter of a few days or even a few hours. Because of their great physiologic and metabolic diversity, fungi render no food exempt from attack.

The other major importance of the fungi, by contrast, lies in their ability to produce valuable substances for the well-being of human beings. The fungi are of great value in the production of alcoholic beverages and many kinds of cheese, and in the aging of steaks. Fungi also are used in the production of valuable chemicals such as succinic, malic, and citric acids, a large variety of antibiotics, and many other valuable substances. Figure 2.19 shows a modern industrial fermenter and a schematic drawing of an older machine.

QUESTIONS TO ANSWER

1. How do microorganisms differ from plants and animals?
2. What are the characteristics of eucaryotic microorganisms?
3. What are the characteristics of procaryotic organisms?
4. What are the distinguishing characteristics of the bacteria?
5. How have the bacteria contributed to the evolutionary development of human beings?
6. How are the bacteria classified?
7. What does human-microbe symbiosis consist of?
8. How do the bacteria cause disease?
9. What is the size range of the bacteria?
10. What is the size range of the viruses?
11. How are viruses like living things?
12. How do the viruses differ from living things?
13. How do the viruses cause disease?
14. What characteristics distinguish the algae from all other living things?
15. How many classes of algae are there? What are the common and scientific names of each?
16. What characteristics distinguish the protozoa from all other living things?
17. What are the four phyla of protozoa? On what basis were these phyla established? What different methods of motility are seen in the protozoa?
18. What are the morphological characteristics of the fungi?
19. What is respiration? Fermentation?
20. What are the distinguishing traits of the mushrooms?

FURTHER READING

Readings from Scientific American. 1985. THE MOLECULES OF LIFE. W. H. Freeman and Co., New York:

Tortora, G. J. 1982. MICROBIOLOGY AN INTRODUCTION. Benjamin/ Cummings Publishing, Menlo Park, California.

Volk, W. A., D. C. Benjamin, R. J. Kadner, and J. T. Parsons. 1986. ESSEN-TIALS OF MEDICAL MICROBIOLOGY, 3d ed. J. B. Lippincott Co., Philadelphia.

Whittaker, R. H. and L. Margulis. 1978. Protist classification and the kingdoms of microorganisms. BioSystems 10: 3–8.

Growth, Dispersal, and Distribution of Microorganisms

3

If thou be'st born to strange sights,
*   things invisible to see,*
Ride ten thousand days and nights
*   till age snow white hair on thee,*
Thou, when thou return'st, will tell me
*   all strange wonders that befall thee.*

                                        Donne

Microorganisms are found everywhere on Earth. Not only do they occupy every possible niche in nature, but they also are capable of growing in hostile environments where no other living thing can exist. The number of microorganisms alive on Earth at any given time is extremely large, and probably uncountable because microorganisms grow and perish at rapid rates. Their populations, therefore, give the semblance of a dynamic system in which the total number in a given place at a given time would represent the equilibrium situation. This is an equilibrium that may shift in either direction at the slightest perturbation, resulting either in massive death or in "blooms" of new microbial cells.

   Microorganisms grow in all the environments of Earth in which plants and animals can exist but, unlike them, microorganisms also grow in hostile habitats. They produce large populations in places as hostile as natural and

Figure 3.1

Bacterial and fungal cells grow on the surfaces and crevices of soil particles. Size of organisms is greatly exaggerated to illustrate the point.

manmade brines, acid pools, hot springs, and frozen wastes. The growth from these varied environments accumulates in the water, the air, and eventually in the soil (see Figure 3.2). The soil contains myriads of microbial cells, **spores**, and **cysts**, bringing some experts to say that the soil is the "natural reservoir" for all microorganisms. Microbial cells are so numerous in the soil that they make up a small but significant fraction of the soil commonly called **soil organic carbon**. Solid evidence indicates that some of these soil organisms may remain alive for periods in excess of a hundred years.

Because microorganisms are so small, they are lifted from the soil and carried aloft by wind currents, with the effect that the air is filled with microorganisms at all times. Many of these, such as the spore-formers and the **mycobacteria**, can survive in the air several months and perhaps even several years. When organisms from the air fall into water or other environments that contain moisture such as foods or the bodies of animals, they grow rapidly, producing large populations that again will dry and rise into the air to repeat the cycle. On the basis of these observations, microorganisms are assumed to grow in the **hydrosphere**, are transported by the **atmosphere**, and accumulate or survive in the **lithosphere**.

Figure 3.2

Microorganisms exist in and are part of the soil.

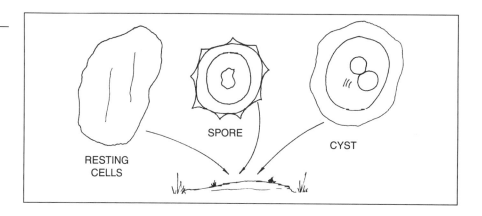

The microorganisms differ from each other in so many fundamental properties that treating them collectively when describing phenomena such as growth, dispersal, and distribution is impossible. To treat each group of organisms separately also would be an extremely lengthy task. The dilemma can be resolved by selecting a model to elucidate the property under consideration. The bacteria traditionally are selected because more is known about them and they provide the best experimental evidence to show the kinetics of microbial growth.

BACTERIAL GROWTH

Bacterial growth and rate of reproduction are essentially synonymous, as cells grow very little before division occurs. In practice, the production of new cells is regarded as growth and the increase in mass of individual cells is regarded as of secondary importance.

All microorganisms including the bacteria reproduce and grow only in the presence of water. Because they are very small, the amount of water required is also slight, and they may well be able to grow in or on substances that appear to be dry. Bacteria require water for growth because they obtain all their nutrients by diffusion through the cell membrane. This means that all nutrients must be dissolved in water before bacteria can utilize them. In addition, the bacterial capsule, cell wall, and cell membrane must be **hydrated** before they can function physiologically.

If sufficient water is available, and if all necessary nutrients are present, growth and reproduction begin almost immediately. Environmental conditions such as temperature, gaseous environment, **pH**, and the presence of nutrients all affect the rate of growth. Because nutrients and oxygen normally are limited and waste products and other toxic **metabolites** accumulate in the growth environment, bacteria seldom exhibit their full capacity for reproduction in nature.

Effect of Temperature on Growth

When the temperature is 0°C or lower and water freezes, bacteria can neither grow nor reproduce as water is then in the solid state and **solvation** and **hydration** phenomena occur only in water in the liquid state. Although many microorganisms are killed by freezing, many others survive quite well. Bacteria and other microorganisms live longer in colder environments and show maximal survival times when in the frozen state. Many bacteria grow well in the temperature range from 0° to 15°C. Called **psychrophilic bacteria**, they are of great importance in the food and beverage industry as they cause spoilage of materials stored in the cold.

Figure 3.3

Effect of temperature on growth rate of psychrophilic (a), mesophilic (b), and thermophilic (c) bacteria.

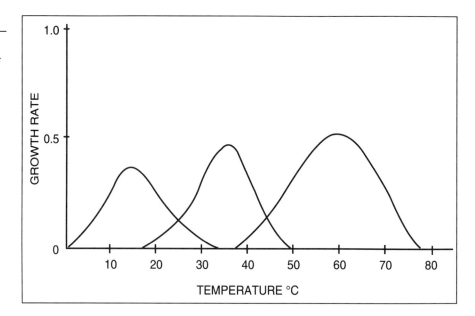

In contrast, microorganisms that grow at room temperatures (in the range from 15° to 45°C) are said to be **mesophilic** while those that grow at temperatures greater than 45°C are called **thermophiles**. Mesophilic microorganisms are responsible for the most damage to foods as they grow faster than the psychrophiles because of the effect of the higher temperature on the rates of chemical reactions.

Finally, thermophilic organisms are significant to the food and beverage industry because the spore-formers among them often produce spores that survive immersion in boiling water for extended periods and some can survive heating even in the **autoclave**. Psychrophilic organisms, which grow best at refrigerator temperatures, also can grow, albeit more sparsely, at higher temperatures. In like manner, the mesophiles can grow in the psychrophilic and thermophilic ranges but, again, not as well as at their optimal temperatures. Organisms that grow only at one temperature interval are said to be obligately psychrophilic, mesophilic, or thermophilic.

Each organism has an optimal growth temperature, the temperature at which its rate of division is highest for a given growth medium and a given set of environmental conditions. Optimal growth temperatures vary as growth conditions change, so in reality each set of growth conditions has an optimal growth temperature. Growth under the given set of conditions will occur at temperatures both higher and lower than the optimal point but not at as high a rate. Figure 3.3 depicts the effect of temperature on growth rate for one set of growth conditions. Some organisms have narrow limits of temperature tolerance; others show great latitude.

Effect of Acidity on Growth

The effect of acidity can be depicted in the same manner, and the optimal condition can be defined in the same way as the effect of temperature on bacterial growth. Acidity, expressed by the symbol pH, is measured as the concentration of hydrogen ions in solution. It is defined as shown in the chemical notation:

$$pH = \log (1 / [H^+])$$

where $[H^+]$ signifies the concentration of hydrogen ions.

The entire pH range is essentially from 0 to 14. Highly acidic solutions such as the liquid in a car battery are said to have a pH less than 1; vinegar, a pH of 4; water, a pH of 7; sodium bicarbonate, a pH of 9; and lye, a pH approaching 14. Water is neither alkaline nor acid but, rather, a condition described as neutral. Acidic solutions have pHs in the range from 0 to 7 whereas alkaline solutions have pHs between 7 and 14.

Most bacteria grow better at neutral pHs (pH 6 to 8), but some prefer an acid pH. These are called **acidophilic**, whereas those that prefer an alkaline pH are said to be **basophilic** microorganisms. Organisms found in sauerkraut are acidophilic. Those found in green beans are basophilic.

Effect of the Gaseous Environment on Growth

The gaseous environment has dramatic effects on microbial growth. The best example of this can be described by the growth of bacteria of the genus *Clostridium*. These bacteria grow well and rapidly in the absence of oxygen, and even small traces of oxygen inhibit their growth completely. Such organisms are called **obligate anaerobes**. Those that must have oxygen to grow are called **obligate aerobes**. A third, and larger, group, which can grow in either situation, are said to be **facultative microorganisms**. A small but important group of bacteria, which cannot grow in the absence of oxygen but are killed by concentrations greater than 0.5% of this gas, are called **microaerophilic**. Many, if not all, bacteria and some fungi, grow better when a small amount of carbon dioxide is added to the gaseous environment. Some anaerobes can grow well only in the presence of hydrogen, but others can grow as well under nitrogen or helium.

Many microorganisms are not affected by gases such as cyanide, carbon monoxide, methane, and hydrogen sulfide. These gases are toxic to all other forms of life.

Effect of Growth Substrate

The amount, but not the rate, of growth is determined by the amount of essential nutrients available to the organism. An essential nutrient is one that makes growth possible. Like all other living things, all microorganisms require nutrients that contain the chemical elements needed for the synthesis of new cells. A source of nitrogen (chemical symbol N) must be present, as the organism needs this for the synthesis of proteins, nucleic acids, amino sugars, vitamins, and other essential cell constituents. A source of carbon (C) must be present for the synthesis of proteins, nucleic acids, lipids, sugars, vitamins, and other cell constituents. In the same manner, oxygen (O), hydrogen (H), sulfur (S), phosphorous (P), iron (Fe), magnesium (Mg), and many other chemical substances essential for building complex molecules must be supplied in adequate amounts to make new microbial cells.

To build molecules for living cells, the organism must have a source of energy. Some, such as the anaerobic photosynthetic bacteria, cyanobacteria, and algae, obtain energy from light. Others, the **chemolithotrophic bacteria**, obtain energy from chemical reactions in which reduced materials are oxidized. These two classes of organisms are called **autotrophic**. The name implies, wrongly, that they feed themselves. On the other hand, bacteria, fungi, protozoa, and all the animals obtain energy from carbohydrates, lipids, proteins, and other energy-containing molecules. These are called **heterotrophic** organisms. This name, also wrongly, implies that these organisms eat anything.

The metabolic steps by which energy is extracted from complex molecules are collectively called **catabolism**, and those used in building new molecules are collectively called **anabolism**. The entire process of converting one kind of chemical substance into another—converting food into new cells—is called metabolism. This sequence of chemical reactions is extremely complex and requires hundreds of enzymes, many vitamins (coenzymes), minerals, and other substances plus a highly specific and rigid set of cell conditions.

MEASURING BACTERIAL GROWTH

Rapid and sustained growth of microbial cells occurs if there are no limiting factors. Bacterial growth is evident to the unaided eye only when the increase is of many orders of magnitude of the cell number during a brief period. If the population doubles in a given time, the event will go unnoticed, as the increase in cell number will not be perceived easily. It will be noted only if the increase in cell number is of the order of a millionfold. A microscope can reveal the increase in bacteria, as pictured in Figure 3.4. The growth of bacteria in a flask, or even in nature, can be measured by several methods, including turbidity, direct count, plate count, and chemical determination.

Turbidity

The amount of light that passes through a test tube full of growth medium can be measured with a **nephelometer**, a **turbidimeter**, or a **colorimeter** see

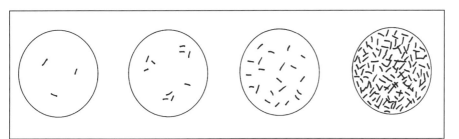

Figure 3.4

Increase in bacterial numbers as seen with the oil immersion lens of the microscope.

Figure 3.5). If a bacterial cell is placed in the growth medium, it will absorb a certain amount of the light passing through the medium, and the difference in light will be proportional to the bulk of the cell. It follows that two cells will absorb twice as much light and ten million cells, ten million times as much. Under the proper conditions, then, the number of cells in a liquid culture can be estimated simply by measuring the amount of light absorbed (turbidity) using the proper instrument and adequate controls. Because dead cells will absorb the same amount of light that live cells absorb, they will be counted as if they were living cells.

Figure 3.5

The turbidity of a culture can be measured with a colorimeter.

Figure 3.6

Petroff-Hausser chamber as seen with the oil immersion lens of the microscope.

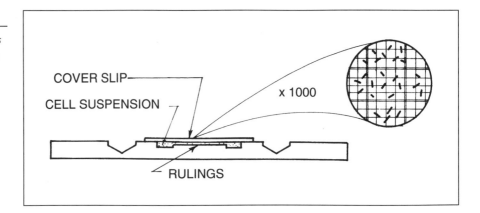

Direct Count

A glass slide marked in a precise pattern and covered with a special piece of glass will hold a known volume of liquid culture. This special instrument, pictured in Figure 3.6, is called a **Petroff-Hausser chamber**. When viewed with the microscope, it can be used to estimate the number of cells in a culture. Like the many other techniques for direct counting of the total number of cells in a bacterial population, it, too, suffers from the same shortcoming: Live and dead cells are counted equally. Even with this fault the direct count of cells is useful, and in some situations it is the only method that can be used.

Plate Count

Liquids that contain microbial cells can be diluted with known quantities of sterile water and the diluted samples spread on sterile nutrient media gelled with **agar** and held in **Petri plates** in such a manner that the exact amount of diluted sample in each plate is known. If the plates are incubated in a suitable environment, viable cells in each dilution will grow and give rise to colonies on the surface of the plate. The assumption is that each colony arose from one cell in the dilution, and from that one can estimate that the number of colonies on a given plate is a measure of the number of cells in the original sample. Calculations used in the plate count method are given in Figure 3.7. This technique has many variations, but they all give the same information.

Chemical Determinations

Bacteria that can grow in a medium free of protein can be counted by measuring the amount of protein that appears in the culture after a given period of incubation. It is assumed that one cell contains a certain amount of pro-

	Sample					
Dilution	**1:10**	**1:100**	**1:1,000**	**1:10,000**	**1:100,000**	**1:1,000,000**
Amount of sample	0.1	0.01	0.001	0.0001	0.00001	0.000001
Amount plated	1 ml	1 ml	1 ml	1 ml	1 ml	1 ml
Count	TM*	TM	TM	131	10	1
Count	TM	TM	TM	147	7	0
Count	TM	TM	TM	120	7	1

$131 + 147 + 120 \div 3 = 133$ cells per 0.0001 ml of sample

$\therefore 133 : 0.0001 = x : 1$

$$x = \frac{133}{.0001} = 133 \times 10,000 = 1,330,000 = 1.33 \times 10^6$$

*Too many

tein and that two cells contain twice as much. Therefore, by measuring the total amount of protein under controlled conditions, the number of cells can be estimated.

Many other constituents of bacterial cells, such as deoxyribonucleic acid, pigments, and lipids, also can be used to measure microbial growth. Other methods must be devised to meet the special conditions of various situations including measurement of the weight of dried cells, measurement of oxygen consumed, measurement of carbon dioxide given off, incorporation of radioactive substances, and production of fluorescent materials.

GROWTH KINETICS

The growth of bacteria is different from that of plants and animals. The individual microbial cell gains little in size or mass before it divides into two equal and identical cells. The two new daughter cells undergo a limited amount of growth before they, too, divide in the same manner (Fig. 3.8). Growth, therefore, is seen as an increase in the number of individuals rather than as an increase in the size of an individual organism.

If the growth in a flask culture is measured by the plate count method and the number of viable cells plotted in graph form, a typical **bacterial growth curve** such as the one in Figure 3.9 will be obtained. The graph in the figure is said to be the typical bacterial growth curve, but actually it is a highly stylized, theoretical description that may bear little resemblance to the growth of bacteria in nature.

Figure 3.8

Growth and division of the bacterial cell are shown in schematic representation.

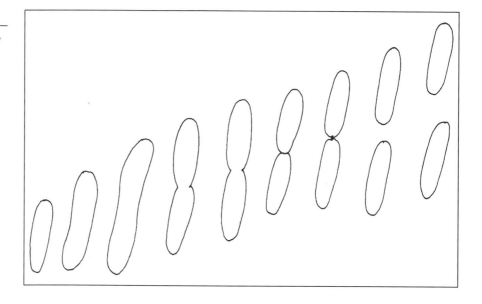

The part of the curve labeled b, the logarithmic phase, is the phase of maximal growth rate and also is the only phase in which the **generation time** is constant. Not only is the generation time constant, but it also is the shortest generation time possible for the organism under the conditions imposed. This part of the growth curve can illustrate the growth potential of the bacteria. The scheme given on the next page shows how the growth potential can be visualized and how the equations for the theoretical growth curve were derived originally.

Another view of bacterial cell division will show that each cell will double itself in a given period. This is the time normally called a **generation**, or better, a generation time. Figure 3.9 shows that the number of cells doubles at each of the times t_1, t_2, t_3, and so on. That is, each generation results in doubling of the population.

The sequence in Figure 3.10 shows that one cell will divide into two in time t_1 and two cells will divide into four in time t_2. It also means that if there is one ton of cells at any given time, there will be two tons of cells in a time period equal to t_1 if the conditions for reproduction are suitable. During the logarithmic phase of growth, the times t_1, t_2, etc. are all equal and represent the maximal rate of growth of the organism—the shortest possible generation time.

This illustration indicates the potential for growth of the bacteria and other microorganisms. In nature, growth normally is limited because of lack of nutrients in the environment, but the growth potential can be measured in laboratory studies. Experimental measurements are close to these

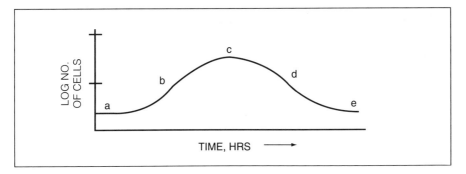

Figure 3.9

Typical (theoretical) growth curve of bacteria.

theoretical calculations, but growth at these rates can be maintained for only brief periods.

Because the growth of a bacterial culture can be measured by counting the number of generations, it is obvious from Figure 3.10 that the total number of cells can be calculated from an equation that describes the relationship between the number of generations and the total number of cells. For example, if the culture starts with one cell, there will be eight cells after three generations. But if the culture starts with two cells and it is allowed to grow for three generations, the total number of cells will be 16, or eight new cells from each original cell. It is obvious, then, that if the number of cells at the beginning (N_0) and the number of generations are known, the number of cells at the end (N_t) can be calculated.

The relationship among these variables can be written as an equation that may be used to find one parameter when the other two are known:

$$N_t = N_0 \times 2^n \qquad \text{(equation \#1)}$$

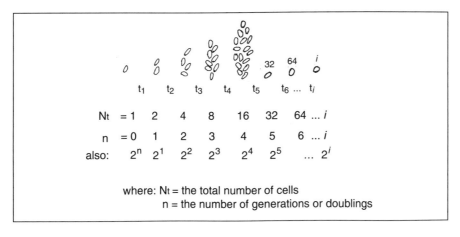

Figure 3.10

Relationships that form mathematical expression of kinetics of bacterial division.

How to Count 123,000,000,000 Bacteria

The growth equation can be used to calculate N_t in the relationship shown above. It is not the most convenient method of estimating bacterial growth, however, when the number of generations becomes large. To use equation #1, ordinary numbers have to be converted to logarithms. To better understand logarithms, we have to understand exponentials. Both of these are fun when approached with an open mind and a pure heart. The relationship between these two methods of counting is given in Table 3.1.

Equation #1 can be converted from the exponential form to the logarithmic form without changing its meaning. That is, it can be written in terms of logarithms without changing any of its properties or altering the relationships expressed in the material just discussed.

$$N_t = N_0 \times 2^n \qquad \text{(equation \#1)}$$

$$\log N_t = \log N_0 + n \times \log 2 \qquad \text{(equation \#2)}$$

The number of cells present in a culture after a given number of generations, n, which take a given amount of time, can be found by solving for $\log N_t$.

$$\log N_t = \log N_0 + n \times 0.301$$

The number of generations n can be determined if the number of cells at the beginning and at the end of the culture time are known.

$$n = \frac{(\log N_t - \log N_0)}{0.301}$$

Equation #2 can be employed in the following manner. If a culture is started with one cell and allowed to grow for seven generations, how many cells will there be at the end of this time?

Using equation #1:

$$
\begin{aligned}
N_t &= N_0 \times 2^n \\
&= 1 \times 2^7 \\
&= 1 \times 128 \\
&= 128
\end{aligned}
$$

Using equation #2:

$$
\begin{aligned}
\log N_t &= \log N_0 + n \times \log 2 \\
&= \log 1 + 7 \times 0.301 \\
&= 0 + 2.107 \\
N_t &= 128
\end{aligned}
$$

If a culture was started with 10 cells and there were 10,000 cells at the end of the culture time, how many generations took place?

Table 3.1 Numbers Expressed by Different Notations of the Logarithm

Number	Multiply 1×10	Exponential	Logarithm of the Number
1	0 times	1×10^0	0
2	0.301	$1 \times 10^{.301}$	0.301
3	0.477	$1 \times 10^{.477}$	0.477
4	0.602	$1 \times 10^{.602}$	0.602
5	0.699	$1 \times 10^{.699}$	0.699
6	0.778	$1 \times 10^{.778}$	0.778
7	0.845	$1 \times 10^{.845}$	0.845
8	0.903	$1 \times 10^{.903}$	0.903
9	0.954	$1 \times 10^{.954}$	0.954
10	1×10	1×10^1	1.000
100	$1 \times 10 \times 10$	1×10^2	2.000
1,000	$1 \times 10 \times 10 \times 10$	1×10^3	3.000
10,000	$1 \times 10 \times 10 \times 10 \times 10$	1×10^4	4.000
100,000	$1 \times 10 \times 10 \times 10 \times 10 \times 10$	1×10^5	5.000
1,000,000	1×10 six times	1×10^6	6.000
10,000,000	1×10 seven times	1×10^7	7.000
100,000,000	1×10 eight times	1×10^8	8.000
1,000,000,000	1×10 nine times	1×10^9	9.000

Using equation #1:

$$
\begin{aligned}
N_t &= N_0 \times 2^n \\
10{,}000 &= 10 \times 2^n \\
9{,}990 &= 2^n
\end{aligned}
$$

the value of n can be found easily by converting the numbers to the form of equation #2.

Using equation #2:

$$
\begin{aligned}
\log N_t &= \log N_0 + n \times \log 2 \\
\log 10{,}000 &= \log 10 + n \times \log 2 \\
4 &= 1 + n \times 0.301 \\
3 \div 0.301 &= n \\
n &= 9.967 \text{ generations}
\end{aligned}
$$

Generations are expressed in units of time. Therefore, if one generation takes 30 minutes, then 9.967 generations will take (30×9.967) 299.01 minutes, or a total culture time of 4.98 hours.

Now assume that a culture of the same organism under the same conditions was started with 460 cells and that it was allowed to grow for 28 generations. What will be the total number of cells at the end of the culture period? The answer can be obtained in various ways, including a graphical solution, but the easiest method is one that employs equation #2:

$$
\begin{aligned}
\log N_t &= \log N_0 + n \times \log 2 \\
&= \log 460 + 28 \times \log 2 \\
&= 2.663 + 28 \times 0.301 \\
&= 11.091
\end{aligned}
$$

and: $N_t = 1.23 \times 10^{11}$ cells

or: $= 123{,}000{,}000{,}000$ cells

The relationship between the number of generations and total growth time is obvious, as the time for each generation is the same during logarithmic growth (Figure 3.8). In the problem above there were 28 generations each of 30 minutes. This means that the total growth time was:

$$
\begin{aligned}
28 \times 30 &= 840 \text{ minutes} \\
840 \div 60 &= 14 \text{ hours}
\end{aligned}
$$

The definition of generation time (G.T.) is:

G. T. = time required for one doubling or cell division

= total growth time (T) ÷ number of generations (n)

$$
\text{G.T.} = \frac{T}{n} \qquad \text{(equation \#3)}
$$

This can be rewritten as:

$$
n = \frac{T}{\text{G.T.}} \qquad \text{(equation \#3a)}
$$

And equation #2:

$$
\log N_t = \log N_0 + n \times 0.301 \qquad \text{(equation \#2)}
$$

Can be rearranged as follows:

$$
\log N_t - \log N_0 = n \times 0.301
$$

$$
n = \frac{(\log N_t - \log N_0)}{0.301} \qquad \text{(equation \#2a)}
$$

As equations (2a) and (3a) show they are both equal to n, they can be combined:

$$
\frac{T}{\text{G.T.}} = \frac{\log N_t - \log N_0}{0.301} \qquad \text{G.T.} = \frac{(T \times 0.301)}{(\log N_t - \log N_0)} \qquad \text{(equation \#4)}
$$

Equation #4 is the growth equation in terms of culture time rather than in terms of the number of generations. In practice, measuring the generations of more than one or two cells at a time is impossible, whereas measuring time is as simple as reading the clock.

Equation #4 applies only during the logarithmic phase of growth and has no meaning during the other phases of the culture. This equation is useful in the laboratory and in the production plant. Its usefulness can be illustrated by the following examples. Suppose there are 2,000 (2.0×10^3) cells in a culture at 8:00 A.M. and 14,000,000 (1.4×10^7) at 1:00 P.M. What is the generation time of the organism?

Using equation #4:

$$\text{G.T.} = \frac{5 \text{ hours} \times \log 2}{\log 1.4 \times 10^7 - \log 2 \times 10^3}$$

$$= \frac{1.505}{7.146 - 3.301}$$

$$= 0.391 \text{ hours} = 23.5 \text{ minutes}$$

The number of generations that took place in the culture described in the problem can be derived using equation #3.

$$n = \frac{T}{\text{G.T.}}$$

$$= \frac{5 \text{ hours}}{0.391 \text{ hours per generation}}$$

$$= 12.79 \text{ generations}$$

A culture of several different bacteria has an average effective doubling time of one hour. The volume of this culture is 25,000 liters, and it is maintained at a constant population level of 1×10^9 cells per milliliter of liquid in the tank by continually adding fresh medium and removing overflow cells. Bacterial cells weigh an average of 1.5×10^{-12} grams each; 5% of each cell is edible, and the rest is water and nonfood components. How much food per day will this culture yield?

Solution:

25,000 liters $\times$ 1,000 = 25,000,000 milliliters   or: 2.5×10^7 milliliters

$$
\begin{aligned}
\text{and:}\quad 2.5 \times 10^7 \text{ ml} \times 1 \times 10^9 \text{ cells} &= 2.5 \times 10^{16} \text{ cells} \\
2.5 \times 10^{16} \text{ cells} \times 1.5 \times 10^{-12} \text{ grams} &= 3.75 \times 10^4 \text{ grams} \\
3.75 \times 10^4 \times 0.05 &= 1.875 \times 10^3 \text{ grams} \\
1.875 \times 10^3 \times 24 \text{ hours} &= 4.5 \times 10^4 \text{ grams} \\
4.5 \times 10^4 \text{ grams} &= 45 \text{ kilograms of food per day}
\end{aligned}
$$

Bacteria in Food

The rapid growth of bacteria described in the foregoing pages is not limited to these organisms. All microorganisms grow rapidly under the proper conditions, and their growth can present huge problems in many parts of the food and beverage industry. These problems often are seen from the production line to the time the food is consumed. In recent times refrigeration, freezing, and canning have largely eliminated this problem and, in essence, reduced it to a nuisance.

In today's view, the strategy for protecting food from the growth of bacteria rests on two major principles. The first depends on sterilization of food and subsequent protection from recontamination. This is best accomplished by placing food in metal cans, glass jars, or plastic containers and sterilizing it by heat or ionizing radiation. Once the containers are sealed, microorganisms will not reach the contents until the container is opened again. Food will remain safe for consumption many years when so treated, but the heat needed to kill all the organisms present brings about many undesirable changes in appearance, taste, aroma, and texture.

The other method relies on preventing bacteria on food from growing. This normally is accomplished by adding chemicals such as the fermentation acids, which kill microorganisms or inhibit their growth. The temperature can be made so cold or so hot that it will prevent growth. Raising the **osmotic pressure** will result in killing or inhibiting bacteria and other microorganisms. Raising the osmotic pressure is, in effect, the same as decreasing the **water activity** below the level required for the growth or survival of microorganisms and results in the protection of foods from microbial attack.

Even though the optimal temperature, pH, nutrition, and other conditions for growth provide the essential requirements for the highest growth rate, substantial growth can occur under other, nonoptimal, conditions. Some organisms lack well defined limits and are highly tolerant of even large variations in growth conditions. This tolerance allows microorganisms to grow under a wide diversity of situations. Some species of the genus *Pseudomonas* can grow at temperatures as high as 55°C and as low as 0°C and in media as rich as animal blood and as poor as distilled water. Of course, growth in a rich blood medium at a temperature of 35°C would be much more rapid than growth in distilled water at 5°C.

Growth of Pathogenic Organisms

Food contaminated with a few pathogenic mesophilic bacteria will not be harmful when preserved by storage in the refrigerator but would be if left at room temperature on a warm summer day. It also would be harmful if left in the refrigerator for a long time, because some of the pathogenic bacteria grow

Figure 3.11

Mature fungal spores such as these will be dispersed by even the slightest air currents. Scanning electron micrograph courtesy of Munhyeong Choi.

in the refrigerator, albeit slowly. In addition, bacteria that produce toxins in food may grow enough to make the food dangerous and possibly lethal.

DISPERSAL

Microorganisms grow where sufficient moisture is present to solubilize nutrients and hydrate cell structures. They either remain at the place where growth occurs or get carried into the air and dispersed. Microorganisms enter the atmosphere as single cells, or as parts of small soil or dust aggregates, or in minute droplets of water or other fluids. Water containing microorganisms may be **aerosolized** by many different actions and forces as it comes in contact with the air. The liquid part of the aerosol might evaporate, leaving one or more microbial cells suspended in the air. Although the vast majority of these organisms die on exposure to air and sunlight, many still survive and remain alive for many days, weeks, and even months.

On the other hand, fungi and bacteria form specialized structures, the spores, (Fig. 3.11) that guarantee their survival even under the most hostile of conditions. Microorganisms that grow in the soil, in water, or in the bodies of animals are introduced into the air in large numbers. These and many other mechanisms are at play constantly and supply the air with vast populations of every kind of microorganism that lives on Earth.

Although many cells die as a result of desiccation, exposure to ultraviolet radiation from the sun, and exposure to atmospheric oxygen, many of them survive. Air is a hostile environment in that it rarely is saturated with water and is highly effective in removing moisture from microbial cells. Air with 50% relative humidity can take large amounts of water from the liquid state and transform it to the vapor phase. In addition, air contains a large amount of oxygen that is toxic to many microbial cells.

Air contains 20% oxygen—that is, 200,000 parts of oxygen per 1,000,000 parts of air—whereas water contains only 8 parts of oxygen per 1,000,000 parts of water. Organisms that develop and grow in water suffer a drastic change when they go from the water environment to the air, and most do not survive the transition. Further, the amount of ultraviolet radiation in sunlight is sufficient to cause lethal **mutations** or even death in small, single-cell organisms. Nonetheless, large numbers of microorganisms survive this hostile environment and reside in the air for long periods.

MICROBIAL LOAD OF THE AIR

Soil particles and other debris, including microbial cells, are stratified in the air (Fig. 3.12) according to size and weight. The larger particles carrying large numbers of organisms will be closer to the ground and tend to settle more rapidly than the smaller ones. Single cells, being the smallest, will remain aloft longer and will rise to great heights. Fungal spores have been found at altitudes as high as 200 kilometers above the Earth's surface. Microorganisms, however, are extremely rare at altitudes in excess of 100 kilometers above the surface of the Earth, and the spores found at 200 kilometers probably represent the limit of the ability of microorganisms to survive at altitudes that approach outer space.

The forces that affect the rate of flow of microorganisms vertically or horizontally and the rate at which microorganisms in the air settle again are extremely complex and in most cases not yet understood. They include particle size, shape, and density; air flow; ground surface characteristics; humidity; solar heat; evaporation of water; transpiration from plants; and probably other, as yet unrecognized, factors. Particles that settle on the ground or on other exposed surfaces are easily picked up again by even slight air movements such as those caused by the movement of animals or human beings.

Much research has been done in this area of study, but the results are not yet complete. It is not known whether the movement of microorganisms in the air follows given patterns or "routes" or if it is an arbitrary, directionless movement. It is well established, however, that air movements introduce large numbers of organisms into the air and that the numbers are larger near the ground than at any other level.

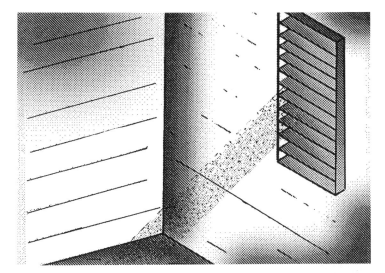

Figure 3.12

The microbial load of the air can be visualized when a strong beam of light enters a dark room.

Pasteur proved that the number of organisms was greater in the streets of Paris than at high altitudes in the mountains. Subsequently, other workers showed that viable fungal spores could travel 1,800 kilometers from the point of release. The number of microbial cells in the air at any time is large, but it fluctuates as conditions change during the course of the day. The major fluctuations in number of organisms suspended in the air are caused by changes in temperature, humidity, rainfall, and variations of the seasons. The load of microorganisms in the air is the result of infinite movement and mixing so that the organisms found at any given site represent more a history of the microbiology of the air at that point than any one occurrence associated with the introduction of organisms into a given sample of air.

Because the volume of the atmosphere is very large, the number of microorganisms present therein also is quite large. Consider a city of 50,000 people, and assume that it lies in a tract of land 20 kilometers long by 20 kilometers wide. Now assume that microorganisms in the air above this town are present to an altitude of 50 kilometers and that the average number of organisms in the air is 50,000 per cubic meter (this is very clean, clear air). The total number can be found as follows:

$$
\begin{aligned}
20 \times 20 \times 50 &= 20{,}000 \text{ cubic kilometers} \\
20{,}000 \times 1{,}000 \times 1{,}000 \times 1{,}000 &= 2 \times 10^{13} \text{ cubic meters} \\
2 \times 10^{13} \times 50{,}000 &= 1 \times 10^{18} \text{ viable organisms} \\
&\quad \text{or } 2 \times 10^{13} \text{ organisms per human being}
\end{aligned}
$$

No one, or even many, of the media used to count microorganisms in the air will allow every viable organism to grow. Therefore, it must be appreciated that the number of organisms such a test detects is much smaller than the actual number in the air.

Because of their very small size, there is no visual evidence whatsoever of the presence of microorganisms in the air. To see them, special microscopes and special techniques are required. Even so, the effect they have on everything on Earth is of major significance. Everything, everywhere is exposed to these organisms as they settle out of the air. Anything or any surface that can be free of microorganisms at any time (sterile) will be covered with them in the next instant. For this reason alone, surgery generally was unsuccessful prior to Pasteur's work, and not until Joseph Lister was able to provide a germ-free theater for surgery did successful operations become a possibility. In the same way, food that was sterilized by cooking at a high temperature for an extended time will become contaminated as soon as its temperature decreases and organisms from the air fall into it anew.

TRANSPORT

The air serves as a transport medium for organisms suspended therein. Imagine the cloud of microorganisms that exists above the hypothetical town described previously. A wind blowing across the land will do two important things: It will carry the organisms from a point of origin to every part of the globe, and it will mix these organisms with those from other places. This movement and mixing of microorganisms is a never-ending process, and the result is that the entire planet represents a homogeneous system in terms of its atmospheric microbial populations.

SURVIVAL

The soil is the reservoir of all microorganisms in nature. Microorganisms enter the soil from the air, the water, decaying vegetation, animal matter, and an untold number of other sources where growth has taken place. However they enter the soil, many cells dry on the surface of soil particles and adhere there as if they were fixed with glue. Large numbers of these will die as a result of desiccation, but a larger number will survive and become part of the soil. Those that survive enter a resting or dormant stage and remain alive for many years. The populations of microorganisms in the soil mix thoroughly by moving about constantly with the coursing of water and air over or through the soil. This is a dynamic system that never comes to rest. The soil also serves as the source of all saprophytic microorganisms in nature.

Many bacteria and fungi, as well as some algae, form spores, whereas other bacteria and protozoa form **cysts**. Both of these structures are survival forms and withstand exposure to high temperatures, high concentrations of

Figure 3.13

Sequence of mineralization reactions by which all matter on Earth is recycled by microbial action.

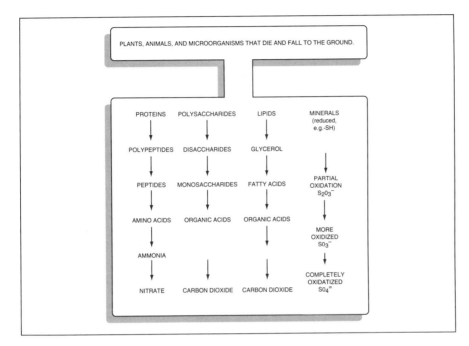

acids and other chemicals, and the effects of radiation. Bacterial spores are the most resistant of all living things. The spores of *Bacillus stearothermophilus* can be placed in boiling water for several hours without affecting their viability. Laboratory studies have shown that the spores of some bacteria can survive desiccation for more than 100 years, and some microbiologists think bacterial spores may survive for thousands of years. Fungal spores are more resistant than the vegetative structures of fungi, but these spores are not as resistant as bacterial spores are. In the same manner, cysts are more resistant than the vegetative cells of bacteria and protozoa, but they are not as resistant to environmental stresses as are fungal or bacterial spores.

Not all organisms that survive for prolonged times in the soil do so by forming spores or cysts. Many survive simply by becoming inactive. These are said to be resting cells, but the mechanisms or structures that permit them to remain alive are not understood yet. Many studies have shown that if soils are stored in the laboratory in sealed containers, the number of living organisms decreases with the passage of time. The rate of decrease is slow, so that after 25 to 30 years many organisms that do not form spores or cysts still will be alive. After this time, non-encysting and non-sporulating organisms will disappear, but the number of spore-formers will remain constant for many more years. Because of the ability of soil organisms to survive for many years, the soil may represent a living "index" of all the organisms that have inhabited the Earth in the recent past.

TOTAL MICROBIAL FLORA OF THE SOIL

The total number of microorganisms in the soil can be only estimated, as no method exists by which all the organisms living there at any one time can be counted. The size of the individual populations depends on many factors, but the most important is soil richness. Rich soils may contain many more than one billion live bacterial cells per gram and many thousands of fungi, algae, and protozoa as well. In general, the total number of organisms in the soil, viable and nonviable, cannot be determined by the methods now available for counting microbial populations. Because of the continual mixing of populations, many experts say that each gram of soil may contain many viable cells of each kind of organism that exist on Earth.

The number and variety of viable organisms reflect something of the history of the soil. Soils exposed to heat such as that of volcanos or to lightning discharges have relatively low populations, and those moistened frequently with rich nutrients, such as sugar mill washings, have extremely high populations. Soil constituents play a strong role in selecting the populations of microorganisms. Soils high in salt content have large populations of halophilic microorganisms. Acid soils contain many acidophilic ones. The root zones of plants, the **rhizosphere**, harbor populations of microorganisms that live from the substances the roots secrete, and organisms that compete with the same plant for specific nutrients may be absent because they cannot compete effectively.

The organisms that populate the soil are inactive only when the soil is dry. When the soil is moist, there is a great surge of metabolic activity, which results in growth and eventually reproduction of all the viable cells living therein. All substances in the soil that may serve as nutrients for microorganisms are taken up at this time and used. These include all soluble substances from dead vegetal and animal matter, as well as the cells of other microorganisms.

Microorganisms utilize all chemical substances that contain energy in one form or another. As energy and chemical constituents are removed from these substances, waste products are released back into the soil (Figures 3.13 and 3.14). The result of this cycling of matter in nature is that populations increase and complex materials are converted into simpler chemical forms. In this complex and dynamic system, populations of a given kind of microorganism wax while those of others wane in a series of unending cycles.

This is the infinite regenerative process of the Earth, which allows life to proceed one eon after another without interruption and without running out of material from which new living forms can be made. The role of the microorganisms as ultimate converters of all matter from the complex to the simpler form cannot be overestimated. In brief, life is possible on the Earth's surface only because of this regeneration and recycling of matter by the bacteria and other microorganisms that dwell in the soil and in the water. All the chemical substances that human beings call food are part of this system.

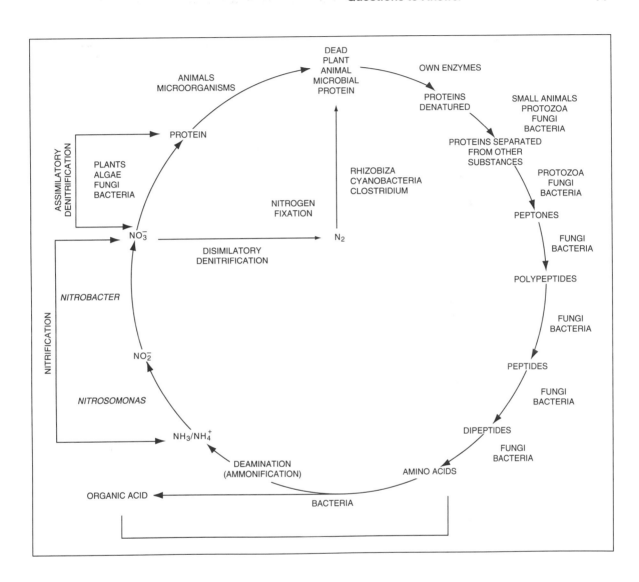

QUESTIONS TO ANSWER

1. What forms of the bacterial cell are adapted specially for survival in dry soil?
2. What is meant by the term "bacterial growth?" How does growth in bacteria differ from growth in metazoans?
3. What are the different phases of the bacterial growth curve?
4. How was the equation $N_t = N_0 \times 2^n$ derived?
5. How was the equation $\log N_t = \log N_0 + n \times \log 2$ derived?

Figure 3.14

Involvement of microorganisms in the conversion of matter; the nitrogen cycle.

6. How was this equation derived?

$$n = \frac{\log N_t - \log N_0}{\log 2}$$

7. How was this equation derived?

$$\text{G.T.} = \frac{\log 2 \times T}{\log N_t - \log N_0}$$

8. What is an exponential?
9. What is a logarithm?
10. Where and how do microorganisms grow?
11. Why does the air contain microorganisms?
12. Why do microorganisms live in the soil?
13. What is the difference between viable cells and total cells?
14. How long do bacterial cells last in the soil?
15. Why do microorganisms degrade complex substances in the soil?
16. Why are microorganisms called recycling agents?
17. What kills many microorganisms in the air?
18. Do all soils contain microorganisms?
19. Does the soil contain viruses?
20. Why does the Earth not run out of food for human beings?

Anderson, D. A. 1973. INTRODUCTION TO MICROBIOLOGY. C. V. Mosby, Saint Louis.

Ehrlich, H. L. 1981. GEOMICROBIOLOGY. Marcel Dekker, New York

Jensen, V., A. Kjrensen. eds. 1986. MICROBIAL COMMUNITIES IN THE SOIL. Elsevier Applied Science Publishers, New York.

Krumbein, W. E. 1983. MICROBIAL GEOCHEMISTRY. Blackwell Scientific Publications, London.

Leadbetter, E. R. and J. S. Pointdexter. 1985. BACTERIA IN NATURE. Plenum Press, New York.

Smith, O. L. 1982. SOIL MICROBIOLOGY: A MODEL OF DECOMPOSITION AND NUTRIENT CYCLING. CRC Press, Boca Raton, Florida.

Stanier, R. Y., E. A. Adelberg, and J. L. Ingraham. 1976. THE MICROBIAL WORLD. Prentice-Hall, Englewood Cliffs, New Jersey.

Tate, R. L. 1986. MICROBIAL AUTECOLOGY. John Wiley & Sons, New York.

Control of Microorganisms

4

If the red slayer think he slays,
Or if the slain think he is slain,
They know not well the subtle ways
I keep, and pass, and turn again.

Emerson

Only during the last hundred years have human beings finally acquired enough knowledge to control the microorganisms effectively. The foundation on which this ability rests was laid in the laboratories of Pasteur, Koch, and the many other 19th-century scientists who first began to understand the nature of microorganisms and develop the techniques by which they could be manipulated. These pioneer bacteriologists represent the highest level at which science can operate, for they not only sought solutions to the scientific problems of their day but also sought to understand the fundamental phenomena that caused those problems.

Because of the intimate relationship between human beings and microorganisms, methods of dealing with these invisible creatures were devised long before their existence was established. In the early days of human existence and even until the end of the 19th century, these methods generally were based on practical knowledge and were designed to solve specific problems under a given set of circumstances. This practical approach to problems caused by microorganisms is still practiced with a great measure of success. The successes, however, tend to be limited to unique situations, and rarely do they provide knowledge applicable to other, even closely related, problems. Modern techniques for dealing with microorganisms were devised long before

Figure 4.1

Depiction of direct incineration of contaminated materials and the contagion they carried in biblical times.

the microorganisms were discovered. Many practices that were developed before knowledge of the microorganisms existed were shown later to coincide perfectly with those based on knowledge of their nature.

EARLY PRACTICES FOR DEALING WITH MICROORGANISMS

Fire

The use of fire to purify objects, although mythical or superstitious in its genesis, was an effective means of controlling the populations of microorganisms in many different situations. Applications of direct incineration are numerous in the daily life of early societies. Biblical dicta regarding the proper method for disposing of the clothing and bedding of lepers was as effective as any method known today, and in many cases incineration is still the method of choice for disposing of items such as these. Figure 4.1 depicts incineration of contaminated material in biblical times. Another example of the pragmatic approach to a problem without understanding the principles involved is the use of cauterizing heat to prevent infection.

Freezing

Probably the first method of controlling the populations of microorganisms on food was to allow it to freeze out of doors. Food obtained in quantities larger than needed at a particular time did not spoil in cold weather as it did during the warmer seasons. It is immediately evident that frozen foods can be thawed and prepared for eating at a later time without loss of quality or flavor. The cache of summer food that could be frozen successfully usually was sufficient to make life possible during the winter, when crops did not grow and when game was difficult or impossible to find. In extreme cold this method works so well that animals killed during the winter may be allowed to freeze without dressing.

Because thawing proceeds from the exterior inward, on thawing the flesh may be removed while the viscera are still frozen, thereby avoiding contamination of the meat. Food placed in the ground in the tundra may be kept frozen for a year or more.

Drying

Other practices undoubtedly developed from simple observations. For example, naturally dry foods such as nuts, cereals, and grains stay fresh for long periods of time, and moist ones such as tomatoes, grapes, and the flesh of animals spoil very rapidly. Attempts at drying moist foods by exposure to the sun, low fire, or smoke (see Figure 4.2) met with rich success and opened the possibility of storing large amounts of food even in tropical lands. The art of drying foods has been carried as far as the imagination has allowed. If dried properly, fruits, meats, berries, and many other kinds of foods can be preserved for long periods.

Figure 4.2

In many communities fish were preserved by drying in the sun or over low fire.

Preserving

Preserving is a practice closely related to drying, as it involves driving off water by heating or boiling. The techniques and methods for saving fruits, vegetables, and meats by preparing them as preserves were all worked out to a high state of technology long before knowledge of microorganisms was acquired. In principle, removing water by evaporation results in greater concentrations of sugar and other solutes to the point of causing a significant increase in osmotic pressure (see later discussion).

In early food technology development, it was a short step from the techniques of preservation by removing water to the method of adding sugar without the need for boiling. The effect is the same: A high concentration of sugar can be attained by adding the purified material.

The addition of sugar to preserves undoubtedly was used first to improve taste and texture, but the resulting preparation also had excellent keeping qualities. In addition, it soon was found that sugar added to any kind of food, including vegetables and meats, improved taste and quality. The result was the origin of families of foods now called conserves, preserves, jams, jellies, marmalades, and candies.

Street vendors in modern Mexico prepare and sell various fruits and vegetables in much the same way as the original inhabitants of that land prepared them thousands of years before. Candy sellers explain that they obtain fruit and vegetables from various sources and prepare them by methods learned from their parents, who in turn had learned them from their parents. The fruits and vegetables are washed well, cut into small pieces, and placed in an uncovered iron "caso" or cooking pot with a large amount of sugar and water. Spices may be added as desired, and the mixture brought to boil while being stirred constantly. When all the water has evaporated, the fruit is stirred rapidly to make sure the melted sugar covers each piece of fruit evenly. The low heat ensures that the hot sugar preparation permeates the interior spaces of the fruit or vegetable. Foods prepared in this manner remain edible for many months or even years.

Some foods remain edible for long periods in their natural condition because they contain antimicrobial properties or structural features that make them impermeable to microbial cells. Examples of the former are oranges and onions, and of the latter, cereal grains and nuts. Figure 4.3 illustrates this point.

Salting

The addition of salt to meat and certain vegetables to preserve them is also a discovery lost in the first mists of our time on Earth. The use of brines to preserve pork, cucumbers, and olives still is done today and, to the credit of

Figure 4.3

Garlic can be preserved for long periods in its natural condition because of its antimicrobial properties.

the ability of early human beings to observe and understand, a method that has not required improvement. When natural freezing was not possible, salting was the primary means to preserve meats. In the prerefrigeration era, in all parts of the world except the Northernmost lands, salt was the universal preservative.

Salting typically was accomplished by rubbing powdered salt on meat and introducing it into all the crevices and spaces of the meat. A layer of salt was placed in the bottom of a container such as a wooden barrel, and then alternate layers of meat and salt were added until the container was full. The sealed container could be transported or stored for extended periods as long as water did not dilute the salt.

The useful life of meat stored in this manner was on the order of 10 to 12 months, but this depended on the conditions of storage. In hot, humid weather, storage may be limited to one or two months, but in cool, dry weather it could be a year or more. Often the duration of sea voyages was determined by the quality of the meat stored in salt. Meat and other foods cannot be stored indefinitely in salt or even in the strongest brines because **halophilic bacteria** and **fungi** eventually develop large populations and putrefaction sets in.

Meat stored in salt cannot be used as is when it is needed. It first must be cleaned to remove all adhering salt and then soaked in several changes of water to remove as much of the salt as possible. Many people find the residual salty taste objectionable, but under any circumstance this taste is not as objectionable as that of putrefying meat.

The preservative properties of sugar, salt, smoking, and drying are all based on the same principle. This principle rests on the increase in osmotic pressure, a phenomenon of nature that will be explained in a subsequent section of this chapter.

Spices

Human beings pioneered the widespread use of spices as preservatives only during the last thousand years. Although the preservative properties of some spices are substantial, many spices first were used to mask the taste and odor of foods that no longer were in the prime of freshness. Spices that were capable of controlling microbial growth and protecting foods from spoilage often were thought to have supernatural qualities. In some early societies this mystic view often took on the same image in food preservation, as exemplified by the mandrake root in female fertility.

Many spices have **antibiotic properties** and inhibit the growth of bacteria and fungi alike. Clove, black pepper, dill, mace, and oregano were used to mask offensive odors and tastes long before they were employed as preservative agents and before they were desirable because of their subtle and enchanting flavors and aromas. Today most spices are used to enhance the taste of food or to create new, and often fascinating, variations in taste and bouquet.

Certain traditional foods, such as the ham preferred by Spaniards, *jamon serrano* (see Figure 4.4), still are preserved by spices. The Spanish divide all ham into two classes: the ham they eat, which they call ham, and the ham that is eaten everywhere else on Earth, which they call English ham because it is prepared by cooking in the English manner.

Figure 4.4

Air curing of ham near Zaragoza, Spain.

Fermentation

One of the most widely used methods for the preservation of fruits and cereals is fermentation of the sugar in these foods with the consequent accumulation of acid or alcohol to a level high enough to prevent further microbial growth. The fermentation is brought about by microorganisms easily selected by simple manipulation of the conditions under which the food is stored. For example, adding 150 grams of salt to 5 liters of water will result in the selection of those bacteria from nature that convert cabbage into sauerkraut. Fermentation has two major benefits:

1. It preserves food.
2. It enhances flavor and palatability.

As the result of these two effects, the food item becomes a more desirable product. Grapes, apples, milk, and hundreds of other foods that will not last long unless they are frozen or dried can be preserved by fermentation. Human beings almost certainly practiced the art of fermentation at least 10,000 years ago. Early human beings mastered a highly developed fermentation technology at least 4,000 years before the discovery of microorganisms. Many experts say that the vintner's product has deteriorated since its foundation was transferred from that of an art to that of a science.

Control of microorganisms in fermented foods is accomplished by acids, alcohols, and other chemicals elaborated in the fermentation liquid. These include formic, acetic, propionic, succinic, lactic, and butyric acids. Various alcohols, aldehydes, esters, ketones, and ethers contribute to the final bouquet of the product. The aroma, taste, and consistency of fermented food products such as vinegar, olives, sauerkraut, buttermilk, and yogurt, plus a fantastic variety of cheeses and leavened breads, are the result of the kinds and amounts of fermentation products.

Fermentation of foods today forms a large part of the food and beverage industry, not only as a means of preservation but also, and perhaps primarily, as a means of providing a variety of foods that otherwise do not exist. This variety of products is possible only because the populations of microorganisms in fermentation vats are kept under constant and rigid control.

However human beings learned the methods of controlling microbial growth without knowing the nature of microorganisms, the discoveries they made were of a universal character. Archeological and historical evidence suggests strongly that all societies, regardless of extent of isolation, have devised methods for drying meat and vegetables, preparing alcoholic beverages, and preserving many foods by acid fermentation.

The methods and products derived from these fermentations are common and well known to all human beings. To a certain extent, fermentation products form the background for the modern food and beverage industry. To understand better the science that undergirds this industry, two important thoughts have to be borne in mind. First, all the practices, methods, and products described previously were discovered without knowledge of the presence of microorganisms, the causative agents of food spoilage and the agents of fermentation. This is a tribute to human intelligence. Second, all the practices, methods, and products described were necessitated by the presence of the microorganisms and their ability to compete with humans for food and other valuable substances. Recognition of this fact indicates the importance of these very small but extremely powerful living things.

MODERN METHODS OF CONTROL

All the methods of controlling the growth of microorganisms described previously are still in practice today. Modern societies, however, are much more complex and require more sophisticated approaches to controlling microbial growth. The newer methods are based on understanding what the microorganisms are and how they function.

The impetus for developing new techniques for removing or eliminating microorganisms is still based on the premise that inspired early humans—the protection of valuable items against attack from microorganisms. A vast industry worth countless billions of dollars is dedicated solely to this purpose. This complex and pervasive industry includes, but is not limited to, all forms of refrigeration, canning, and packaging of foods, freezing, fermentation, and all other means of preserving foods and beverages.

Lest human beings become proud of their ability to control and exploit the microorganisms, it must be borne in mind that approximately a fourth of all foods produced for human consumption become inedible as a result of microbial contamination. One also must recognize that even slight disturbances, such as the failure of electric power, will result in immense losses of stored food products. Significant human effort is required simply to hold the microorganisms at bay and victory is still far off for either side.

PHYSICAL METHODS

Incineration

Burning of contaminated materials, as in the earlier biblical example, is still the method of choice in many modern practices. Sterilization of air and other materials by flame or electric coil to prevent the entry of organisms is a common process of modern technology. These methods require fairly simple equipment and can be adapted to many different situations.

In the simplest technique, an open flame is used to kill microorganisms on the surfaces of fireproof objects, much like in the laboratory, where the nichrome wire inoculating loop is heated in the Bunsen burner before and after bringing it into contact with bacterial cultures (see Figure 4.5). Direct

Figure 4.5

Sterilization of the nichrome inoculating loop by direct incineration in the laboratory.

incineration frequently is the method of choice for removing microorganisms from the air because it is cheap, clean, and effective.

Air that is sterilized by incineration may have to be cooled by allowing it to stand or by refrigerating it. In some operations the heated air may be used to maintain the temperature of the working environment or as a means to preheat materials before use. Often, all the air may not have to be sterilized, as sterilizing only part of it will keep the number of organisms at some acceptable level (Figure 4.6).

Air sterilized by direct incineration also will be free of all suspended particles that burn, but the air, in exchange, will contain the combustion products produced in the incinerator. Control of microbial populations is essential in food processing areas because introducing large numbers of organisms into food and beverages hastens spoilage. This is especially important when the products are not given a final heating treatment, as in the production of milk, cheese, ice cream, and other dairy products. The size of air populations is also of major concern in meat lockers, vegetable storage bins, and display refrigerators, because it determines the shelf life of products stored therein.

Dry Heat

Dry heat is the method of choice for sterilizing all materials that are not damaged by heat and that will not evaporate when heated. This method is applied most often to the treatment of glass or metal objects such as the containers used in canning and bottling foods and beverages. Dry heat for sterilizing or for reducing microbial populations usually is accomplished in an oven almost identical to that of modern household stoves.

Sterility can be accomplished by heating objects to 160°C for 60 minutes. Timing does not begin until the item to be sterilized reaches the desired temperature. Often this cannot be ascertained by any means other than trial and error. The lowest temperature and time of exposure that results in sterility of the desired material is the combination of choice, as additional heating is of no value and may even damage the materials being sterilized.

Moist Heat

When microorganisms are heated in the presence of water or steam, they are killed at temperatures much lower than those needed in dry air. Water, however, cannot be heated to temperatures higher than 100°C because water evaporates at this temperature and the steam carries the excess heat away with it. Many bacteria form spores that are not killed by exposure to temperatures of 100°C even in the presence of water. These spores can be killed only by a 60-minute exposure to dry heat at 160°C or exposure to moist heat at 121°C for 15 minutes. The latter is attained in an **autoclave**, a device in which steam

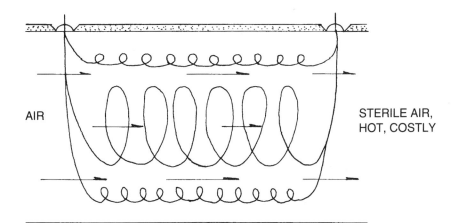

AIR

STERILE AIR, HOT, COSTLY

can be trapped and heated to high temperatures. As heat is applied to steam, the pressure inside the autoclave builds up and the temperature increases. Figure 4.7 is a schematic diagram of a modern autoclave. The common home pressure cooker is an autoclave of fairly simple design.

When the pressure inside the autoclave is zero pounds of steam per square inch above standard atmospheric pressure (psi), the temperature of steam in the autoclave is 100°C at sea level. At a pressure of 5 psi, it is 109°C; at 10 psi it is 116°C; and at 15 psi it is 121°C. Steam at 121°C will kill all living things and the spores of all bacteria except those of some strains of *Bacillus stearothermophilus.*

Autoclaves used in industry for sterilizing canned or bottled foods are called **retorts**. These often are large enough to sterilize thousands of cans or jars at one time. Steam is the most common method of preserving food, for it represents a compromise among the various parameters of cost, effort, effectiveness, time, and undesirable changes in the product.

Pasteurization

Pasteurization is the name given to a method that Louis Pasteur devised to keep wines from going sour. In its essential form the method consists of gentle heating to destroy a specific population of microorganisms. Today this method is used universally for the treatment of all milk products because the organism that causes tuberculosis in human beings, *Mycobacterium tuberculosis,* can be killed without much change in the taste, appearance, or consistency of the product.

Pasteurization of milk is accomplished by heating to 60°C for 20 minutes. This is the traditional method and is called the **holder method** because

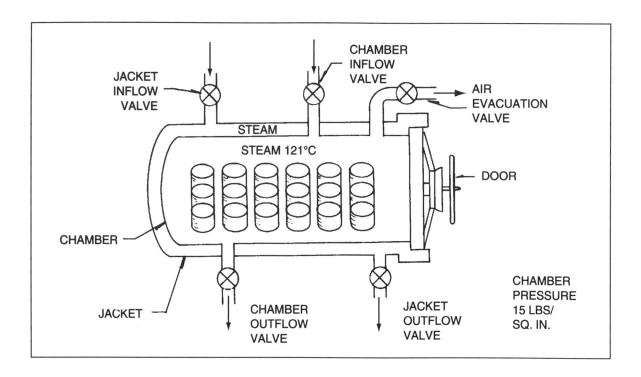

Figure 4.7

Schematic diagram of a modern autoclave.

pasteurized milk is held at 10°C after pasteurization. Today, the **flash method** is used typically. The product is heated to 72°C for 15 seconds, followed by rapid cooling to 10°C. Other dairy products, such as cheese, ice cream, yogurt, as well as vegetable and fruit juices, also are pasteurized to remove harmful bacteria and extend shelf life.

Osmotic Pressure

Drying, smoking, salting, and conserving are all methods of food preservation that depend on the increase in **osmotic pressure** to inhibit microbial growth. Osmotic pressure increases when water is removed from food and water-soluble chemicals in the food become concentrated. If a microbial cell falls into this environment, the water in the cell will be extracted as a result of the high osmotic pressure of the system. The effect is to inhibit growth or cause death of the cell.

This is called a **hypertonic system**. The opposite happens when a high concentration of chemical substances exists inside the cell, as is normally the case, and a low concentration exists outside of the cell. In this case water from the environment will enter the cell and, if sufficient, the cell will become engorged and rupture. It can be seen easily when raisins or other dehydrated

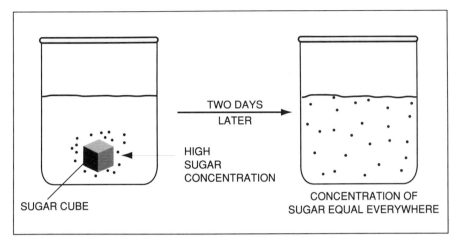

Figure 4.8

Schematic depiction of osmosis; dots indicate sugar molecules in solution.

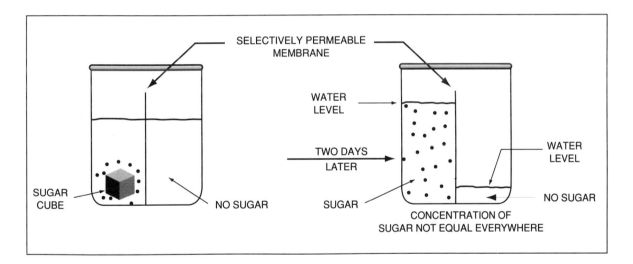

Figure 4.9

Schematic diagram representing effect of differentially permeable membrane; dots indicate sugar molecules in solution.

foods are placed in water, swell, and eventually fill with water until they burst. This is called a **hypotonic system**.

If the amount of dissolved chemicals is the same inside of the cell as outside of the cell, water does not leave or enter the cell. This is called an **isotonic system**.

These three phenomena—hypertonic, hypotonic, and isotonic—are explained in graphic form in Figures 4.8 to 4.10. The underlying natural phenomena are as follows:

1. A solute moves from an area of high concentration to one of low concentration until no difference in concentrations exists (see Figure 4.8). This is one of the fundamental laws of nature.

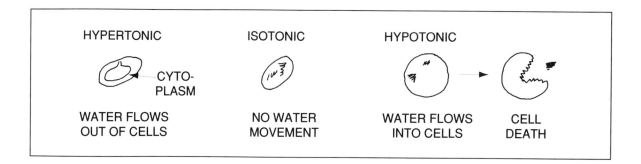

HYPERTONIC

CYTO-
PLASM

WATER FLOWS
OUT OF CELLS

ISOTONIC

NO WATER
MOVEMENT

HYPOTONIC

WATER FLOWS
INTO CELLS

CELL
DEATH

Figure 4.10

Water moving into and out of bacterial cells as result of changes in osmotic pressure of liquid in which they are suspended.

2. If a **differentially permeable** membrane separates two parts of a liquid such as water, and if one part contains a solute that does not pass through the membrane and the other contains water only, water will move to the side of the membrane that contains the solute. Water will continue to move across the membrane until the weight of the water is equal to the force that draws the water across the membrane (Figure 4.9). This force is called **osmotic pressure** and is of great importance in food preservation (Figure 4.10).

A phenomenon closely related to osmotic pressure, **water availability**, is indicated by the symbol a_w. In all aspects of microbial life, water is the solvent with which all nutrients, waste products, and cell parts are associated. Collectively, the state or extent of solution is called **solvation**. This term implies that a certain amount of water is required to dissolve a certain amount of solute and that the water, therefore, is not available for the solvation of other substances. The values of water availability are given as ratios, and their effect on cell growth is shown in Table 4.1. Figure 4.11 pictures osmophilic organisms, which grow well in conditions of high osmotic pressure and low water availability.

Filtration

Microorganisms, with the exception of viruses, range in size from 0.3 to 100 μm. They can be removed from fluids, including air, by passing the fluid

Table 4.1   Values of Water Availability Obtained with Glucose and Their Effect on Microbial Growth

a_w	Effect on Microbial Growth
0.99 to 0.90	Range for bacterial growth
0.99 to 0.85	Range for yeast growth
0.99 to 0.75	Range for mold growth
0.85 to 0.60	Range for osmophilic organisms

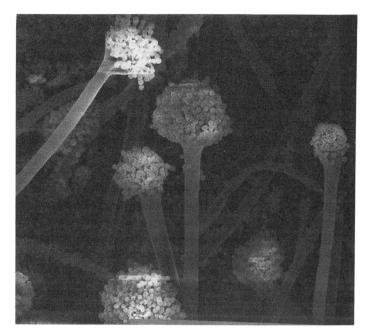

Figure 4.11

Osmophilic organisms grow well in environments with high osmotic pressure and low water availability.

through filters that have pores on the order of 0.22 μm in diameter. Liquids that must be sterilized but that will be destroyed by heating can be sterilized effectively by filtration. Medicines, drugs, cosmetic preparations, beer, and even gasoline have been protected successfully from microbial action by filtration.

The air in hospital operating rooms, nuclear submarines, medicine preparation areas, and food preparation areas is sterilized most conveniently by filtration. Because the temperature of the air treated by this means is not changed as it is in heat sterilization systems, the volume of air treated remains constant, making engineering considerations of air movements less complicated. Nevertheless, the cost of filters and the energy required to move the air through the filters make this method more expensive than incineration. Figure 4.12 depicts plastic membranes used to remove microorganisms from liquids and air.

Another method of air filtration depends not on filter membranes but, rather, on liquids, such as water, to trap the cells of microorganisms suspended in the air. Although other liquids often are employed, water is the most common, as it is inexpensive, easy to handle, and quite effective. Air cleaned by sparging water will emerge saturated with water but almost free of microorganisms and dust particles. The air can be dried easily, either by refrigeration or by chemical means.

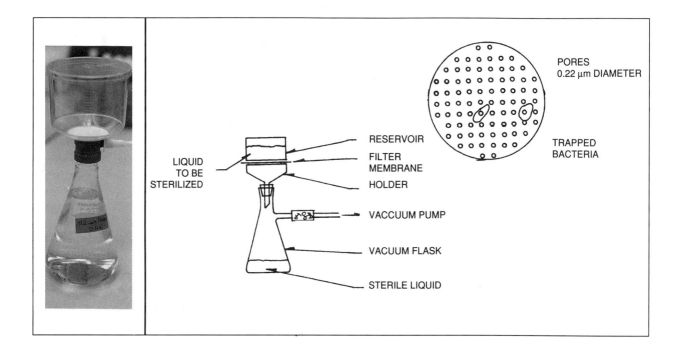

RESERVOIR

FILTER MEMBRANE

HOLDER

VACCUUM PUMP

VACUUM FLASK

STERILE LIQUID

LIQUID TO BE STERILIZED

PORES 0.22 µm DIAMETER

TRAPPED BACTERIA

Figure 4.12

Plastic membranes of known pore size are used to remove microorganisms from liquids and air.

Exclusion

Many modern products are sterile in their natural condition but must be maintained sterile lest they spoil. They may have properties such that they cannot be heated, filtered, or otherwise sterilized. The resulting dilemma can be approached only from the point of view of protecting the product from contamination. This protection can be accomplished only by excluding all microorganisms from the environment of the product.

Few examples of this problem can be given that would be of interest in the food and beverage industry other than that represented by the egg. Except when the hen is infected with *Salmonella enteritidis*, the yolk and white of the egg are sterile and will remain so as long as microorganisms are kept from reaching the interior. The shell and its protein covering maintain the substance of the egg in a sterile condition as long as neither is damaged but, once disturbed, microbial invasion occurs almost immediately.

Reference to blood transfusions and organ transplantations should suffice to establish the principle and need for exclusion of microorganisms. At present, sterilizing eggs, blood, or body tissues is impossible once they are contaminated. Protecting them from contamination is the only way of maintaining their usefulness.

Ionizing Radiation

After World War II, radioactive materials became readily available in large enough quantities to make high-intensity radiation fields possible. With this innovation, sterilization by radiation became an accepted practice. **Cold sterilization** is most desirable when heating causes detrimental changes in the materials treated. Because radiation, usually gamma rays, easily penetrates most commercial containers in which foods and other products are sold, the material to be treated can be sealed in cans, jars, bottles, or hermetic packages and then sterilized. All living things can be killed by radiation, and microorganisms are no exception. Although radiation offers a simple and effective way of protecting almost any kind of material from microbial contamination, it is an expensive process and one with major hazards. Also, excessive doses of radiation bring about some changes in taste and consistency, but these changes can be controlled by various means.

Non-Ionizing Radiation

Ultraviolet (UV) radiation is an important agent in the control of microbial populations. In the food industry it is used most often to reduce the number of microorganisms in the environment where food is prepared. This is accomplished by conveniently placing UV "lights" with maximal energy output at 2650 **Angstrom units (Å)** in locations where human beings will not be exposed but where microorganisms in the air will receive maximum exposure. A choice location is near the ceiling with adequate shading so all the radiation goes upward to the ceiling and none reaches the work stations of personnel in the area.

This arrangement is shown in simplified form in Figure 4.13. If air now is made to move across the UV tube, organisms so exposed will be killed with high efficiency. The microbial load of the air can be reduced significantly by this method.

In installations where table surfaces, floors, and working utensils have to be decontaminated, UV sources are directed downward toward those areas and activated by door switches designed to protect workers from UV radiation. These switches turn on the UV sources when workers leave the area and turn them off when workers enter. The objective is to have the UV radiation on when no one is in the room and to have it off when workers are present and to accomplish this with maximal safety for workers.

Part of the UV spectrum, the lower wavelengths, are weakly ionizing and cause ionization of both air and water in the air, resulting in the formation of small amounts of ozone. Ozone is extremely toxic for human beings, so great care is required when workers are present. Ozone also is a strong oxidizing agent (it reacts with most materials), which produces extensive damage to working equipment.

Figure 4.13

Schematic arrangement of UV radiation tubes in food preparation area.

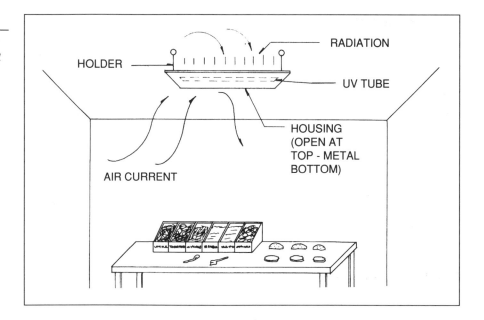

CHEMICAL METHODS

As described early in this chapter, human beings discovered long ago that certain chemical substances, such as alcohol and fermentation acids, are excellent preservatives for many kinds of food. Recently, many processes have been discovered by which foods and other substances can be protected from microbial action. These are based on specific knowledge of the characteristics of microorganisms.

The techniques of modern microbiology have been used to examine the scientific principles involved in many traditional methods of microbial control. In many cases, these principles have been amended to give better results in terms of efficiency, control of quality, and predictability. Some of the chemical methods by which populations of microorganisms can be controlled effectively with chemical substances are described next.

Acids

Many bacteria and fungi are able to live without air. These anaerobic microorganisms exist by a type of metabolism called fermentation. Anaerobes degrade carbohydrates and other large molecules and excrete complex molecules, such as acids and alcohols, into their environment. Other kinds of chemical substances also are produced, but they are not as critical to the preservation of foods and beverages as are acids and alcohols. This process is described in simplified chemical notation in Figure 4.14. In contrast to the anaerobes, aerobes degrade the same kinds of carbohydrates but excrete only carbon dioxide and water.

When foods are attacked by bacteria and fungi a large variety of fermentation products occurs naturally. If the concentration of acids and other fermentation products is large enough, the food may change significantly in taste and texture, but it also will be protected from further attack because fermentation acids prevent the growth of almost all other organisms. The most common, and most desirable, fermentation acids are acetic, butyric, and lactic, but many others also are produced. The acids and other products and the relative amounts present are determined by the kind and number of microorganisms involved, by the nutrients available, and by the conditions under which the fermentation takes place. In some situations one or two acids may predominate, but normally a mixture of several will be found. All fermentations are self-limiting in that the number and kind of microorganisms that survive in the fermentation liquid are determined by the acids produced therein.

Because microbial populations are controlled effectively by acids, it does not matter whether these acids are produced by fermentation processes, as just described, or added to the food as preservatives. The pickling of such foods as cucumbers, peppers, mixed vegetables, and meats such as pig's feet and sausages may be accomplished by adding acetic acid to the food item. Although the extent of protection is the same as if the acid had been produced by natural fermentation, the lack of flavor and bouquet is noticeable immediately.

Any acid will inhibit microbial growth and can be used to control the development of large populations of microorganisms, but most acids cannot be used in the food and beverage industry because of their toxicity or the detrimental effects they have on the taste and structure of foods. A few acids other than those produced by fermentation also are employed in preserving food and drink because of their ability to inhibit microbial growth. Among these are sorbic acid, benzoic acid and its derivatives (**methylparaben** and **propylparaben**), caprylic acid, and sulfur acids in the salt form. The latter include sulfur dioxide, sulfite, metabisulfite, and bisulfite.

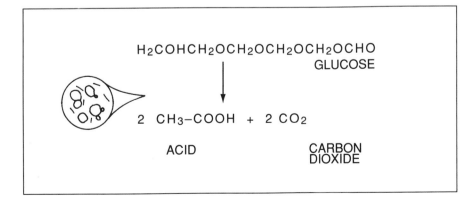

Figure 4.14

Glucose molecules are converted to acetic acid and carbon dioxide by anaerobic organisms.

Figure 4.15

A curve showing the effect of acidity on bacterial growth.

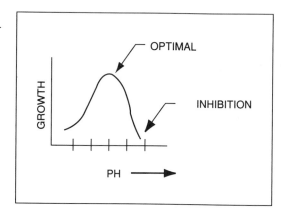

The effect of acids on microbial growth can be expressed in terms of the optimal pH required for the organisms present to grow (Figure 4.15). On the other hand, if the salt form is active in controlling growth, the effect must be attributable to other chemical properties of the substance in question.

Alcohols

As mentioned previously, another method of protecting foods and beverages from microbial activity requires the presence of alcohol. The source of the alcohol is of no consequence. It may be produced by fermentation of carbohydrates in the food substance itself, or it may be added in the chemically pure form.

Alcoholic fermentations form the basis for one of the major industries of modern technological societies. They are of such great importance that they often comprise a separate chapter in the literature of the microbiology of foods and beverages. Thus, the alcoholic fermentations are treated separately in a different section of this text.

Other Chemical Methods

Many chemical substances affect the growth and survival of microorganisms. The food and beverage industry utilizes some of these successfully as preservatives, but most are of no value in this regard because they are toxic, leave noxious tastes and odors, or in other ways affect the quality of food adversely. Among the substances that can be used to preserve food, those that enhance taste and food quality, such as acetic acid, are preferred over those that leave traces of undesirable odors and tastes, such as caprylic acid.

In modern industrialized societies like the United States, the use of preservatives is regulated by government agencies whose objective is to pro-

Table 4.2 Partial List of Approved Preservatives and Their Uses

Preservative	Concentration	Use
Antibiotics	0.0007%	Poultry, fish, cheese
Butyl hydroxyanisole	0.02%	Nuts, crackers, chips
Ethylene oxide	changing	Fruits, spices
Propionic acid	0.1%	Bread, bakery products
Ethyl formate	0.02%	Nuts, spices, dried fruits
Methyl paraben	vary	Fruit drinks, syrups, beer
Propyl paraben	vary	Dried fruits, salad dressings
Propylene oxide	changing	Dry spices, fruits, starch
Smoke (aldehydes, phenols, etc.)	saturation	Hams, beef, poultry
Sodium benzoate	0.1%	Soft drinks, relishes
Sodium diacetate	0.30%	Bread, pastries, fruit
Sorbic acid	various	Beverages, many foods
Sulfites	0.1%	Relishes, sauerkraut

tect public health and well-being. Manufacturers and purveyors must show proof that the preservatives and other additives used in the preparation of food are safe and that they accomplish the objective claimed. All the substances listed in Table 4.2 have been certified on both of these criteria, so they are assumed to be safe for general consumption.

CLASSIFICATION OF CHEMICALS

Chemical substances used in the food and beverage services are classified according to their use and to their effect on microorganisms. Some of these classifications, listed in Table 4.3, are incorporated in the laws prescribing certain practices in many states, counties, and communities in the United States.

OVERVIEW OF MICROBIAL CONTROL

In protecting valuable substances such as foods and beverages from microbial attack, whether the microorganisms are killed or simply rendered ineffective matters little. In terms of microbiological science, both results imply that the growth of microorganisms will be prevented and, consequently, that the product is safe from microbial attack. Many chemical substances pre-

Table 4.3   Classification of Various Chemicals Used in Food Services to Control Microbial Growth

Sanitizing agent	Substance used to reduce the number of microorganisms on implements, utensils, and work areas. Examples: soap, detergent, ammonia, chlorine, ethylene oxide, phenylphenol
Disinfectant	Substance used to eliminate all organisms capable of causing infection or disease. Examples: same as above
Disinfestant	Substance used to remove stomach worms, liver flukes, intestinal worms, body lice, and other animals from the body
Antiseptic	Substance used to reduce the number of microorganisms on the skin and other body surfaces. Examples: soap, merthiolate, alcohol, hexylresorcinol
Sterilizant	Substance used to destroy all forms of life. Examples: mercuric chloride, iodine.
Germicide or microbicide	Substance that kills microorganisms but not their spores.
Sporicide	Substance that kills bacterial spores. Examples: chlorine, ultraviolet radiation, hydrogen peroxide, mercuric chloride
Fumigant	Substance used to eliminate insects such as ticks, lice, mites, flies, ants, cockroaches, and animals such as mice and rats

vent the growth of microorganisms without killing them. These chemicals are called **microbistatic agents**, in contrast to **microbicidal agents**, which work by killing microorganisms. The suffix in both cases is applicable to any prefix in that it can be used as **bacteristatic** and **bactericidal**, **fungistatic** and **fungicidal**, **virustatic** and **virucidal** to indicate growth inhibition and killing, respectively.

Often substances that are microbistatic at low concentrations are microbicidal at high concentrations. Also, an agent that is microbicidal against a specific organism under one set of conditions may well be microbistatic against the same organism under a different set of conditions.

THE DEATH OF BACTERIA

Growth and reproduction of all microorganisms can be controlled through the methods described above. Each type of control affects each kind of organism in a specific and predetermined way. In addition, a vast array of variables affects each case, making generalizations extremely difficult or even impossible. Because examining even a small number of variables in detail is impractical, the few generalizations that are possible are considered here. Also, because the bacteria are understood better than the other organisms, they are used as the example, with the caveat that what affects one organism under one set of conditions may not apply to other organisms or even the same organism under different conditions.

The mechanics of bacterial death are not yet understood completely, but the kinetics of the death of many kinds of bacteria have been well worked out. When a suspension of bacterial cells is heated to a moderate temperature and then a small sample is removed and analyzed for survivors, it is found that some of the cells have died and others survived even though all were exposed to the same temperature for the same length of time. If the surviving cells again are exposed to moderate heat again, some perish and others survive.

This procedure can be repeated many times with the same result. Some cells are killed by a brief exposure to heat, and others survive repeated exposures. The cells that perish and those that survive have no physiologic differences. Therefore, it is said that bacterial cells die in a statistical manner—by random chance.

In a more common experiment that also depicts the death of bacteria, cells are exposed to a moderate amount of heat for increasing periods of time. The results of these experiments may be as those shown in Table 4.4. From a practical point of view, the data obtained from experiments such as these yield valid results only if the experimental procedure enables the researcher to assess the death of individual cells. Because bacterial death is assessed by colony formation and many bacteria exist in clusters, chains, or clumps, special procedures must be used to assess properly the results obtained when bacteria are employed in experiments of this type.

Even cursory examination of these data shows that some cells survived heating for 12 minutes and others did not survive 2 minutes. The surviving cells were not more resistant to heat than were the ones that died early, as shown by the first experiment. Survival of bacterial cells is based on statistical probability, not on the extent of exposure to the killing agent. The pattern of death follows the same kinetics as a first-order chemical reaction; that is, the same proportion of cells dies at each increment of heating.

This relationship between the time of exposure and the number of cells killed leads microbiologists to say that bacterial cells die because the heat causes a specific, as yet unknown, chemical reaction to take place in the cell. This hypothetical reaction is thought of as a **critical target** in the cell. The relationship is **linear** and can be used to predict the number of cells that will be killed in a given time of exposure to the specific killing agent. The number of cells that will die in any succeeding increment of time can be predicted and, from this, the amount of heating necessary to kill the last surviving cell in any population of bacterial cells.

The relationship is shown in the graph in Figure 4.16. The mathematical relationship between the logarithm of the surviving number and the time of

Table 4.4   The Death of Single-Celled Bacteria Exposed
               to High Temperatures (Explained in Narrative)

Minutes Exposure to 60°C	Number of Cells at Start	Number Killed	Percent Killed	Number of Survivors
0	1,000,000	0	0	all
2	1,000,000	900,000	90	100,000
4	100,000	90,000	90	10,000
6	10,000	9,000	90	1,000
8	1,000	900	90	100
10	100	90	90	10
12	10	9	90	1
14	1	1	90	0

heating, the **coefficient of killing**, can be used to predict both the time and the temperature necessary to kill any number of the bacterial cells present in any population. This knowledge is vital in the canning and bottling industries and also is useful in all aspects of food and beverage preservation.

Either of two methods can be applied to measure the effect of heat on bacterial cells. One, called the **thermal death time**, is defined as the time required to kill a given number of cells at a given temperature. The other, the **thermal death point**, is determined by measuring the temperature required to kill a given number of cells in a given amount of time. The thermal death time can be derived from the data in Figure 4.16.

The pattern of bacterial cell death, **death kinetics**, observed as a result of exposure to heat is the same as that seen upon exposure to radiations of various kinds, to chemicals, and even to mechanical forces such as osmotic pressure and ultrasound waves (see Figure 4.17). A coefficient of killing for each of these agents can be obtained by the same means described for heat, but one has little relation to the other because each kind of organism reacts differently to different agents and different conditions.

THE PHENOL COEFFICIENT

A special method was developed to compare the efficiency of chemicals in killing bacterial cells or their spores. As the name implies, the chemical to be tested is compared to phenol. The test procedure is shown in Figure 4.18.

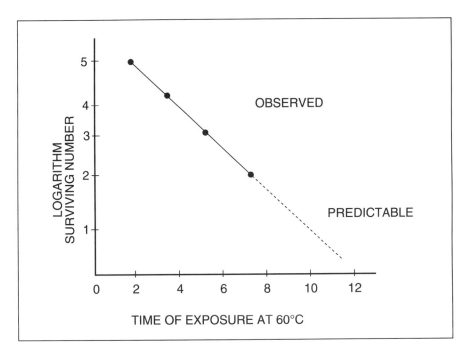

Figure 4.16

Effect of heat on survival of single-celled bacteria. The dotted line is an extrapolation (prediction) based on the data.

If another chemical is tested in the same way and a phenol coefficient of, say, 33 is obtained, the second substance is said to be 10 times better than Brand X. In purchasing cleansers and disinfectants, the phenol coefficient is important in determining price. In the example given here, if Brand X costs $35 per liter and another brand with a higher phenol coefficient costs $200,

Figure 4.17

Effect of X-rays (a) and salt (b) on bacterial survival.

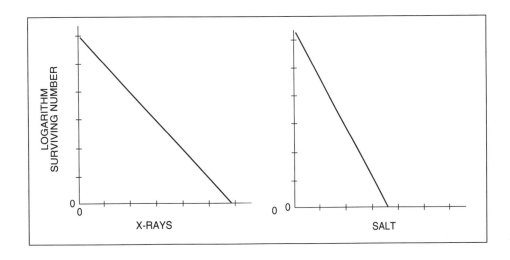

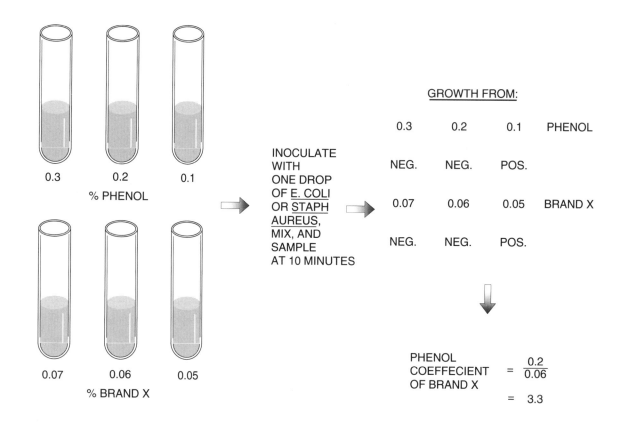

Cost of phenol   = $18.00 per liter
Cost of Brand X = $35.00 per liter

Disinfection equivalent; phenol = 1
                      Brand X = 3.3

Cost per disinfection equivalent; phenol = 18.00 / 1 = 18.00
                              Brand X = 35.00 / 3.3 = 10.61

Brand X is 18.00 ÷ 10.61 = 1.7 times cheaper than phenol

Figure 4.18

Determination of the phenol coefficient and its use in purchasing antimicrobials.

the other may be the best buy if all other considerations, including color, appearance, smell, residue, keeping qualities, ease of handling, and availability, are equal. The benefit in purchasing the other brand lies in the fact that a smaller quantity will have greater killing power than Brand X.

QUESTIONS TO ANSWER

1. How did early humans develop methods for preserving food?
2. What are five early methods of food preservation?
3. Why does rice have to be stored in a dry place?
4. What are five physical methods by which food is preserved today?
5. What are five chemical methods by which food is preserved today?
6. What does "logarithmic death of bacteria" mean?
7. Why is the death of bacteria said to be like a first-order chemical reaction?
8. What temperature and time are required to sterilize dry items?
9. What temperature and time are required to sterilize moist items?
10. Does boiling water render it sterile? Explain.
11. How can one sterilize a sugar solution without heating it?
12. Can ionizing radiation be used to sterilize food? Explain.
13. Can ultraviolet radiation be used to sterilize canned foods? Explain.
14. How do pasteurization and sterilization differ?
15. Can gasoline be sterilized? Explain.
16. Can human blood for transfusion be sterilized? Explain.
17. What is an isotonic system?
18. What is a hypertonic system?
19. How are marmalades protected from microbial growth?
20. Why does sauerkraut not spoil readily?

FURTHER READING

Banwart, G. J. 1979. BASIC FOOD MICROBIOLOGY. AVI Publishing, Westport, Connecticut. See Chapters 10, 11, and 12.

Hayes, P. R. 1985. FOOD MICROBIOLOGY AND HYGIENE. Elsevier Applied Science Publishers, New York.

Jay, J. M. 1986. "Microbial Spoilage Indicators and Metabolites" in: FOOD-BORNE MICROORGANISMS AND TOXINS: THEIR DEVELOPING METHODOLOGY. eds: M. D. Pierson and N. J. Stern. Marcel Dekker, New York.

Volk, W. A. and M. F. Wheeler. 1980. BASIC MICROBIOLOGY, 4th. ed. J. B. Lippincott Company, Philadelphia. See Chapters 12 and 13; for better understanding read Chapters 14 and 15 as well.

Microorganisms as Agents of Disease

5

Some cultivate in broths impure
The clients of our body—these,
Increasing without Venus, cure,
Or cause disease.

Horace

(translated by Kipling)

Diseases caused by microorganisms are called **infectious diseases** because they can be transmitted from one individual to another. Further, these diseases occur only when a specific organism is present in or on the body of a susceptible host. The **germ theory of disease** holds that disease is caused by invasion of the body by microorganisms and their growth therein. This theory implies that these microorganisms must have the ability to overcome the substantial defense mechanisms of the host to penetrate into the interior of the body.

Microorganisms also must be able to establish themselves in the tissues of the host in spite of the many severe forces in the body that act diligently to kill or inactivate invading microorganisms. Only a few organisms are capable of overcoming all the obstacles and defenses the host presents and of parasitizing it. Even when invasion and colonization have been accomplished, the outcome may not be the production of disease.

Figure 5.1

Bacteria Shigella dysenteriae, *which causes bacillary dysentery in human beings.*

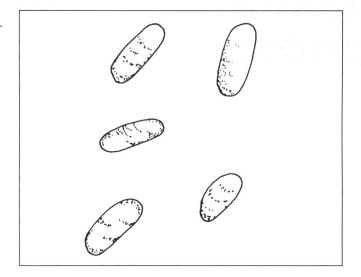

To cause illness, the organism must bring about some change in the host's cells and tissues. Many properties of microorganisms have detrimental effects on the host that harbors them. Organisms are said to be **pathogenic** if the total effect of these detrimental reactions results in illness of the host. If, on the other hand, the presence of the parasite has no detrimental effect on the host's body, the parasite is said to be **avirulent**. Finally, disease is defined as the effect of the relationship between host and parasite on the host. Within this definition exists a spectrum ranging from severe to benign disease.

The vast majority of microorganisms live in nature as **saprophytes**, lacking the power to invade the cells of living hosts. Only occasionally and under special conditions do saprophytes turn to parasitism as a means of existence. **Obligate parasites**, however, can live only at the expense of a host and cannot exist independently in nature. All the viruses are obligate parasites, and so are many bacteria. The viruses cause many human diseases, including poliomyelitis, hepatitis, mumps, and AIDS. Bacteria are the cause of typhoid fever, dysentery, and undulant fever, among many other diseases. Figure 5.1 depicts the bacteria *Shigella dysenteriae*, which causes bacillary dysentery in humans.

Many fungi, protozoa, and bacteria are classified as **opportunistic parasites**. These organisms normally live as saprophytes but under certain special circumstances will invade the bodies of susceptible hosts and cause disease. Organisms that invade the body and establish the initial focus of infection are called primary pathogens. Secondary and subsequent pathogens are those that enter the body through the lesion made by the primary pathogen.

All living things are subject to invasion or infection by other living things. Disease is universal in nature. Each organism has its particular parasites that cause specific diseases. Microorganisms that cause diseases in human beings do not normally cause diseases in plants, and those that cause disease in plants generally do not cause disease in insects. This kind of **host-parasite specificity**, encompasses human pathogens, insect pathogens, pathogens of cattle, pathogens of horses, pathogens of cats, and so on.

In a few cases the **host spectrum** of a given pathogen may be broad enough to cause disease in several kinds of hosts. The best example is seen in the virus that causes rabies, as the same virus can cause the same disease in animals as different from each other as human beings, dogs, cattle, and rabbits. In other cases the pathogen may cause one disease in a given host and quite a different disease in another. An example of this phenomenon is seen in the cowpox virus, which causes a severe skin infection or pox in the cow but only a single, small, and insignificant skin lesion in human beings. Still other pathogens, such as *Neisseria gonorrhoea*, have limited host ranges. *Neisseria gonorrhoea* causes disease in human beings only, primarily in males. Its specificity is accentuated further in that it can attack only the mucous membranes of genital tissues, eyes, and mouth.

SEARCH FOR THE CAUSE OF INFECTIOUS DISEASE

Current thought on the nature of infectious disease is based on the germ theory of disease and depends on understanding the nature of microorganisms and the effect they have on human bodies. Of the vast millions of different microorganisms that inhabit the Earth, only a few hundred are capable of causing disease in human beings.

Before considering the classifications of disease-causing organisms, it must be established that disease is the result of the interaction between parasite and host. In this regard, then, an element of individuality is present in every illness. Of three individuals infected by the same virus, one may have a severe illness, the other a mild illness, and the third none at all.

The presence and extent of illness depend heavily on individual resistance to the effects of the given parasite. Individual resistance is known to be affected by many personal factors including age, sex, state of health, predisposing disease, socioeconomic status, employment, and many other, still unrecognized factors.

Most of the work associated with finding the causative agent of human disease at this time can be divided into two categories. The first is concerned with describing the **etiology** of diseases that appear *de novo* in the population. In the last decade this includes acquired immunodeficiency syndrome, legionellosis, Kuru, Lyme disease, and outbreaks of a deadly disease caused by the Ebola virus. Although some claim that legionellosis occurred as early as 1946, the claims are ambiguous. With the aid of computers capable of handling large amounts of data, it is now possible to examine millions of medical records at one time. As a result, diseases that formerly went unnoticed can be detected.

Comprising the second category are diseases that occur with such low frequency that they have gone unrecognized for years. Listeriosis is a case in point. Often this work depends on extensive searches through old medical records with high-speed computers in efforts to find the few recorded cases among the millions of other syndromes.

The search for the cause of infectious disease started with the philosophical speculations of Lucretius in the first century B.C. From that time to the 19th century, when Pasteur and Koch proved that the germ theory of disease was tenable, many thousands of scientists and as many charlatans sought to explain how small, invisible, living things traveled through the air from one sick person to another. So intense was this work that claim upon claim stated that the causative agents of disease were in effect minuscule beasts of fantastic shape and unimaginable ferocity.

In this morass of fevered imagination and rampant distortion, Fracastoro de Verona published a book in 1546 entitled *De Contagione*, which explained different forms of transmission that are still valid today. Kircher, Jesty, Jenner, Bassi, Hunter, Davaine, Ryer, and Henle published experimental findings that eventually focused the search for causative agents of disease. After Pasteur and Koch proved that anthrax in sheep was caused by a specific bacterium, other researchers quickly followed to find the etiologic agents of many common diseases.

CLASSIFICATION OF DISEASES

Tables 5.1, 5.2, 5.3, and 5.4 outline different systems of classification of disease: by symptoms and clinical picture, by characteristics of the **etiologic agent**, by clinical syndrome, and by route of transmission, respectively. All methods of classification have little or no basis in logic or in science. The categories presented here are devised simply because of the desire for order and for the convenience of physicians, epidemiologists, and the keepers of vital statistics.

Diseases also are classified in traditional or folk ways. The most common is that imposed by the observations of ordinary village folk. These designations generally are based on direct experience, often mixed with myth, legend, and superstition. The naming and classifying of skin infections into a family of poxes during the middle ages was a noteworthy accomplishment based on acute observation and later was accepted into other, scientifically determined classifications. In this system, cow pox, small pox, and the great pox were distinguished one from the other by the size, color, and structure

Table 5.1 Classification of Diseases According to Symptoms and Clinical Picture

Class	Symptoms
Nonclinical	Infection with no symptoms
Malaise	Illness with indefinite symptoms
Clinical	Illness with definite symptoms requiring medical attention
Acute	Rapidly developing syndrome
Fulminating	Development of syndrome abbreviated into short time; prognosis bad
Chronic	Clinical disease that develops slowly and persists for long time
Localized	Disease process limited to certain tissues or organs
Disseminated	Disease process involving many parts of the body
Systemic	Infection involving all parts of the body
Convalescence	Infectious agent under control or eliminated but some symptoms remaining

Table 5.2 Classification of Infectious Diseases According to Characteristics of Etiologic Agent

Class	Etiologic Agent

I. Gram-positive, anaerobic, spore-forming rods	A. *Clostridium perfringens*
	  1. Gangrene
	  2. Peritonitis
	  3. Wound infection
	  4. Food poisoning
	B. *Clostridium botulinum*
	  1. Food poisoning
	  2. Wound botulism
	  3. Infant botulism

II. Gram-positive, aerobic, spore-forming rods	A. *Bacillus cereus*
	  1. Food poisoning
	B. *Bacillus anthracis*
	  1. Cutaneous anthrax
	  2. Pulmonary anthrax
	  3. Systemic anthrax

Table 5.3 Classification of Human Diseases
According to Clinical Syndrome

Class	Clinical Cause
I. Diarrhea	A. *Salmonella spp.*
	B. *Shigella spp.*
	C. *Campylobacter jejuni*
	D. *Bacillus cereus*
II. Dysentery	A. *Entamoeba histolytica*
	B. *Shigella dysenteriae*
	C. *Escherichia coli*

Table 5.4 Classification of Infectious Diseases
According to Route of Transmission

Class	Transmission Route
I.	Direct body contact
II.	Sexual contact
III.	Biological vectors
IV.	Respiratory contact by droplets
V.	Fomites
VI.	Food- and water-borne
VII.	Air-borne
VIII.	Special cases such as blood transfusions, organ transplantations, and contaminated syringes used in intravenous injections

of the skin lesions. Today, these diseases are named cowpox, smallpox, and syphilis, respectively.

The nomenclature of infectious diseases is of prime importance since properly designated illnesses can be recognized and studied without confusion. In the absence of a valid system of names and classification, the ordering and accumulation of data, knowledge becomes an impossibility. It is almost the same as if one tried to run a bank without the names of the account holders.

CHARACTERISTICS OF THE ETIOLOGIC AGENT

The study of the biological characteristics of pathogenic microorganisms makes one of the most dramatic chapters in the study of microbiology. Even while the germ theory of disease was still in the process of verification and

Table 5.5 Classification of Microorganisms According
to Their Ability to Cause Disease

Class	Definition
Pathogen	Organism with genetic potential to cause disease
Secondary	Pathogen or saprophyte that cannot invade alone but will cause infection when another organism establishes a primary infection
Opportunist	Organism that normally does not cause disease but will under special conditions
Virulent	Organism with genetic potential to cause clinical disease
Avirulent	Pathogenic organism that has lost its ability to cause disease

confirmation, Roux and Yersin, two French scientists, proved that diphtheria in human beings was caused by a **toxin** produced and excreted by the bacteria associated with this disease, *Corynebacterium diphtheriae*. As a result of their work, along with that of many other investigators, diphtheria was brought under control in the 1920s.

The period from 1880 to the 1920s is regarded as the **golden age of bacteriology**. During this period Pasteur, Koch, Ehrlich, and many others showed that bacteria produce disease only when they possess certain properties inimical to the host. A given organism was shown to produce disease when it elaborates the toxin but totally harmless if it loses the ability to produce toxin. Pasteur exploited this phenomenon to produce **attenuated cultures** of the causative agent of fowl cholera and the rabies virus, enabling him to induce immunity to these two diseases in animals and in human beings. In the case of diphtheria, **toxigenic** strains of the organism are pathogenic but nontoxigenic strains are not.

In addition, some strains of a given pathogen were shown to produce powerful toxins and others to produce weak toxins. These are called highly virulent and weakly virulent organisms, respectively. It also was learned that many organisms have the ability to change from the pathogenic form to the nonpathogenic form in response to environmental conditions or by mutation. Table 5.5 provides a synopsis of microorganisms according to their ability to cause disease.

The genetic characteristics that contribute to pathogenicity can be described simply, although their discovery forms a long and complex chapter in microbiology. The discovery, description, and explanation of the mechanisms by which bacteria cause disease opened new horizons to the understanding of how the human body is structured and how it functions.

Some bacterial elements, such as the bacteriophages (bacterial viruses) presented problems in understanding pathogenicity that could not be solved for more than a hundred years. It was well known that only human beings who harbored *Corynebacterium diphtheriae* developed diphtheria, but not all who harbored the organism suffered from the disease. The organisms isolated were identical in all respects save the fact that some produced diphtheria toxin and others did not. Microbiological convention of the time explained this by stating that some strains of the organism were toxigenic and others were not. Although the explanation was logical, it failed to elucidate the differences that existed among the toxigenic and nontoxigenic strains of *C. diphtheriae*.

Not until the 1960s did Gorman show that bacteriophages that parasitize *C. diphtheriae* carry the genes responsible for producing toxin. Only those strains of *C. diphtheriae* infected with bacteriophages were capable of producing toxin. The phenomenon whereby a nontoxigenic organism is changed into a toxigenic one by infection with a bacteriophage is called **phage conversion**. From this description it is evident that the bacterium is not pathogenic and that the bacteriophage, acting through the bacterium, is what possesses the genetic potential for producing disease.

ATTRIBUTES OF PATHOGENICITY

The mechanisms of pathogenicity of many of the parasitic microorganisms among the viruses, bacteria, fungi, and protozoa still are not understood completely, although it is well established that many pathogenic organisms produce substances known to damage and even destroy host cells. Many of these microbial substances also have the ability to kill various animals when the substance is isolated from culture fluids and injected into the animal. Although pathogenicity cannot be explained entirely on the basis of toxins and other closely related substances, knowledge of their presence and activity explains many previously misunderstood observations.

Undoubtedly many other substances that contribute to pathogenicity exist, but only the ones listed in Table 5.6 have been described adequately. Modern microbiologists assume that, when all the substances and cell constituents contributing to pathogenicity have been identified and characterized, the work that Pasteur, Koch, Roux, and Yersin started will be finished. At this time, however, much work remains to be done.

EPIDEMIOLOGY

Knowledge of the patterns by which disease is transmitted throughout a population is essential to understanding the nature of infectious diseases. Without knowledge of the manner in which diseases are spread from one host to another, controlling dissemination is practically impossible. Epidemiology is the study of the mechanisms and forces that play a part in the spread of dis-

Table 5.6 Major Agents of Pathogenicity

Agent	Description
Capsules	Complex carbohydrate polymers that envelop and protect bacterial cell; inhibit phagocytosis
Toxins	Complex molecules produced by some pathogens; affect cell function
Enzymes	Protein molecules that destroy chemical structure of host cells
Adherence factors	Chemical areas on surfaces of cells that allow them to adhere to host cells and grow there
Bacteriophages and plasmids	DNA molecules that carry genes for toxins and for antibiotic resistance; infect bacterial cells and cause them to produce toxins

ease. A pathogenic organism that can live only in the cells of human beings, such as *Neisseria gonorrhoea*, must be present in a given number of individuals in a given population lest everyone get cured and the germ become extinct. This is called the **maintenance rate of infection**, and the presence of disease in the population then is said to be **endemic**. If the disease spreads rapidly so that many cases appear at the same time, the disease is said to exist in **epidemic** form (see Table 5.7).

Epidemiologists use the data obtained from studies on the nature of endemic and epidemic disease to predict the appearance of disease in new populations, to predict the number of cases that will appear, and to predict the involvement of still other populations. The techniques of epidemiology include surveillance, records, and analysis. Each of these is discussed briefly in the following pages.

Surveillance

Surveillance requires constant observation of the population for the appearance of disease and for movements of disease within the population. Many

Table 5.7 Classification of Infectious Diseases According to Epidemiological Criteria

Class	Criteria
Sporadic	Disease appears at irregular intervals
Endemic	Present in the community at most times
Epidemic	Sudden increase in number of cases
Pandemic	Worldwide epidemic
Zoonotic	Disease of animals transmissible to man
Nosocomial	Disease acquired in a hospital

pathogenic organisms, such as *Staphylococcus aureus,* are found in the nose and throat of healthy individuals even when no cases of infection are present in the community. Assume that in a given population 20% of persons tested harbor the organism but few cases of staphylococcal disease are found in that group of people. If the number of individuals harboring the organism increases to 40% and cases begin to appear in epidemic form, epidemiologists then can predict appearance of the disease by monitoring the population to establish the number of individuals who harbor the organism in their nose and throat.

Epidemiological surveillance also can be done by testing for the presence of antibodies in the bloodstream of selected members of the community, as presently is done in tracking AIDS. Many other methods of surveillance are used, and the principal question is always the same: How many cases or potential cases of a given disease exist in the population at a given time, and how can this information be applied to predict the appearance of epidemic disease?

Records

Record keeping requires that the information obtained from surveillance be preserved accurately for immediate and future study. Records of every case of the disease, including time and place of infection, state of immunity of the person before infection, clinical symptoms, type of therapy, result of therapy, and final outcome of the case, must be meticulous and contain a maximum of relevant information. Records must be centralized and prepared in a form suitable for dissemination to all individuals and authorities concerned. All records must be maintained permanently and within easy access of health practitioners and researchers.

The spread of infectious disease in a population can be characterized by the number of cases involved, by the etiologic agent, by the characteristics of the population affected, by route of transmission, and by many other criteria. The availability of these data is what enables medical personnel to forecast and avert epidemic disease. Figure 5.2 documents the prediction of a cholera epidemic in Brazil.

Analysis

Analysis encompasses all knowledge available about the organisms that cause disease, about the subject population, and about the patterns of dissemination. The data available must be examined for unusual or unexpected patterns that may yield information essential for preventing further spread of the disease. This includes genetic variation of pathogens and degree of immunity of various segments of the population.

1st Rio Case of Cholera raises fears

The Associated Press

RIO DE JANEIRO, Brazil - The first confirmed case of cholera along Brazil's populous southeast coast has health officials in Rio de Janeiro fearing an epidemic similar to ones that have ravaged other Latin American cities.

Dr. Meri Baran, a physician with Secretary of Municipal Health, predicted up to 200,000 cholera cases and 1,200 deaths could occur in the Rio de Janeiro area during the next

Figure 5.2

Prediction of cholera epidemic including morbidity and mortality. From Arkansas Democrat-Gazette, *Dec. 1, 1991, p. 13A.*

The value of proper analysis of epidemiological data has been demonstrated recently by the discovery of several "new" diseases. These diseases have such a low incidence that each case previously had been considered as unique and not representative of a definite syndrome. By analyzing large numbers of reports of unrecognized diseases, enough cases were accumulated to reveal similarities in clinical signs and other necessary criteria.

Legionellosis, Lyme disease, Kuru, and campylobacteriosis were discovered by analyzing numerous previously documented medical records that had been classified as **disease of unknown origin**. Some experts claim that these and other rare diseases could have been identified only because of the availability of high-speed computers capable of assimilating and analyzing large amounts of data. The data, of course, were obtained and recorded accurately by careful and intelligent epidemiologists. Figure 5.2 illustrates this point.

TRANSMISSION

The spread of disease can be prevented or, in most cases, inhibited if detailed knowledge of the mechanism of transmission is known. Several essential

Figure 5.3

Sketch showing route of transmission of polio-myelitis; dots emitting from second individual on the right represent saliva droplets that may contain viruses.

factors in transmission must be satisfied if the transfer of pathogenic micro-organisms from one human being to another is to be accomplished success-fully. Anything that interferes with transmission prevents transfer.

To follow the movements of disease among a population, the means by which the disease is transmitted from one individual to another must be well established. The epidemiology of air-borne diseases is very different from that of water-borne infections. Only after establishing the route of transmission can adequate surveillance be put into place so the rate at which new cases appear may be determined. In addition, the time at which cases appear in sur-rounding communities can be observed only if the route of transmission is known. The information obtained from these observations also can be used to predict the time, and force, with which the disease will strike a distant com-munity.

For instance, in tracing the spread of AIDS across the United States, checking the number of rats per household would have been confusing and futile because rats have nothing to do with the transmission of this disease. On the other hand, the epidemiology of plague cannot be studied without know-ing the extent of rat populations, and knowledge of the sexual habits of humans will be useless in this context. Figure 5.3 illustrates the route of trans-mission of some viruses.

Population Characteristics

The susceptible population for any disease has to be identified so effort is not wasted surveying those who are not involved in the epidemic. Human populations can be classified by many different criteria. The first division generally is made by gender. Some diseases, such as vaginitis, have predilections for women for obvious reasons. Other predilections are not understood as easily, as is the case for stomach tuberculosis in various races.

Other divisions are made on the basis of age, occupation, economic status, nutritional condition, and personal behavior. These categories are not mutually exclusive and, indeed, often are used to "type" a given segment of the population to distinguish it from another. For instance, the group described as "middle-aged, low-income, alcoholic" is distinguishable from the group described as "young, affluent, athletically active." The distinction can be made not only according to sociological criteria but also according to the microorganisms found with highest frequency when illnesses occur.

Immunity

The state of immunity of the population must be established. A population immunized to a given disease is very different from one that has no immunity. The best example is found in the contrast between smallpox and gonorrhea. In the former, international law for many years required that everyone be immunized, and smallpox has been eradicated as a consequence. On the other hand, because there is no immunization for gonorrhea, the number of cases increases constantly.

Education

Often, one of the key aspects of a successful epidemiological study is an informed community. The first steps in education are to make the public aware of the danger, how to avoid risk, and how to report new cases of the disease to public health authorities. The epidemiological report can be enhanced by an intelligent and well informed public, whereas an uneducated public often will misinform and confuse researchers.

Identity

Epidemiological studies can succeed only if the nature of the etiologic agent is known. To gain this knowledge, it must first be established that all cases of the disease are caused by the same organism. Failing to establish this may lead, as it has many times in the past, to confusion of two or more different epidemics that may be taking place at the same time in the same population.

Even if the identity of the causative agent is determined, the specific strain causing the epidemic has to be known. More than 2,000 different strains of *Salmonella typhi* exist, and knowing that this organism is causing an epidemic of disease is of little value if the specific strain is not known. The knowledge is indispensable for identifying the epidemic and for applying control measures. Because all strains of salmonella are closely related, however, strain variations are of little value in managing ill individuals or in determining clinical procedures.

Variation

Genetic variations in the etiological agent are important in epidemiology. In the case of influenza, a pandemic generated by the virus will cause much of the world's population to become immune, and as a consequence the disease will subside until only a few cases exist. At approximately 10-year intervals the organism undergoes a mutation, and the immunity that human beings had developed is no longer protective because the organism is changed. The result is another pandemic with another cycle of immunity, mutation, and epidemic. This situation has existed for a long time, and it will continue into the future for many more generations of both human beings and the influenza virus.

REPORTABLE DISEASES

The information gathered by epidemiologists is of value only if it is assembled and reported properly. The Centers for Disease Control (CDC) in Atlanta, Georgia, compiles a list called the index of reportable diseases in the United States (see Figure 5.4). When any illness listed in the index of reportable diseases is diagnosed or even suspected anywhere in the United States, it must be reported to the CDC with sufficient information to permit epidemiological evaluation. The CDC in turn publishes a *Morbidity and Mortality Weekly Report* (see Figure 5.5), which is available to all who are concerned with public health. Many states have their own periodic reports dealing with local problems. As a result of the work of these agencies, the effects of transmissible diseases on the community can be assessed at a glance. Some of the reportable bacterial diseases in the United States are:

Anthrax	Leptospirosis	Shigellosis
Botulism	Listeriosis	Tetanus
Brucellosis	Lymphogranuloma	Tuberculosis
Chancroid	venereum	Tularemia
Cholera	Meningitis	Typhoid fever
Diphtheria	Pertussis	Typhus fever
Granuloma	Plague	Yersiniosis
inguinale	Psittacosis	
Legionellosis	Rheumatic fever	
Leprosy	Salmonellosis	

505 Cigarette Smoking among Public High School Students — Rhode Island
507 Deaths due to Chronic Obstructive Pulmonary Disease and Allied Conditions
510 Antigenic Variation of Recent Influenza A(H1N1) Viruses
517 ACIP: Monovalent Influenza A(H1N1) Vaccine, 1986-1987

Printed and distributed by the Massachusetts Medical Society, publishers of *The New England Journal of Medicine*

Figure 5.4

Weekly report of number of reportable diseases in U.S. population.

Perspectives in Disease Prevention and Health Promotion

Cigarette Smoking among Public High School Students — Rhode Island

From July 1983 through December 1984, as part of a health-risk survey, information was obtained from 11,657 Rhode Island public high school students about their cigarette smoking practices. Overall, 22.3% of these students reported that they smoked cigarettes. Cigarette smoking increased by grade and was more common among females (26.5%) than among males (17.5%). The difference between females and males was due primarily to a larger proportion of females who reported smoking less than one pack per day (Figure 1).

During this period, 19 (63.3%) of the 30 public high schools in Rhode Island took part in

Figure 5.5

Publication of Centers for Disease Control.

TABLE I. Summary—cases specified notifiable diseases, United States

Disease	32nd Week Ending			Cumulative, 32nd Week Ending		
	Aug. 9, 1986	Aug. 10, 1985	Median 1981-1985	Aug. 9, 1986	Aug. 10, 1985	Median 1981-1985
Acquired Immunodeficiency Syndrome (AIDS)	371	155	N	7,631	4,632	N
Aseptic meningitis	354	349	366	4,188	3,780	3,873
Encephalitis: Primary (arthropod-borne & unspec.)	28	28	35	543	617	644
Post-infectious	3	-	2	67	86	62
Gonorrhea: Civilian	18,527	16,871	19,424	527,166	530,774	543,763
Military	259	285	575	10,145	12,785	14,891
Hepatitis: Type A	385	393	426	13,244	13,228	13,228
Type B	463	444	489	15,754	15,430	14,397
Non A, Non B	65	58	N	2,168	2,511	N
Unspecified	91	110	126	2,877	3,495	4,383
Legionellosis	21	21	N	380	437	N
Leprosy	-	23	3	168	241	155
Malaria	21	18	23	591	595	595
Measles: Total*	83	75	29	4,847	2,236	2,172
Indigenous	81	54	N	4,615	1,878	N
Imported	2	21	N	232	358	N
Meningococcal infections: Total	31	16	35	1,705	1,612	1,916
Civilian	31	16	35	1,703	1,606	1,912
Military	-	-	-	2	6	9
Mumps	124	14	25	3,153	2,089	2,335
Pertussis	156	85	61	1,746	1,327	1,254
Rubella (German measles)	8	19	10	368	466	738
Syphilis (Primary & Secondary): Civilian	564	518	662	15,877	16,266	18,411
Military	2	-	4	107	113	232
Toxic Shock syndrome	2	11	N	214	245	N
Tuberculosis	546	325	448	13,241	12,819	14,116
Tularemia	7	6	7	76	107	141
Typhoid fever	16	5	9	171	205	229
Typhus fever, tick-borne (RMSF)	29	24	34	450	398	688
Rabies, animal	85	113	136	3,338	3,257	3,898

Because the substances of the human body, like those of the bodies of all other living things, are an attractive habitat and source of food for many, if not all, microorganisms, life is like a poised equilibrium in which parasites tend to enter and body defenses to repel. In the end (i.e., death of the individual) the microbes win and nothing will change this.

The optimist, of course, can point to the example of smallpox and wait for the day when all diseases have been eradicated. For every optimist, however, a pessimist laments the fact that smallpox has disappeared only to be replaced by AIDS.

QUESTIONS TO ANSWER

1. What is a pathogenic organism?
2. What is an obligate parasite?
3. What is an opportunistic parasite?
4. What is the definition of *host-parasite specificity*?
5. What is meant by "host spectrum?"
6. What is virulence?
7. What is an attenuated organism?
8. What are three methods by which diseases can be classified?
9. What are bacteriophages?
10. How do bacterial capsules affect pathogenicity?
11. What is the definition of epidemiology?
12. What is meant by surveillance of disease in a community?
13. What is an endemic disease?
14. What is a nosocomial disease?
15. What are five "reportable diseases" of human beings?

FURTHER READING

Baron, S. and L. M. Alperin (eds). 1982. MEDICAL MICROBIOLOGY. Addison-Wesley, Menlo Park, California.

Davis, B. D., R. Dulbecco, H. N. Eisen, and H. S. Ginsberg. 1980. MICRO-BIOLOGY, 3d ed. Harper & Row, Hagerstown, MD.

Finegold, S. M. 1982. Bailey and Scott's DIAGNOSTIC MICROBIOLOGY, C. V. Mosby Co., St. Louis.

Milgrom, F. and T. D. Flanagan (eds). 1982. MEDICAL MICROBIOLOGY. Churchill Livingstone, New York.

Bacterial Diseases Transmitted by Food and Water

6

Well and wisely said the Greek
Be thou faithful, but not fond;
To the altar's foot thy fellow —seek
The Furies wait beyond.

Emerson

The objectives of this text will be better accomplished by limiting further discussion of communicable disease to illnesses that are transmitted by food and water. This distinction, like the classification of transmissible diseases, is an artificial one in that organisms that generally are associated with enteric infections may, under special conditions, cause other diseases. As an example, the protozoan *Entamoeba histolytica* normally causes dysentery and is transmitted by the feces-to-food route. Under appropriate conditions, however, it can be transmitted by direct contact and causes severe skin ulcerations.

In transmitting microorganisms by food and water, the fecal matter of an infected person or animal finds its way into the food or drink of a healthy, susceptible person. The normal route of transmission works by two principal mechanisms. The first is by direct incorporation of fecal matter containing pathogenic bacteria into food and drink. This form of contamination happens in many different ways but most often when hygienic conditions become compromised. Eating with utensils soiled with fecal matter, contaminated water,

Figure 6.1

Thoroughly washing hands with soap helps control the spread of disease.

dirty hands, and direct contamination of food are all major avenues of transmission (see Figure 6.1).

Contamination of food at the point of production frequently is encountered in areas where raw sewage is used in the irrigation of ground crops such as lettuce, radishes, and strawberries. Because these foods are consumed without cooking, the bacteria present on leaves and fruit become part of the food.

The second mechanism involves some intervening agent, such as food or beverage, where the bacteria may grow and produce large populations—an amplification of the number of organisms. The number of pathogenic organisms introduced into the food or beverage may have been below that needed to cause infection, but after a period of growth, many times more than the requisite number of organisms may be present.

In another aspect of this kind of transfer, organisms that are introduced into food products, which then are shipped under refrigeration or after being frozen, may affect people in places far from the point of origin. Because microorganisms remain viable for prolonged periods under refrigeration or after freezing, the effects of contamination may not be observed for many months or even years.

During World War II some small lots of eggs contaminated with salmonella were mixed with large lots of uncontaminated eggs, which gave rise to epidemics of salmonellosis. These egg mixtures were dried over gentle heat and shipped to armed forces installations all over the world. Because the directions for using dried eggs stipulated placing the dried egg powder in milk or water and letting it sit during the night, the bacteria present had sufficient time to develop large populations. If the eggs were cooked thoroughly, the bacteria were killed and no infections resulted. On the other hand, all items coming in contact with uncooked egg mixture, including the cook's hands, were contaminated with salmonella. Also, if the eggs were consumed "easy"

Table 6.1 Intestinal Disorders of Human Beings Caused by Bacteria

Organism	Disease	Mode of Transmission
Brucella abortus	Undulant fever	consumption of unpasteurized dairy products
Brucella melitensis	Undulant fever	consumption of unpasteurized dairy products
Escherichia coli	Summer diarrhea	contaminated food and water
Salmonella typhi	Typhoid fever	contaminated food and water
Salmonella spp.	Salmonellosis	contaminated food and water
Shigella spp.	Dysentery	contaminated food and water
Vibrio cholera	Cholera	contaminated food and water
Yersinia enterocolitica	Yersiniosis	unpasteurized dairy products
Campylobacter jejuni	Campylobacteriosis	contaminated food and water

or "soft-cooked," large numbers of bacteria survived the cooking and caused infection. Thousands of cases of salmonellosis resulted from eating these eggs before the source of contamination was discovered. The same thing happened aboard U. S. Navy ships, as well as in the navies and armies of other countries employing the same methods of food preparation.

FOOD AS A VEHICLE FOR MICROBIAL TRANSMISSION

With the exceptions of only a few foods such as sauerkraut, pickled vegetables, and salted meats, all the food that human beings consume is a perfect growth medium for the bacteria that cause intestinal disorders. If the bacteria are introduced into foods by any of the mechanisms described previously, they will remain there, and perhaps increase in number, until the food is consumed. Although foods that have been cooked thoroughly can be consumed safely, this usually is not the case. Because bacterial death follows **first-order kinetics**, the number of organisms present determines the time and temperature required to kill enough bacteria to make the food safe for eating. If the number is large, even overcooking may not be sufficient to kill all the organisms present. This often is the case when extensive growth has taken place.

Food is an excellent medium for transporting bacteria to the intestinal tract in that it protects them in the stomach. The digestive juices in the stomach easily kill most bacteria. Food, however, neutralizes stomach acid and binds digestive enzymes, allowing large numbers of live bacteria to enter the intestinal tract. The number of salmonella mixed in beef broth needed to cause an infection is approximately ten thousand times less than that needed when the organisms are present in water. The difference between the two levels represents the protective effect of the beef broth.

The third point of significance that makes food an excellent vehicle for transmitting bacteria resides in its variability. This means that, for any organism and any set of conditions, one or more kinds of foods will harbor the bacteria during transfer from the feces of one individual to the intestinal tract of another. In the final analyses however, certain foods do not support the growth of some bacteria. Salmonella and other enteric pathogens will not survive even brief exposure to acid foods such as tomatoes, sauerkraut, and mustard.

WATER AS A VEHICLE FOR MICROBIAL TRANSMISSION

The most common method of dissemination of enteric infections is by ingestion of water contaminated with enteric pathogens. Household sewage that carries fecal matter containing enteric pathogens also has a rich and varied load of organic chemicals. These chemicals are a rich growth medium that support the growth of many enteric microorganisms including those that are pathogenic. These organisms can survive and even grow in the sewage, and often large populations result. Food, water, utensils, insects, and the bodies of animals and human beings alike become contaminated when they come in contact with sewage.

Until this century all human beings lived in fear of the water they consumed. Although never certain of how water carried the intestinal diseases they feared, people always assumed that water played a role of some kind. In many communities water from various wells or fountains was rated good or bad according to the number of users who became ill after drinking it. Today, any community wishing to provide water that is safe to drink may do so with little effort or expense, as the quality of the water can be ascertained precisely by microbiological and chemical analyses before it is used.

Water purification was practiced in many households as long ago as the beginning of the Christian era and perhaps even longer. Water was purified largely for aesthetic purposes rather than from knowledge of the presence of microorganisms. The ancient Arab custom of adding lime to water and filtering it through clay or unglazed porcelain vessels to make it sparkling clear (see Figure 6.2) also had the effect of removing microbial cells and rendering the water pure and safe to drink. Although municipal water purification first began in Germany in the early part of the 19th century, it did not become a universal practice until the 20th century. Even today, vast segments of the world's population drink water that has not been filtered and is neither pure nor safe.

Under normal circumstances, water that is not purified may be the bearer of epidemics of typhoid fever, dysentery, cholera, poliomyelitis, hepatitis, and other enteric disorders. Although most industrialized countries have succeeded in eradicating cholera and poliomyelitis and in controlling epidemics

Figure 6.2

Traditional Arabic water purifier; lime is added to water, which then is filtered through unglazed porcelain.

of other enteric diseases, many emerging countries exist today as they did one hundred or even one thousand years ago in that they still have no effective water purification system. Cholera, typhoid, dysentery, and the other epidemic diseases still decimate those populations frequently. In spite of the efforts of the World Health Organization, international philanthropists, and local governments, increasing populations in many communities tend to increase more rapidly than the available supplies of pure drinking water.

THE ENTEROBACTERIACEAE

Most bacteria that cause enteric diseases transmitted by food and water are placed in a group called the **Enterobacteriaceae**. Unlike almost all other microorganisms, these have the ability to pass through the stomach without being killed. Once past the stomach, they begin to reproduce in the small intestine and in a few days produce large populations capable of causing disease. Some do this by invading the cells of the intestinal tract. Others produce

Figure 6.3

Mixtures of enteric bacteria in the intestinal tract; bar represents 10 μm. Scanning electron micrograph by Munhyeong Choi.

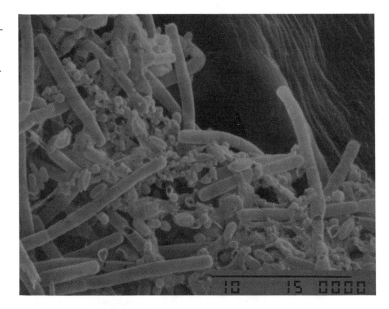

toxins that affect cells adversely. Some bacteria penetrate intestinal tissues and gain entry into the body. Enteric pathogens produce disease by several different mechanisms, but all possess **endotoxin** and affect the host in somewhat the same way through the action of this toxin.

Most enteric infections are self-limiting. The disease lasts from two days to several weeks. Some infections give rise to the carrier state, in which the pathogen becomes a member of the **normal flora** of the host and the host becomes immune to the pathogenic effects of the organism. These **carriers** serve as a natural source of infection in the population.

The most celebrated carrier of salmonellosis was Mary Mallon, better known as Typhoid Mary. A European maid-cook-chef who worked in and around New York City at the beginning of the 20th century, she was the person in which the carrier phenomenon was first recognized. She had entered New York harbor in a German vessel ridden with typhoid fever. Although most of her fellow steerage passengers were stricken with fever and many died, she was not afflicted. Typhoid Mary was employed as a maid and worked until her employers died of typhoid fever. She moved from one house to another and later, when she became a cook, from one restaurant to another. Some 18 years after her career began, public health authorities identified her as the source of typhoid. Before she was apprehended and confined, it is said she caused the deaths of an estimated 5,300 individuals. She remained a symptom-free carrier until she died.

All the enteric bacteria are Gram negative, small, rod-shaped organisms (see Figure 6.3) that contain a **lipopolysaccharide (LPS)** in the cell wall. Some

Table 6.2   Genera of Family Enterobacteriaceae and Their Roles in Nature

Genus	Role
Escherichia	Found in the intestinal tract of all humans beings, most of the great apes, and many other animals
Enterobacter	Found in the intestinal tract of animals, in the soil, and in water
Hafnia	Very similar to *Enterobacter*
Edwardsiella	Occasionally cause diarrhea in human beings; normally found in soil and water and as a pathogen on some plants
Citrobacter	Opportunistic pathogen in human beings
Klebsiella	Cause pneumonia in human beings and hospital-acquired burn and urinary tract infections; common opportunistic pathogens
Salmonella	Cause typhoid fever and salmonellosis in human beings
Shigella	Cause bacillary dysentery in human beings
Yersinia	Enteric pathogen of farm animals, cats, dogs, and occasionally human beings
Serratia	Soil and air bacteria; opportunistic pathogens
Erwinia	Major plant pathogens
Morganella	Cause nosocomial pneumonia and urinary tract infections; opportunistic pathogens
Providencia	Same as *Morganella*

forms of LPS are endotoxin, and others are simply cell wall constituents with no toxic activity. The LPS also serves as a family of **antigenic determinants** that can be used to identify many different strains within the same genera and species of enterobacteria. The enterics are a group of similar organisms placed in the genera listed in Table 6.2.

The enteric bacteria are easy to separate into species and strain classifications. They are **peritrichously flagellated** when motile, **facultative**, and capable of growing on simple laboratory media. Some have the capacity to ferment lactose with production of acid and gas, and some produce hydrogen sulfide from proteins. They grow readily in the laboratory in different media, and a preliminary identification can be made by the series of biochemical tests shown in Table 6.3.

Additional tests using a variety of media and biochemical reactions must be performed to make distinctions among the remainder of the species. Strains within species, as well as some species distinctions, are made using purified **strain-specific antisera**. These tests often are automated and the results transmitted directly to a computer for identification.

Table 6.3   Separation of Genera of *Enterobacteriaceae*
by Reactions on Two Laboratory Test Media

Organism	Triple Sugar Iron Agar				Lysine Iron Agar		
	Slant	*Butt*	*Gas*	*H₂S*	*Slant*	*Butt*	*H₂S*
Escherichia	Acid	Acid	+	−	Alk	Alk	−
	Alk	Acid	−	−	Alk	Acid	−
Enterobacter	Acid	Acid	+	−	Alk	Alk	−
Hafnia	Alk	Acid	+	−	Alk	Alk	−
Edwardsiella	Alk	Acid	+	+	Alk	Alk	+
Citrobacter	Alk	Acid	+	+	Alk	Acid	+
	Acid	Acid	+	−	Alk	Acid	−
Klebsiella	Acid	Acid	+	−	Alk	Alk	−
	Alk	Acid	+	−	Alk	Acid	−
Salmonella	Alk	Acid	−	+	Alk	Alk	+
	Alk	Acid	+	−	Alk	Alk	−
Shigella	Alk	Acid	−	−	Alk	Acid	−
Yersinia	Alk	Acid	−	−	Alk	Acid	−
Serratia	Alk	Acid	+	−	Alk	Alk	−
	Acid	Acid	−	−	Acid	Acid	−
Morganella	Alk	Acid	+	−	Red	Acid	−
Providencia	Alk	Acid	−	−	Red	Alk	−

Alk = alkaline
+   = positive, gas or hydrogen sulfide present
−   = negative, gas or hydrogen sulfide not present

Salmonella

Salmonellosis is the name given to any infection caused by bacteria of the genus *Salmonella* whether the infection is enteric or **extraenteric**. The salmonellas are a large and fairly diverse group of bacteria that cause various diseases in human beings and in other animals as well. Of the diseases caused by these bacteria, typhoid fever is the most severe and paratyphoid is a distant second in severity. The most common salmonella infection is gastroenteritis, and the most unusual is *salmonella septicemia*. Gastroenteritis also is known by older names, such as enteric fever, intestinal fever, and food poisoning, and a new name, **tourist diarrhea**, recently has appeared. The genus *Salmonella* contains several species (see Table 6.4), and at least 2,500 **antigenic variants**.

Table 6.4   Species Distribution of Salmonella Infections in Texas in 1986

Species	Number of Cases	Percentage
S. typhimurium	368	23.1
S. newport	175	11.0
S. heidelberg	133	8.3
Group B	90	5.6
S. javiana	78	4.9
S. enteritidis	63	3.9
S. infantis	47	2.9
S. agona	46	2.9
S. montevideo	39	2.4
Others	494	31.0

Gastroenteritis   After an incubation period of 10 to 36 hours, progressive nausea, intestinal distress, and vomiting ensue. As the symptoms progress, the cells of the intestinal tract become infected, and diarrhea, fever, chills, and acute abdominal pain develop. Symptoms continue for 3 to 15 days, and often the disease disappears spontaneously. The death rate from gastroenteritis is approximately 3% of untreated cases in normal, healthy individuals but is much higher in **medically compromised** individuals. Often the bacteria persist in the intestinal tract and are shed in the feces even after all symptoms of infection disappear. If antibiotics are employed in controlling the infection, the salmonellas that persist in the intestinal tract usually are antibiotic-resistant. Figure 6.4 shows the three antigen-antibody specific reactions in the test tube: (a) positive, (b) negative, (c) and (d) controls.

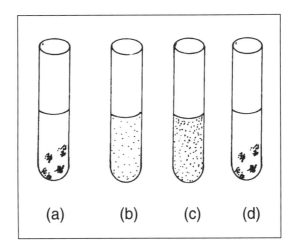

(a)   (b)   (c)   (d)

Figure 6.4

Antigen-antibody specific reactions, (a) positive test, (b) negative test, (c) negative control, and (d) positive control.

Table 6.5   Antigenic Formulae of Some Serotypes of *Salmonella enteritidis*

Organism	O Antigens	H Antigens Phase 1	Phase 2
S. enteritidis	1,4,5,12	b	1,2
	1,4,5,12	i	1,2
	6,7	c	1,5
	6,8	e,h	1,2

Etiology   The most common causative agent of salmonellosis in the United States is *Salmonella enteritidis*. Other organisms that also cause salmonellosis are *S. typhimurium* and *S. schottmuelleri*. The few species of the genus *Salmonella* contain hundreds of antigenic variants. The differences among variants or **serotypes** are subtle but important in diagnosis and in epidemiology. The genus is divided into serotypes on the basis of the antibodies produced when the bacterial cell is injected into laboratory animals. Serotypes can be used to identify more than 2,500 distinct antigenic variants within the genus *Salmonella*.

Serotypes, or antigenic types, are determined by the presence of **somatic antigens**, the lipopolysaccharide fraction of the cell wall, and **flagellar antigens**. Flagella are made of antigenic proteins called flagellins. The former are called O antigens and the latter H antigens. The H antigen is of two types: H1 and H2. The serotype designations for several serological variants of *S. enteritidis* are given in Table 6.5. The organism described in line one in Table 6.5 is written as follows; *S. enteritides* 1,4,5,12; b; 1,2. The one on the second line is written *S. enteritidis* 1,4,5,12; i; 1,2. These are two closely related organisms identical in every respect except some slight differences in the proteins of their flagella.

Epidemiology and Control   The natural reservoir of all the salmonellas is the intestinal tract of birds, reptiles, all domestic animals, and human beings. Many animals harbor large populations of salmonella without being sick, and they regularly excrete the pathogenic bacteria in their feces. Household pets are a constant source of infection for children and adults when the pets come into close contact with the family. Baby turtles, some with painted shells, sold in stores during the 1970s often carried salmonella, which became established in the aquarium water. Children and others who played with the animals or the water acquired salmonellosis in epidemic proportions. An estimated 325,000 cases of this infection per year prevailed until the sale of baby turtles was prohibited by law.

Salmonellosis transmitted from one individual to another via the feces-to-food route accounts for the endemic presence of the disease in all populations. Epidemics of salmonellosis occur only when the bacteria are spread by water and, to a lesser extent, when spread by contaminated food.

Farm workers, abattoir (slaughterhouse) personnel, and butchers are considered at high risk because poultry, hogs, and cattle often carry salmonellas. Household vermin and insects are effective intermediaries in the spread of these microorganisms. The bacteria, along with contaminated material, are carried on the mouths and other body parts of ants, cockroaches, and flies. Insects also transfer viable bacterial cells to food and utensils, which then become potential sources of infection. These carriers of contagion are called **mechanical vectors**, in contrast to biological vectors. The latter are described as insects that acquire microorganisms which then multiply in the insect's body before being transmitted to another host, such as humans.

Some special cases of transmission, although unnatural, illustrate the full potential for infection. Contaminated ice cream, for example, has been shown to yield viable *Salmonella* after many months of storage in the freezer. Federal regulations that dictate proper labeling, including production lot numbers, are designed to make it possible to find contaminated foods once they have been identified as a potential source of disease.

Epidemics of salmonellosis can be averted by separating fecal matter from food effectively. Because the routes by which bacteria from the intestinal tracts of humans beings and other animals enter the food preparation area are often subtle and diverse, separation may become an onerous task at best and impossible at worst. Scrupulous hygiene at both the personal and the municipal level is fundamental to the control of salmonellosis.

Personal hygiene entails washing the hands before preparing food, cooking all foods properly, as none of the enteric bacteria are resistant to even mild heating, and avoiding uncooked foods that may have been contaminated when harvested or in transportation and preparation. When there is reason to suspect that food may be contaminated, it should be treated as if it were. When water or milk is suspected to be contaminated, it should be avoided, but if this is not possible, heating to a slow boil for a few seconds will render these items safe. Figure 6.5 summarizes carriers of *Salmonella* to the family.

At the municipal level the best means of avoiding salmonella epidemics is an effective water purification system. This normally consists of filtration and chlorination or, much better, **ozonization**. These methods are more effective when coordinated with constant surveillance of the system and the water. Surveillance is based on an analytical laboratory in which trained microbiologists work together with field inspectors. The latter gather samples for testing in the laboratory and work to eliminate sources of infection. Surveillance must be done continuously if epidemics are to be averted.

Extraenteric Salmonellosis Infections of wounds, burns, surgical lacerations, and pneumonia caused by salmonellas are not unusual. These infections

Figure 6.5

Carriers of Salmonella *to the home.*

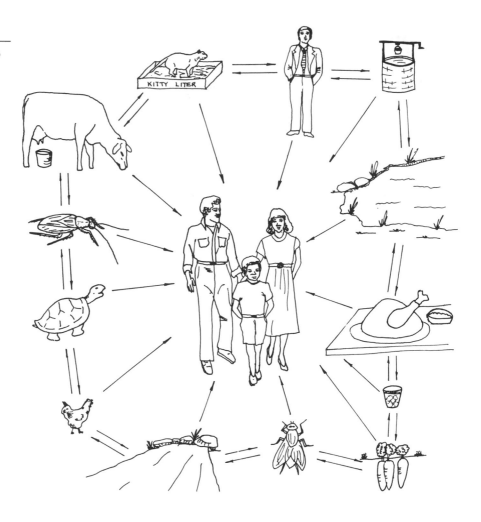

usually are acquired by contamination of the body tissues with fecal material during the **septicemic episode** of the disease. They also may originate externally from contact with contaminated materials including the feces of human beings or that of other animals.

Typhoid Fever

History   Salmonellosis caused by *Salmonella typhi* is a more serious disease than that caused by the other bacteria of the genus *Salmonella*. The disease has two forms: an early enteric form, followed by a prolonged febrile episode. The disease is classified as a fever because of the latter, more severe, syndrome.

Typhoid fever has been one the most common and most serious diseases in humans throughout the ages, as it spreads in epidemic form through contaminated water and food. Its spread can be checked only by modern methods of sanitary engineering supported by good microbiological practice. In the absence of these modern methods of control, epidemics of typhoid fever are inevitable. *Salmonella typhi* has affected the history of humanity in many ways, and the very nature of human beings has been shaped by the interaction between this bacterium and humankind.

I believe that modern humans' overly developed taste for alcohol has its genesis in the ancient fear of typhoid fever and other enteric disorders. As early as the time of Christ, personal health could be guarded best during times of enteric disease by avoiding the local water and drinking only beer or wine. One can argue that, in this way, individuals who did not have the capacity to drink only beer or wine for long periods would be more likely to perish from enteric disease than those who were able, or perhaps only too willing, to take all their liquid requirements in the form of beer and wine.

In evolutionary terms, this process obviously would serve to enrich future populations, modern human beings, with genes that would enhance a tolerance, or perhaps exaggerated appreciation, for alcoholic beverages. Although wine and beer can be contaminated with enteric organisms, this can be done only under laboratory conditions involving the presence of alcohol and acids in fermented materials. The difference between water and beer was expressed in the following manner by some appreciative but anonymous soul.

> Microscopic lens doth show
> That water teems with insects queer
> Oh! What a comfort it is to know
> There are no wiggling things in beer.

Infection   The incubation period of typhoid fever is 1 to 3 weeks, during which time the bacteria grow into large numbers in the small intestine. The death of large numbers of Gram-negative cells in the intestinal tract results in the release of large quantities of endotoxin. This toxin affects the ability of the cells of the intestinal lining to regulate the flow of water from the bloodstream into the intestinal canal, and severe diarrhea results. The first syndrome of typhoid fever, diarrhea represents the intestinal phase of the disease. It lasts several days to several weeks and varies in severity. During this phase of the disease, *Salmonella typhi* can be isolated from stool specimens but from no other body secretions.

In the latter part of the enteric phase, bacteria invade cells that line the small intestine, and from here they enter the lymphatic circulation and then the bloodstream. At this point body temperature rises, abdominal pain occurs, diarrhea lingers, and the intestinal wall becomes ulcerated. Ulceration is

Figure 6.6

Flagella stain of Salmonella typhi.

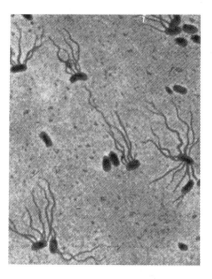

significant because perforation of the intestinal lining results in peritonitis. Blood from the intestinal wall circulation carries the bacteria to the lymph glands, spleen, liver, gallbladder, skin, lungs, and all other organs of the body.

At this time the symptoms may include lassitude, high fever, pneumonia, mental confusion, and prostration. Bacterial growth may break through the outer skin, giving rise to skin lesions on the abdomen called **Rose spots**. The febrile syndrome lasts 1 to 3 weeks and is self-limiting. During this phase of the disease, *S. typhi* may be isolated from feces, urine, blood, and skin lesions.

In untreated cases of typhoid fever, the death rate may be as high as 50%. In those who survive, a long period of convalescence begins on the fifth to seventh week after infection, and the carrier state is common. Bacteria may continue to grow in the intestinal tract, urinary bladder, gallbladder, or other site in the body without further discomfort to the carrier. **Serological typing** makes it possible to establish whether the surviving bacteria are the same strain that caused the primary disease or if the person has acquired a new infection.

Etiology   The causative agent of typhoid fever is *Salmonella typhi* (Figure 6.6). This organism differs from the other salmonellas in that it produces a more severe infection than any of the others and also in that it causes typhoid fever almost exclusively in human beings. Apes and other animals that suffer infection of *S. typhi* show only diarrhea or asymptomatic infection. *S. typhi* has many serotypes, but all are of similar virulence.

The natural reservoir of this organism is the intestinal tract of human beings, but it also may be found in animals. The major means of transmission is the feces-to-food route, as with other salmonellas. The bacteria can

Figure 6.7

The story of Typhoid Mary.

At the turn of the century, approximately 25,000 deaths per year resulted from typhoid fever in the United States. Today, with a much larger population, the number rarely reaches 20 per year.

Mary Mallon came to the attention of the New York City Health Department in 1906 following an investigation made by George Soper, an official of the Department. He found that Mary had worked as a cook for three families in Long Island homes when, consecutively, typhoid fever broke out in these homes. Soper found Mary working as a cook in New York City, but she fled after threatening him with a large carving knife. She hid in and around New York until 1908, when she was arrested and forcefully examined. All tests performed showed that she carried large numbers of *Salmonella typhi* in her intestinal tract, but she refused the treatment available. Mary was held in jail until 1910 but released for lack of legal grounds on which to detain her.

By now known as Typhoid Mary, she disappeared until 1915, when she again was traced by several typhoid outbreaks in the area near New York City. She was returned to the city and held as a hospital patient under legal restraint until her death in 1938. At the time of her death, Mary was still a carrier of typhoid fever bacteria. Ironically, the sulfa drugs needed to cure her had been available since 1932.

survive in sewage and polluted waters many days. This route of transmission is one of the most common sources of epidemics.

The organism can be easily isolated from the feces, blood, urine , and sputum of infected or convalescent individuals and from the feces of carriers, because all strains grow luxuriantly in selective media containing **selenite** or an iodine and sulfur complex called **tetrathionate**. Using selective media and enrichment culture techniques even a few salmonellas can be cultured from feces containing many billions of other bacteria per gram of sample.

Epidemiology and Control   Epidemiology and control are the same as for the other salmonellas. Approximately 500 cases of typhoid fever occur in the USA each year, but deaths occur rarely. The story of Typhoid Mary, reiterated in Figure 6.7, reminds us of the prevalence and fear of typhoid fever a century ago. Despite all the measures taken to avert epidemics, they still happen all too often. Currently, epidemics appear in the United States only when travelers or emigres arrive during the incubation period and subsequently contaminate the water or food of the local population. In evolving countries where public water supplies are not treated adequately, epidemics of typhoid fever are common and loss of life may reach levels similar to those that existed before the advent of modern medicine. Although vaccines are not available, trials are being held with preparations that seem to reduce the incidence of epidemic cases of typhoid fever by approximately 80%. These vaccines are being tested in certain areas of India and other Eastern communities.

Shigella

Unlike salmonellosis, shigellosis is a disease primarily of human beings and is not found in most animals. Only the higher apes harbor shigellas, and these often do so without symptoms of enteric distress. Bacillary dysentery and amebic dysentery were not recognized as separate diseases until 1897, when Kiyoshi Shiga (1870–1957) discovered the bacteria that cause bacillary dysentery. Both kinds of dysentery often are found in epidemic form where large groups of people gather under less than ideal sanitary conditions.

Historically, bacillary dysentery is a disease of pilgrims, prisoners, dwellers in asylums, and armies. It is said that more battles have been lost to dysentery than to arms. During the battle for Granada in 1492, the Spanish monarchs Fernando and Isabela found that dysentery had struck the Arab armies several days before it was found in the Spanish lines. They attacked immediately, taking advantage of the knowledge that the Arab army would have fewer effective soldiers than the Spanish army, and they won a decisive victory over a much larger force. This historic victory occurred just before the Spanish army itself was struck severely by dysentery.

Infection   Dysentery is a highly infectious disease acquired by ingesting food or water contaminated with the causative organism. The incubation period is 1 to 5 days, followed by the sudden onset of acute abdominal cramps, diarrhea, and a low-grade fever. The organisms remain in the intestinal tract and do not invade the interior of the body. They attack and destroy the cells lining the colon, resulting in severe irritation of the tissues. The toxins liberated by the bacteria give rise to an abnormally large flow of water from the blood into the intestinal tract, resulting in copious diarrhea often containing flecks of mucus and blood. The diarrheic stool contains large numbers of *Shigella* but few other organisms are normally seen in the intestinal tract.

Excessive loss of fluids results in dehydration of the body, and loss of **electrolytes** from the blood induces shock. If electrolytes are not restored rapidly to their normal levels and balance, death may follow. The infection rate may approach 100% in vulnerable populations, and the mortality rate may be as high as 40% in untreated cases. Mortality rates often approach 100% during epidemics in prisoner-of-war camps. Such mortality rates were observed in prison camps during the Civil War in the United States and in some Japanese camps during World War II. A recent epidemic caused by antibiotic-resistant *Shigella* in Central America caused the death of 25% of those infected. Dysentery is said to be a disease of **high morbidity** and **high mortality**. Mortality rates tend to be much higher in infants, in the aged, and in the medically compromised than in healthy adults.

The shigellas normally do not cause septicemia as the salmonellas do. Consequently, the former is an intestinal illness and not a systemic infection

	Glucose	Mannitol	Ornithine decarboxylase	β– Galactosidase
S. dysenteriae	A	–	–	±
S. flexneri	A	+	–	–
S. boydii	A	+	–	±
S. sonnei	A	+	+	+

A = acid but no gas
– = no reaction
+ = positive reaction
± = variable reaction

Figure 6.8

Species of Shigella *can be identified by biochemical reactions.*

like typhoid fever. The shigellas produce lipopolysaccharide endotoxin and also a recently discovered exotoxin.

Etiology   Bacillary dysentery is caused by several species of bacteria of the genus *Shigella.* They differ from the salmonellas in that they ferment glucose with the production of acid but not gas. Shigellas are not motile and, as a consequence, they have no H (flagellar) antigens, although, like all other enterobacters, they do have O antigens. All shigella infections produce the symptoms of LPS enterotoxin, including high temperature, abdominal pain, diarrhea, and electrolyte imbalance.

The most common isolate from cases of dysentery in the United States is *S. boydii*, whereas *S. dysenteriae*, *S. sonnei*, and *S. flexneri* are found only rarely. The most severe cases of dysentery are caused by *S. dysenteriae*, an organism found most frequently in tropical and semitropical climates. *Shigella dysenteriae* is the principal causative agent of bacillary dysentery in Japan and other Asian countries.

Each species has a few serotypes but not the variety seen in the salmonellas. The number of carriers is small in comparison to salmonellas, but convalescents continue to shed the organisms for many weeks after all symptoms of the disease have disappeared. These are not true carriers, however, as the organism does not become established permanently in the intestinal tract as part of the normal flora of the individual.

When dysentery is well established and the symptoms are evident, isolation of the etiologic agent is simple, because predominance of the organism in the stool water almost guarantees success. When the stool is normal, the shigellas can be isolated on selective laboratory media such as salmonella-shigella agar with little difficulty. Once in pure culture, the species can be ascertained by employing several fermentation tests, and identity can be confirmed with specific antisera. The scheme in Figure 6.8 shows this process.

Epidemiology and Control   In 1986 Texas had 2,454 cases of shigellosis, an incidence of 14.6 per 100,000 population. Of these, 47% were *Shigella sonnei*; 16%, *S. flexneri*; 3%, *S. boydii*; 1% *S. dysenteriae*; and 34%, unidentified. The highest incidence (62 cases per 100,000) was observed in children between ages 1 and 4 years. Newborn children had an incidence of 25 per 100,000, and 5- to 9-year-olds, 26 per 100,000. Texans of Hispanic origin had an overall incidence of 24 per 100,000, African Americans 11 per 100,000, and Anglo Americans 8.5 cases per 100,000 population.

Bacillary dysentery is a disease primarily of primates, but the bacteria may appear as transients in the intestinal tracts of foxes, dogs, and wolves. To a lesser extent they also have been found in farmyard fowl including chickens, ducks, and geese. Transmission is generally by the feces-to-food route via contaminated food and water and also by contamination of comestibles by vermin and insects.

The number of natural sources of *Shigella* is more limited than that of the salmonellas because shigellosis is almost exclusively a disease of human beings. The number of organisms required to cause an infection is small, so the disease can transmit easily from one individual to another. The infective dose for dysentery is approximately 150 live cells, compared to approximately one billion live cells for typhoid fever.

Bacillary dysentery can be controlled by the same means described for salmonellosis, except that only the human-to-human route has to be guarded. Several immunization programs are being tested at this time, and at least one possibly may be successful. At present there is no effective immunization program.

Escherichia coli

Escherichia is the name of a genus with only one species, *coli*, although in the past, the genus was divided into several species. These bacteria (Figure 6.9) are members of the Enterobacteriaceae and share many similarities with pathogenic organisms such as salmonella and shigella. *Escherichia coli* has more than 1,000 serotypes, some of which are highly virulent primary pathogens of warm-blooded animals, including human beings. The nonpathogenic *E. coli* are so prominent in the fecal matter of human beings that they are considered symbionts of humans. Because they are found with such regularity in all animals capable of harboring salmonella and shigella, they have come to be used as indicators of the contamination of foods, milk, and water with the feces of human beings or other animals.

Coliforms   The presence of *Escherichia* and closely related enteric bacteria, the **coliforms** (*E. coli*-like organisms), in food and water is universally accepted as incontrovertible evidence of fecal contamina. This also means that the absence of *E. coli* and coliforms assures that food and drink are free of

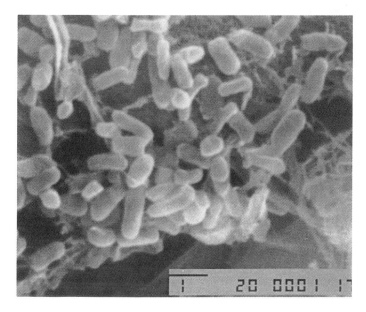

Figure 6.9

Scanning electron micrograph of pure culture of Escherichia coli.

enteric pathogens. This rule has some exceptions, but these are beyond the scope of this text. In Europe *Streptococcus faecalis* is used as the indicator organism of fecal contamination, as it is found only in the intestinal tracts of primates including human beings.

All milk, water, and food intended for human or household pet consumption must be examined by a procedure such as that depicted in Figure 6.10. This coliform test often yields positive results when only soil and water bacteria are present in the sample tested. Because many members of the enterobacteriaceae are found normally in soil and water, interpreting the results of these tests may be in error at times.

The possibility for error may be reduced by identifying *Escherichia coli* as shown in Figure 6.10, or, alternatively, by a separate test called the fecal coliform test. The test for fecal coliforms is accomplished using the protocol given above but with an incubator temperature of 44.5°C. *Escherichia coli* grows well at this temperature, but the coliforms do not. It is important to ascertain the fact that a positive coliform test is based on the finding of *E. coli*, as mixed cultures of soil and water organisms may yield spurious results. This is accomplished by streaking plates of Eosin Methylene Blue Agar from positive lactose fermentation tubes grown at 44.5°C and incubating the plates, also at 44.5°C.

The "confirmed test" for the identity of *E. coli* is performed with the IMViC tests. The I stands for the production of indole from tryptophan; M for methyl red positive, which means that acid was produced from glucose; Vi for the Voges-Proskauer test, which indicates the production of acetylmethylcarbinol; and C for the utilization of citrate as a sole source of carbon.

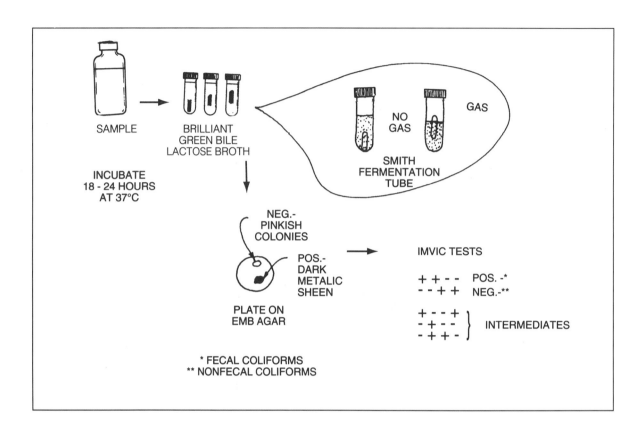

Figure 6.10

Standard coliform test procedure using IMViC tests.

As shown in Figure 6.10, *E. coli* is positive for the first two tests and negative for the last two, making the results ++− −. *Enterobacter*, an organism that may be found in the intestinal tract of human beings or animals, as well as in soil and water, is − −++. Coliforms give other "intermediate" results.

Figure 6.11 pictures an automatic Gram stainer. This instrument is used when large numbers of cultures have to be examined for Gram staining reaction. It gives consistent results and can be operated at a saving in money and labor in any operation where several hundred slides have to be stained weekly. Use of this machine can be justified as the Gram stain not only is the most commonly used stain in the microbiological laboratory but also is often the only staining procedure employed. It can be used successfully in identifying *E. coli*, especially when a known culture of *E. coli* and a Gram positive organism such as *Staphylococcus aureus* are used for comparison on each slide prepared.

Pathogenic Escherichia coli   *Escherichia coli* causes at least three well defined diseases as a consequence of the ability of these bacteria to acquire genetic factors, genes, from pathogenic bacteria. Because these organisms are related closely to the salmonellas and the shigellas, they can acquire deoxyri-

Figure 6.11

*Automatic Gram stainer.
Photo courtesy of TOMTEC,
Orange, Connecticut.*

bonucleic acid (DNA) from these two great families of pathogenic bacteria. Genetic elements responsible for the production of endotoxin and other enterotoxins can be transferred by **transformation**, **transduction**, **conjugation**, and **plasmid infection**. Figure 6.12 depicts the transfer of a plasmid from one cell to another. Although some strains of *E. coli* can become pathogenic as a result of genetic contact with pathogens, others seem to be constituitively pathogenic. One strain of recent notoriety, *E. coli* O:157:H7, has been found in a variety of foods including beef, hamburger patties, chicken, and apple juice. This strain has caused numerous deaths.

Enteropathogenic Escherichia coli   Some controversy exists regarding the pathogenicity of enteropathogenic *E. coli*, but they almost certainly do produce disease. The clinical syndrome is called **epidemic infantile diarrhea** or **infant summer diarrhea** because it is found predominantly, but not solely, in infants. This is a disease encountered where poor hygienic conditions predominate, and it may appear in epidemic form as a result of water and food contamination. The infection also is known to break out spontaneously in nurseries, kindergartens, refugee camps, and other places with large populations of infants and small children.

In untreated cases the death rate may be as high as 50% of infected individuals, but treatment with antibiotics reduces this number dramatically. Toxins produced by enteropathogenic *E. coli* are similar to those found in salmonella, and it is assumed that they were acquired by one of the methods of DNA recombination.

Figure 6.12

Transfer of plasmid from one cell to another.

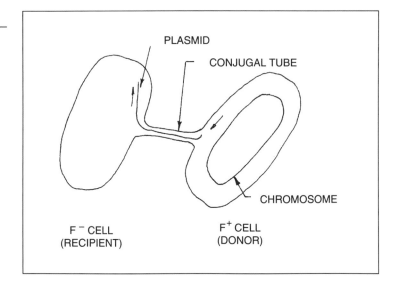

Enterotoxigenic Escherichia coli   Some strains of *Escherichia coli* are infected with plasmids, which carry genes that endow the organism with the ability to produce enterotoxins. These strains of *E. coli* are called **enterotoxigenic** *E. coli* because they cause gastroenteritis. The varying degrees of virulence of gastroenteritis are determined by the number and nature of the pathogenic genes the organism carries.

Two enterotoxins that have been identified are called LT and ST. The designations mean **heat labile toxin** and **heat stable toxin**, respectively. These cause secretion of water from the blood supply of the intestinal wall circulation with resulting severe diarrhea. Other symptoms include dehydration and loss of electrolytes, shock, and prostration. The clinical disease, often called **watery diarrhea**, closely resembles the diarrhea of bacillary dysentery. This diarrheic disease is especially devastating to the newborn and to medically compromised individuals.

Enteroinvasive Escherichia coli   Many strains of *E. coli* produce a shigella-like diarrheal disease in children and adults. It is an intensive diarrhea with flecks of mucus and blood similar to that of the dysentery caused by *Shigella dysenteriae*. Symptoms of the disease vary from case to case, and at least part of the syndrome depends on the toxin genes the organism carries. The disease is localized in the large intestine, where the bacteria invade the cells of the lining, causing extensive tissue damage.

Symptoms may be limited to a brief episode of mild diarrhea, or they may be those of an acute disease indistinguishable from dysentery. Infection with the highly virulent forms of enteroinvasive *E. coli* often results in death. The disease is found primarily in situations where poor hygienic conditions prevail and where children suffer from malnutrition. This is one of the diseases that falls under the general name of **turista**, **Montezuma's revenge**, or more commonly, **traveler's diarrhea**. It is associated primarily with poor personal hygiene; the bacteria are found in water, ice, uncooked vegetables, fruits and other foods. Local residents, having grown up with the infecting strains, have immunity.

Extraenteric E. coli Infections   Most strains of *E. coli* are not pathogenic for human beings when confined to the intestinal tract, but in other parts of the body they cause serious illnesses. Urinary tract infections are caused by many different bacteria, but the organism encountered most frequently in these extraenteric infections is *E. coli*. This is because of the proximity of feces to the urinary meatus in part and in another part to the ability of these bacteria to grow in many different environments. The infection is common in young women who are beginning to have sexual intercourse regularly because the inexperienced male can introduce fecal matter from the perianal region into the genital tissues so easily. In a previous, more romantic age, the syndrome was called **honeymoon cystitis**.

Once it is established in the urinary tract, *Escherichia coli* also may invade the kidneys causing the syndrome called **pyelonephritis**. Invasion of the circulatory system, usually from intestinal wounds, results in bacteremia, which also may give rise to infection of the various organs of the body including the liver, spleen, gallbladder, pancreas, and even the brain under certain conditions. Contamination of the peritoneal tissues as a result of fecal contamination from a wound or surgical perforation of the intestines generally results in massive peritonitis with a high probability of death.

Escherichia coli is one of the most common causes of meningitis in small children, and it is also common in adults. It often is found in nosocomial infections and is one of the most common infections acquired from medical apparatus such as respirators, catheters, and specula.

Epidemiology and Control   The spread of pathogenic *E. coli* is by the feces-to-food route, either by direct transmission or by indirect routes such as contaminated water, milk, or food. Some of the diseases these bacteria cause are associated with poor hygienic conditions. The phrase "poor hygienic conditions" implies the presence of flies, cockroaches, and filth in the living environment, plus lack of adequate personal hygiene. Contamination of food and water by pathogenic *E. coli* causes epidemics in much the same manner as those caused by salmonella and shigella. In like manner, the same measures that are effective in controlling other enteric diseases are adequate for controlling *E. coli* epidemics.

Before it was established that some strains of *E. coli* are pathogenic, the assumption was that enteric disorders not attributable to salmonella, shigella, or cholera were attributable to viral infections. Some microbiologists referred to the toxigenic organisms isolated as "lactose fermenting" shigella and also as "atypical shigella."

Yersinia

The genus *Yersinia* contains three species: *pestis, enterocolitica,* and *pseudotuberculosis. Yersinia pestis* causes plague and will not be discussed further here because it is not associated with food and drink. The other two organisms are found primarily in the intestinal tracts of human beings and other warm-blooded animals. Transmitted by contaminated food and water, they cause intestinal disorders when ingested in large numbers.

Symptoms that accompany infection are limited to acute diarrhea that appears suddenly and at full intensity. The organisms form abscesses in the intestinal tissues and therefore are shielded from chemotherapeutic agents. Yersinias are isolated easily from fecal specimens, and identification in the laboratory presents no serious problem. Diarrhea caused by *Y. enterocolitica* can become chronic and debilitating, but the more serious threat is concern for secondary infections that may lead to complications.

Klebsiella

Klebsiellosis is caused by an organism historically called *Friedlander's Bacillus* but now is named *Klebsiella pneumonia.* It is found commonly in the intestinal tract of human beings and other warm-blooded animals but also can be isolated from water and soil. These organisms are associated with epidemics of diarrhea and seem to become pathogenic only when they acquire genes from pathogenic strains of *E. coli.*

Epidemic diarrhea results from the transmission of pathogenic klebsiellas by the feces-to-food route. In nonenteric disease, klebsiella is classified as an opportunistic pathogen and often is found as the primary cause of pneumonia, septicemia, and urinary tract infections. Klebsiellas also are found as secondary invaders in respiratory tract infections, in burns, and in traumatic injuries to the skin. Klebsiellas are common in nosocomial infections, where they are the second most common cause of urinary tract infections. They also cause 3% of all cases of pneumonia.

OTHER DISEASES TRANSMITTED BY FOOD AND DRINK

Many enteric diseases are caused by organisms that do not belong to the family of bacteria named Enterobacteriaceae. They belong to other, diverse families of bacteria but share a common habitat with the Enterobacteriaceae. They are frequently found in the intestinal tract of human beings and other animals. Because they are passed from one individual to another by the feces-to-food route of transmission, they are of prime importance to this discussion. The diseases they cause include cholera, campylobacteriosis, and listeriosis.

Cholera

Cholera is an intestinal disorder that has afflicted people from the beginning. Descriptions of the disease can be recognized in the earliest writings of many ancient societies, leading modern scientists to realize that symptoms of the disease have not changed in approximately 5,000 years. Cholera is a diarrheic disease caused by *Vibrio cholerae*, an organism with cellular morphology resembling an orthographic comma when viewed with the microscope. It belongs to the group of bacteria called **spirilla** because the cells are in the form of a spiral or, more accurately, a corkscrew. *Vibrio cholerae* survive for prolonged periods in heavily contaminated water, but not in clear water, and produce **cholera zones** throughout the world.

Cholera is thought of as an Asian disease because it always has been endemic in India and other Asian countries. It is a severe epidemic disease of high morbidity and high mortality. Approximately 60% of clinical cases result in death when treatment is not available. Transmission is by food and water, but control is not difficult. Cholera is found almost exclusively in human beings, and epidemics take hold of the entire community where water and food are normally contaminated by human excrement.

Symptomatology   After ingestion of *V. cholerae* in food or water a 2- to 5-day incubation period follows. Because stomach acid kills the organisms easily, the number of cells ingested has to be large if any are to survive passage through the stomach. Just a few live cells need be present to cause an infection when a large meal is consumed, but large numbers are necessary when water only is consumed. Once the bacteria enter the small intestine, they begin to grow rapidly and the population of vibrios becomes predominant in the small intestine.

Symptoms such as nausea, vomiting, diarrhea, and intestinal distress appear suddenly. The diarrhea persists, and large volumes of body water are lost in a short time. The diarrhea of cholera has a characteristic odor and appearance. Called **rice water stool**, it contains numerous flecks of mucus but little fecal matter. Microscopic examination reveals large numbers of epithelial cells and large numbers of bacterial cells with a predominance of *V. cholerae.*

The main characteristic of cholera is the volume of diarrheal discharge. This may reach volumes of 12 to 15 liters of fluid per day and in extraordinary cases may reach 30 to 40 liters per day. Major effects on the body are

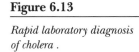

Figure 6.13

Rapid laboratory diagnosis of cholera.

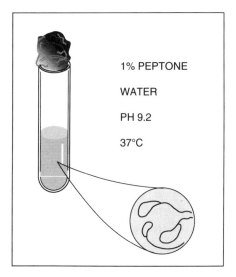

dehydration and loss of electrolytes. The death rate ranges from 35% to 75% of untreated cases, but simply replacing liquid and electrolytes can reduce the rate to about 2%. This replenishment can be accomplished in the field with sodium bicarbonate solution in a volume of water equal to that lost in the diarrhea. In a medical setting, introducing antibiotics, glucose, and electrolytes in an amount of water sufficient to replace the volume lost in diarrhea can bring the death rate down to less than 1%.

Etiology   Cholera is caused by organisms that belong to the genus and species *V. cholerae*. They are small, comma-shaped, pointed, Gram-negative bacteria motile by one polar flagellum. They grow well in simple laboratory media but do not tolerate carbohydrates well. They prefer alkaline media, not acid conditions. Isolation from diarrheic stools is easy, and diagnosis of cholera can be accomplished in 24 hours (see Figure 6.13). Cultures in 1% **peptone water** at a pH of 9.2 and temperature of 37°C yield essentially pure cultures of cholera vibrios when these are present in the diarrheic stool. Four strains of *V. cholerae* have been found: **Hikojima**, **Inaba**, **Ogawa**, and **El Tor**.

Epidemiology   *Vibrio cholerae* is found only in humans, and the intestinal tract of human beings is thought to be its natural habitat as it is not normally found in animals. It is introduced into the environment by the feces of infected human beings. The organism does not survive out of the body for long periods unless the water is heavily contaminated with organic matter. Samples of drinking water inoculated with fecal matter from cholera victims often do not yield *V. cholerae* even though they give positive coliform tests. This is due to a rapid die-off of the cholera organisms in drinking water. In areas

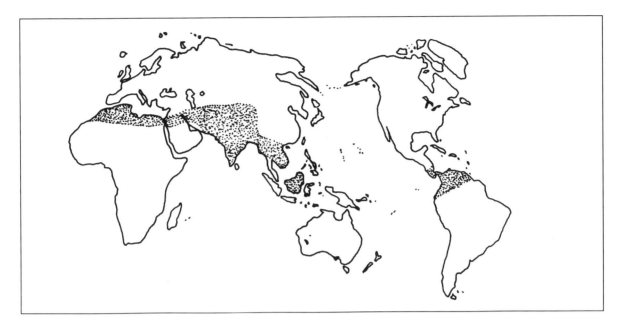

where cholera is endemic, monitoring of food, water, and milk is essential in efforts to control the spread of cholera.

The historic home of cholera has been India and the countries surrounding it. Epidemics traditionally have started in this area of the world and moved westward, sweeping the entire globe. Seven major pandemics have been recorded since 1818, the last one ending in 1975. The most recent epidemic in the United States took place during the years 1911 to 1913. Since that time only one death from native cholera has occurred in the continental United States. The epidemic now under way in South America and Mexico reached El Paso, Texas, in August of 1993. Several cases have been reported in laboratory workers but none from natural sources. The cases of cholera in Western, industrialized countries generally are regarded as travelers' cholera because they often originate in areas where the disease is endemic. Prevention of epidemics requires rapid detection, diagnosis, and isolation of infected individuals.

Control   To control the spread of cholera food and water must be protected from sewage contaminated with human feces. Maintaining a pure water supply and good personal hygiene also are essential in preventing the spread of cholera. Introduction of vibrios directly into prepared food by handlers or servers is one of the main routes of transmission in epidemics. Sanitary measures have effectively eradicated the disease from many countries, but it is still present in both endemic and epidemic form in India, Bangladesh, Indochina, the Philippines, parts of Africa, and tropical America. Figure 6.14 shows the areas of the world in which epidemics of cholera are occurring.

Figure 6.14

Areas of the world where cholera is endemic.

Figure 6.15

Growth medium and incubation conditions for isolation of campylobacters.

CAMPYLOBACTER MEDIUM

BRUCELLA AGAR BASE
10% SHEEP RED BLOOD CELLS
7% LYSED HORSE RED BLOOD CELLS
VANCOMYCIN
POLYMYXIN B
CEPHALOTHIN
AMPHOTERICIN B

C. JEJUNI	C. INTESTINALIS
42°C	25°C
5% O_2	5% O_2
10% CO_2	10% CO_2
85% N_2	85% N_2

Other Vibrios   Other organisms that cause cholera-like disorders are *V. parahaemolyticus, V. mimicus, V. fluvialis,* and *V. vulnificus.* The routes of transmission of these vibrios have not been described completely, but they are known to be associated with the consumption of fish and shellfish that have not been cooked properly. Although most of the cases of diarrhea from these species of vibrio have been observed in India, the causative bacteria have been isolated from shellfish originating in many areas of the world. This serves as a warning that the potential for epidemic disease is present in many parts of the world, not only in "cholera countries."

Campylobacter jejuni   Bacteria of the genus *Campylobacter* have been of great interest to veterinarians since the 1930s because *C. fetus* causes a venereal disease in cattle that results in abortion and sterility. The importance of campylobacters in enteric disease in humans was not discovered until the late 1960s. The exacting requirements of campylobacters for low concentrations of oxygen, and the fact that they grow slowly in the laboratory, made them impossible to detect before this time (see Figure 6.15). We know now that *C. jejuni* and *C. intestinalis* cause several different intestinal disorders in human beings and that they are a prevalent cause of diarrhea in almost all populations of the world.

After an incubation period of 2 to 4 days, diarrhea becomes well established. The disease is characterized by fever and may include bacteremia, endocarditis, and meningitis. The diarrheic episode may be of short duration, but the other, more serious, syndromes may persist and even lead to death.

Campylobacters are not a part of the normal bacterial load of human beings. They are found in many animals—chickens, turkeys, pigs, sheep, goats, and cattle—and animal pets such as cats and dogs. It is fairly well established that the campylobacters make up part of the intestinal microflora of these animals and that this is the source of infections for human beings. Campylobacter diarrheas result from consumption of foods contaminated by animal feces. Unpasteurized milk is a common source of diarrhea, as is water from rivers and streams used by animals. Handling of infected farm animals and pets also has been linked to infection. People at highest risk are farm workers, veterinarians, and children.

Control measures consist of protecting food and drink from contamination with the fecal matter of farm and domestic animals. Fluids such as saliva, blood, and milk from infected animals are known to carry the bacteria and also may be avenues of contamination. Although less common, the infection can be transmitted from human to human either by the feces-to-food route or by direct contact.

Listeriosis

Listeria monocytogenes is a small, Gram-positive, rod-shaped bacterium normally found in soil, water, barnyard manure, and the fecal matter of healthy mice, rats, poultry, sheep, goats, cattle, and human beings. This organism is considered an opportunistic pathogen of human beings and a primary pathogen of animals. It produces infectious illness in cattle, sheep, and goats; and laboratory animals such as rats, mice, guinea pigs, and chickens can be infected with pure cultures of the organism. The main concern to veterinarians is the epidemic spread of listeriosis in cattle because untended cases often lead to abortion. The milk of infected animals yields large populations of the organism, making it unsafe for human consumption. Listeriosis recently has been elevated from that of a rare and not very important human enteric infection to that of a major menace. Although the disease always has been present in rural communities, it now appears to be more an urban problem.

Before pasteurization and analytical microbiology, listeria infections probably were prevalent but unrecognized. The first epidemic of listeriosis in human beings, reported in 1929, was associated with unpasteurized milk from cows known to have bovine listeriosis. Since that time, listeriosis has appeared sporadically in the United States and in many other countries as well.

The first major outbreak of listeriosis appeared in Boston, Mass. in 1979, associated with consumption of raw vegetables that may have been exposed to infected dairy cattle. The total number of cases was not established definitively, but it was more than 20, of these twenty confirmed cases, 5 died, giving a mortality rate of 20%. In 1981, 41 people who ate coleslaw in Halifax, Canada were stricken and 18 died, giving a death rate of 43%. In 1983, 49 individuals acquired listeriosis from animals with active infections of listeria and 14 died. Two outbreaks involving 164 individuals and a total of 81 deaths were blamed on pasteurized cheese. The mortality rate is approximately 35% of all cases recorded. Pasteurization was adequate but, as *L. monocytogenes* is found inside white blood cells in the milk of infected cows, some of the bacteria were not killed by heating at 73°C for 3 seconds. In view of this, pasteurization temperatures and conditions are being reconsidered. New legislation probably will be enacted to address this problem.

Human listeriosis is a systemic infection with varied syndromes. It most frequently manifests itself in meningitis, but septicemia, endocarditis, pneumonia, and abortion also are encountered. Congenital infection arises from placental transfer or vaginal infection at birth, and symptoms of the disease are developed fully in the newborn 24 to 48 hours after birth. The most usual syndrome of congenital listeriosis in human beings is pneumonia.

QUESTIONS TO ANSWER

1. What is the feces-to-food route of disease transmission?
2. Why is ice cream a good vehicle for transporting salmonella?
3. What are the names of 11 genera of enteric bacteria?
4. What is the difference between *Escherichia* and *Enterobacter*?
5. What is salmonellosis?
6. What is typhoid fever?
7. What is shigellosis?
8. What is bacillary dysentery?
9. What are the keys to the serotypes of *Salmonella typhi*?
10. How are the species of *Shigella* differentiated from each other?
11. What are coliforms?
12. How is the coliform test performed on hamburgers?
13. What are fecal coliforms?
14. What are the three types of pathogenic *E. coli*?
15. What is meant by the term "extraenteric infection?"
16. How are the yersinias associated with enteric disease?
17. What is cholera?
18. What is *Vibrio parahemolyticus*?
19. What is *Campylobacter jejuni*?
20. Why do bacteria cause enteric disorders?

FURTHER READING

American Academy of Pediatrics. 1986. RED BOOK: REPORT OF THE COM-MITTEE ON INFECTIOUS DISEASES, 20th ed. American Academy of Pediatrics, Elk Grove Village, Illinois.

Center for Disease Control. 1983. MORBIDITY AND MORTALITY WEEKLY REPORT: ANNUAL SUMMARY 1983. Massachusetts Medical Society, Waltham, Massachusetts.

Elliot, R. P., D. S. Clark, K. H. Lewis, H. Lundbeck, J. C. Olsen, and B. Simonsen. eds. 1988. MICROORGANISMS IN FOODS: THEIR SIGNIFICANCE AND METHODS OF ENUMERATION. University of Toronto Press. Toronto.

Gahan, C. G. and J. K. Collins. 1990. LISTERIOSIS BIOLOGY AND IMPLICATIONS FOR THE FOOD INDUSTRY. TRENDS IN FOOD SCIENCE AND TECHNOLOGY. 2:89–93.

Milgrom, F. and T. D. Flanagan. 1982. MEDICAL MICROBIOLOGY. Churchill Livingstone, New York.

Mims, C. A. 1982. THE PATHOGENESIS OF INFECTIOUS DISEASE. Academic Press, New York.

Riemann, H. and F. L. Bryan. 1979. FOOD-BORNE INFECTIONS AND INTOXICATIONS. Academic Press, New York.

Volk, W. A., D. C. Benjamin, R. J. Kadner, and J. T. Parsons. 1986. ESSENTIALS OF MEDICAL MICROBIOLOGY. J. B. Lippincott, Philadelphia.

Enteric Diseases Caused by Protozoa

7

He gathered all that springs to birth
From the many-venomed earth;
First a little, thence to more,
He sampled all her killing store.

Housman

Protozoa are either single-celled, eucaryotic, complex microorganisms with a large array of subcellular **organelles** or they are very small, simple, single-celled animals. Of the 50,000 or more species that have been identified, only a few are capable of producing disease in human beings. Of the 25 to 30 diseases of human beings caused by protozoa, only five or six are transmitted by food and water. All of these are passed from one human being to another by the feces-to-food route, generally by contamination of food and water.

The protozoa were discovered by Antonie van Leeuwenhoek. Upon examining his own stool with his lens-microscopes, van Leeuwenhoek noted that these microorganisms may be the cause of the diarrhea that afflicted him at the time. He made these observations in the 1670s, but not until the end of the 19th century were van Leeuwenhoek's pointed speculations proven by scientifically valid experiments.

PATHOGENIC MECHANISMS

The pathogenic properties of some of the protozoa are well established, but those of others are only assumed because researchers have not presented definitive data. It is established beyond doubt that *Entamoeba histolytica* is pathogenic, but the role of *Dientamoeba fragilis* in pathogenic processes is only suspected. Although toxins such as those of the bacteria have not been observed in the protozoa, many protozoa excrete enzymes such as the **proteases** and **phospholipases** which are capable of destroying tissue cells. Some protozoa produce **glycoproteins** and **indoles**, which are mild toxins active on certain cells of the body.

EFFECT OF ANTIBODIES ON PROTOZOA

Antibodies produced by the host's body in response to the presence of invading protozoa are essentially the same as those produced against fungi, bacteria, and viruses. These antibodies are not effective when the protozoa remain in the lumen of the intestinal tract, but they are effective against organisms in the body tissues. The effect of antibodies against protozoa is often limited to controlling further spread of the parasite but is generally not sufficient to eliminate it, as is the case in bacterial infections. As a result, there are no usable immunization procedures against protozoa. Meanwhile, great loss of life occurs yearly worldwide as a result of protozoan disease.

Part of the ability of many protozoa to invade a host's body successfully lies in the fact that protozoa can become **immunologically invisible** to the defense mechanisms of the host. **Antigenic mimicry** is a phenomenon in which the protozoan parasite becomes enshrouded with substances from the body fluids of the host, thereby disguising the fact that it is a foreign substance in the body. The **blocking** reaction is one in which the surface of the parasite becomes covered by antibodies that have little or no effect on the protozoa but prevent further attack by more aggressive or more effective antibodies. Still other protozoa are capable of escaping the host's immune defense system by living inside of cells, where they are out of reach of circulating blood and lymph. Finally, some protozoa avoid immunologic confrontation by constantly changing **antigenic specificity** and thus staying one step ahead of the host's antibody defenses.

Because the protozoa are closely related to the animals, biologically and physiologically, antibiotics and the other **chemotherapeutic agents** that are effective in controlling and eliminating bacteria are not useful in treating protozoan infections of human beings and other animals. On the other hand, no chemical substance can be used effectively in the prevention, control, and eradication of protozoan infections. These chemotherapeutic agents depend on specific toxicity for the protozoan pathogen but lack the sweeping antimicrobial effect that penicillin and other antibiotics have on bacteria. Chemotherapeutic agents that are effective against protozoa are:

Amphotericin B

Chloroquine

Dehydroemetine

Diiodohydroxyquin

Furazolidine

Isethionate

Metronidazole

Paramomycin

Pentamidine

Quinacrine hydrochloride

In spite of this unsatisfactory state of affairs, research on effective chemotherapy against protozoa is quite intense. Spanish church history reveals that the age of modern chemotherapy against pathogenic protozoa began in 1638 with the finding of chemicals capable of killing the protozoan parasites that cause malaria without injuring the human host. **Pelletier** and **Caventou** succeeded in isolating **quinine** from **Jesuit powder** in 1820. Figure 7.1 recounts a brief history of Jesuit powder. Quinine was the sole means of

When the Countess of Chinchon, wife of the Spanish viceroy to Peru, contracted malaria in 1638, she was treated with an Indian remedy. The medicine was extracted from the bark of a tree that grew only in Peru. Wishing to share this miracle drug with her countrymen, the countess sent large quantities of the bark to Spain in the care of Jesuit priests.

From that time until 1820, when the drug was identified by Caventou and Pelletier, the bark was known as Jesuit bark, or, when powdered, as Jesuit powder. In that 200-year span of time, the monopoly on sale of the bark or powder rested in the hands of Spanish and Portuguese Jesuits. These religious zealots, it is said, made the powder available to everyone, at little or no cost to Catholics but extremely high cost to others unless they renounced their religion and accepted baptism in the Catholic Church.

Jesuit bark was the only effective treatment for malaria during this period making it highly desirable. Nevertheless, Protestants claimed that it was no remedy at all but, rather "the Pope's evil brew." A British physician capable of obtaining large quantities of bark mixed it with wine and used it as a secret potion among English Protestants. It is assumed that many of these people knew but were willing to accept the secret potion explanation rather than do without the medicine.

Figure 7.1

Brief history of Jesuit bark.

effective prevention and treatment of malaria from 1638 until the middle of World War II.

SARCODINA

Amebic Dysentery

Any consideration of protozoan diseases has to begin with amebic dysentery (also spelled amoebic dysentery), as this is the most serious disease produced by protozoa in human beings and also one of the most common diseases in all parts of the world. Amebic dysentery is found throughout the world, but the higher rates of infection are in the warmer climates.

The disease historically known as dysentery can be traced back to the very beginnings of chronicled history. The name comes from the Greek language and means malfunctioning enteron or sick gut (dys = not well + enteron = gut). Dysentery as a clinical entity can be caused by many different agents, both animate and inanimate. The former include protozoan, bacterial, and fungal toxins, viruses, and even animal parasites. Natural products and chemical substances also cause dysentery and diarrhea when ingested in large amounts. Examples of the former are prunes and raisins, and magnesium hydroxide and mineral oil are examples of the latter. When mineral oil or magnesium hydroxide is administered in a sufficiently large amount, it can induce a diarrhea indistinguishable from that caused by *Entamoeba histolytica.* This property is what makes magnesium hydroxide the drug of choice for inducing limited, well-controlled diarrhea in patients suffering from constipation.

History   The causative agent of amebic dysentery, *Entamoeba histolytica* (Figure 7.2) was discovered in 1875 by the German physiologist **F. Losch.** Discovery of the relationship between dysentery and the specific organism was complicated by the fact that the designated amebas were not present in all cases of dysentery and also by the fact that not everyone who harbored amebas was dysenteric. The dilemma was solved in large part in 1898 when Kiyoshi Shiga found that dysentery in patients who were free of amebas was caused by bacteria. With this work, the designations **amebic dysentery** and **bacillary dysentery** were established by the early part of the 20th century. The remainder of the dilemma was resolved with the finding that many individuals harbor *E. histolytica* but are immune to its pathogenesis, and also that other amebas that closely resemble this pathogen exist but belong to genera and species that are not pathogenic for human beings.

Conclusive proof of the etiology of amebic dysentery was not obtained until 1913, when **Walker** and **Sellards** succeeded in isolating sufficient cysts of *E. histolytica* from the feces of dysenteric individuals to carry out a properly

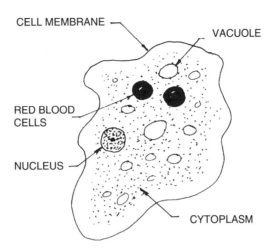

Figure 7.2

Structure and anatomy of Entamoeba histolytica.

designed experiment. The cysts were fed to 20 healthy volunteers known to be free of *E. histolytica* and to not shed the cysts of this organism in their feces. In a short time, 18 of the volunteers began to show signs of infection and to shed cysts of the pathogen in their feces. Of the 18, four developed dysentery, confirmed by clinical syndrome and laboratory diagnosis. Both kinds of proof were conducted by independent diagnosticians.

Amebic dysentery is endemic in many parts of the world (Figure 7.3) and has had a significant role in defining the history of the lands where it exists.

Figure 7.3

Areas of the world where amebic dysentery is endemic.

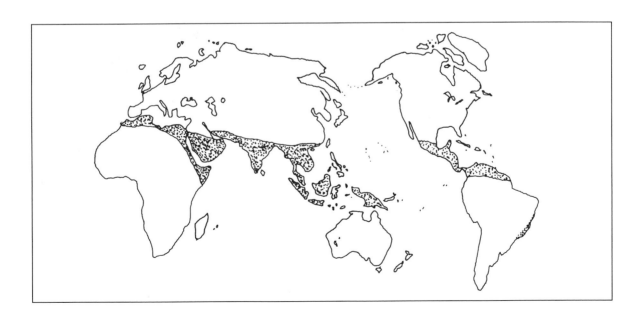

Individuals who are exposed to *E. histolytica* early in life develop a remarkable tolerance that does not exist in those not equally exposed. As a consequence, entering lands where amebic dysentery is endemic is a risk not everyone will undertake lightly. This bit of common wisdom has been as valid for conquerors and their armies throughout the centuries as it is today for tour guides and their charges.

Infection   Symptoms of the infection appear some 10 to 120 days after ingesting the cysts of *E. histolytica* in food or water. The **trophozoites** are not infective, as they are killed in the stomach. In contrast, the cysts survive passage through the stomach and indeed become **activated** by the digestive action of the stomach fluid. **Excystment** begins in the small intestine and ends with the release of four nubile amebas by the time the upper part of the small intestine is reached. New amebas immediately begin to feed, enlarge, reproduce, and become established in the intestinal tract. Many trophozoites pass out of the body with the fecal material; others adhere to the walls of the intestinal lining and feed on the cells and blood of these tissues.

Colonies of many hundreds or thousands of trophozoites form in small crevices, causing ulceration and tissue **necrosis**. Perforation and ulceration may lead to secondary bacterial infection, opening the door to a whole new catalogue of complications. Some of the trophozoites penetrate the tissue barriers and enter the body; others fall off and are carried away by the digesta. As they move toward the colon, the trophozoites encyst and exit the body in this dormant, protected form. In the presence of diarrhea, the organisms are expelled from the body in the trophozoite form because they do not have enough time to encyst. Even so, many of these trophozoites encyst outside of the body provided the fecal matter remains warm for an hour or more. Those that enter the body tissues move about on their own or are carried by the bloodstream to all parts of the body. Infection of the liver, spleen, pancreas, and even the brain are common in untreated cases.

Symptomatology   Amebiasis is generally a subclinical infection that lasts several weeks and then disappears spontaneously. In human beings the disease may manifest itself in any of three different syndromes:

1. Mild intestinal discomfort, mild diarrhea or constipation, and pronounced flatulence.
2. All the symptoms given in item (1) but in more pronounced form, and in addition, headache, fever, chills, acute abdominal pain, and dysentery, which shows strands of mucus and blood.

   The symptoms for both syndromes may become established and last several weeks before remitting spontaneously. Death is quite rare from either syndrome, but the latter requires skillful medical attention for successful management. The great majority of cases of amebic dysentery elicit one or the other of these two syndromes. Cases that lead to severe disease and death of the patient, invariably fall in the third syndrome.

3. Symptoms as above but a more serious attack on the intestinal lining, leading to severe ulceration and invasion of the inner tissues of the body. This is followed by **peritonitis** in approximately 20% of cases, **hepatitis** in approximately 50% of cases, lung tissue involvement (abscesses) in approximately 20% of cases, and invasion of the brain in approximately 10% of cases. Almost all the tissues of the body can become involved, yielding a variety of clinical syndromes including skin rashes and ulcerations, pancreatitis, infection of the spleen, and renal failure. Movement of the ameba to the alveoli of the lungs may lead to efflux of liquid and impairment of breathing. The mortality rate is approximately 80% in untreated cases and becomes 100% when multiple complications occur.

Etiology　The name *Entamoeba histolytica* is derived from the Greek as follows: *ent* = within + *amoeba* = change and *histo* = tissue + *lytico* = loosening or undoing. Therefore, the name describes an organism that changes its internal form and has the ability to destroy tissues. This organism belongs to the class *Sarcodina*; i.e., it is motile by the extension of **pseudopodia** (see Figure 7.4) and very active in the **trophozoite** stage. The organism is found in soil and water in the dormant state, but its natural habitat is the intestinal tract of human beings. Trophozoites measure 10 to 50 µm and cysts 10 to 20 µm in diameter.

Several **races** of *E. histolytica* can be distinguished on the basis of size, form, and **relative pathogenesis**. Pathogenesis varies greatly in these protozoa in that some races are extremely virulent and others are nonvirulent. The basis for this variation is not yet understood.

Epidemiology　*Entamoeba histolytica* is transmitted by food and drink under a wide variety of conditions. In nature, *E. histolytica* is found in the intestinal tracts of human beings, other primates, cattle, swine, dogs, cats, and many other mammals. No clinically identifiable disease is produced in any of the

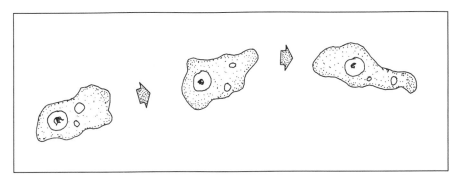

Figure 7.4

Trophozoites of Entamoeba histolytica *move by extension of pseudopodia.*

Figure 7.5

Life cycle of E. histolytica *in human beings.*

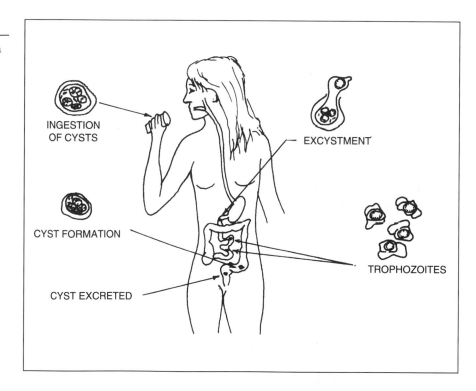

animal hosts, although some of the higher primates may suffer a low-grade diarrhea when infections are heavy.

Although it is more prevalent in tropical and semitropical areas, amebiasis is found in all parts and climes of the world. Amebiasis is more consistently associated with improper sewage removal and impure water supplies than with geographic locations. Epidemics of amebic dysentery occur in communities located above the Arctic Circle and also in those located at the equator. In many communities, the presence of *E. histolytica* is associated with the use of human or animal waste for irrigation of food crops that are consumed raw as are lettuce, radishes, and strawberries.

The cysts of *E. histolytica* are resistant to harsh environmental conditions and survive in the soil for 15 days or longer, depending on conditions of moisture and temperature. Cysts can be carried from the soil, from human and animal manure, and from contaminated water by flies, cockroaches, mice, and other vermin. These animals serve as **mechanical vectors** in carrying viable cysts to food or food preparation areas. When cysts are included in food and water, infection usually results. Approximately 1,000 viable cysts must be present to cause amebiasis in the average human being.

Figure 7.5 illustrates the life cycle of *E. histolytica*. The stages in the life cycle are related to the areas of the body affected at the various stages.

The incidence of infection varies from country to country according to the degree of sophistication of the sanitary facilities in use. In some areas of the United States, the incidence of infection is less than 0.1%; in others it is as high as 20% of the population. By contrast, in some areas of Colombia the incidence of infection is less than 1%, and in some rural areas of that country, it may be as high as 80% of the population. In the United States the major route of transmission is the feces-to-food route among human beings who prepare food and those who consume it. In Colombia the major route of transmission is contaminated food and water, but the other methods of transmission also are in evidence. The incidence of clinical amebiasis is only slightly higher in Colombia than in the United States because much of the population of Colombia is highly tolerant to the presence of these pathogens.

A method of transmission from person to person that had not been documented previously has been discovered recently. A comprehensive study showed that of each 100 homosexual men examined, 25 were infected with *E. histolytica* whereas the population in general showed an infection ratio of fewer than one per 100. The incidence of infection also is high in women who practice anal intercourse. In view of this, amebiasis may have to be considered as also transmissible by sexual contact. The number of cases available for study of this method of transmission is small, and some time may be required to determine the importance of sexual transmission of amebiasis.

Control   The most effective means of control for the spread of amebiasis resides in effective removal of human and animal waste from the environment where food and water are prepared for human consumption. This control is best accomplished by municipal sanitation facilities where fecal matter is disposed of in a manner that makes it unavailable to insects and vermin. Personal hygiene and personal habits also are important in that household pets, especially dogs, can become infected and bring the pathogens into the house or into the environment from which insects and vermin then may carry the organisms onto food and drink.

At this time, no vaccines are available to protect the population from infection by *E. histolytica*, and none appear imminent. No method of therapy exists that is not, to some extent, also harmful to the host's body. Some of the drugs used in the most successful therapeutic regimens, such as chloroquine, metronidazole, and diiodohydroxyquin, also are used as preventives and, as such, they work quite well in controlling amebiasis.

Other Amebas

Other amebas commonly found in the intestinal tract of human beings are shown in Figure 7.6. These are not pathogens but nevertheless are important, as they have to be positively identified when observed in the fecal matter of human beings lest they be confused with *E. histolytica*. Individuals

Figure 7.6

Amebas commonly found in the intestinal tract of human beings.

Organism	Size in μm	Trophozoite	Cyst
Dientamoeba fragilis	5–20	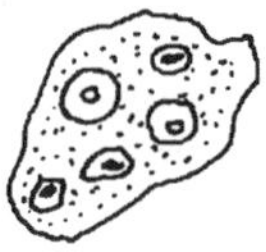	
Endolimax nana	5–15		
Entamoeba coli	10–50	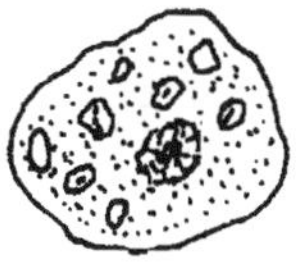	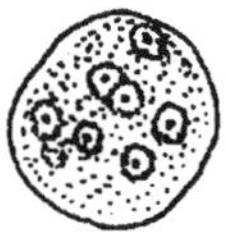
Entamoeba hartmanni	5–12	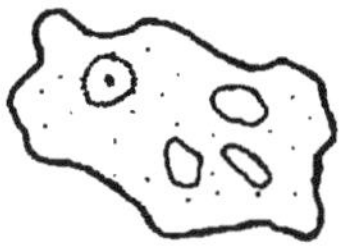	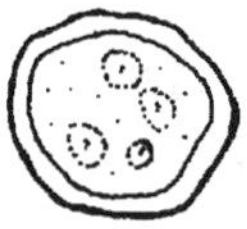
Entamoeba histolytica	10–60		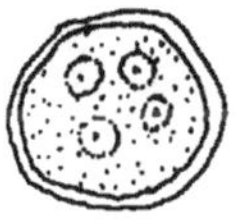
Iodamoeba bütschlii	5–25		

carrying *E. histolytica* who are diagnosed as carrying nonpathogenic amebas then are in the position to infect family members and all others around them without ever knowing they are doing so. The opposite case also is important. That is, if individuals were to be diagnosed and treated as if they had *E. histolytica* while in reality they harbored only a nonpathogenic ameba, they would be subjected unnecessarily to a severe chemotherapeutic regimen when it was not necessary.

At any rate, a disease such as amebiasis can be diagnosed only by definitive, unambiguous identification of the causative organism. This implies that *E. histolytica* and the other amebas commonly found in human beings must be distinguished and identified unambiguously.

DIENTAMOEBA FRAGILIS

This protozoan agent is cosmopolitan in its distribution, infecting approximately 5% of the world's population. High rates of infection are common in communities with underdeveloped sanitary systems. These amebas often are found in half of the populations in communities where water supplies are not filtered or otherwise adequately treated. *Dientamoeba fragilis* is associated with a low-grade diarrhea of indefinite symptoms and no known pathology. It often is found in individuals who do not show clinical signs of diarrhea or other intestinal discomfort, and when present in epidemic form, only some of those infected will have diarrhea. For these reasons, its status as a pathogen is not firmly established.

When diarrhea is present and the ameba are found in the stool in the absence of known pathogens such as *Shigella*, *Salmonella*, *Campylobacter*, and *E. histolytica*, it is said that *D. fragilis* is the agent causing the diarrhea. In the same situation, if *D. fragilis* is not found, the diarrhea is said to be of unknown etiology.

Dientamoeba fragilis is transmitted by the feces-to-food route. It also may be transmitted by direct fecal contamination through contact between individuals who share bedding, clothing, and other objects that come in contact with fecal matter. Because *D. fragilis* does not produce cysts, and because the trophozoites of this protozoan are killed by acids much weaker than stomach acid, the method of transmission was unknown until the 1970s. The mystery was solved when the eggs of pinworms, *Enterobius vermicularis* were found to contain the trophozoites of *D. fragilis*. When the eggs are swallowed, the trophozoites survive passage through the stomach, protected by the shell structures of the worm egg. The eggs of *E. vermicularis* hatch in the small intestine of the host and release live, healthy trophozoites in addition to the worm larvae. Whether this is or is not the complete answer to this old mystery is not yet confirmed, but the phenomenon described offers the only tenable explanation of the mechanism of transmission.

Solving one mystery often reveals a new one. The current question is: How do the trophozoites of *Dientamoeba fragilis* get into the eggs of the pinworm? Do they enter the body of the worm, go to the ovaries, and move into the egg before the outer integument is laid down? Or do they burrow through the integument of the mature egg and then repair the hole through which they enter?

Figure 7.7

Structure of Giardia lamblia.

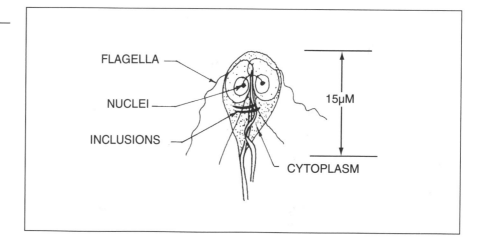

MASTIGOPHORA

Giardia lamblia

Etiology   *Giardia lamblia* is a protozoan parasite of the class Mastigophora. The adult form is found in the lumen of the small intestine of human beings and other animals. These protozoa enter the body in the cyst stage, pass through the stomach as such, and excyst in the upper part of the small intestine, the duodenum. The emerging trophozoites begin to feed immediately, attach to the intestinal wall, and begin to reproduce. Mature trophozoites detach from the intestinal wall and move downward with the digesta until they reach the colon. Here they change into the cyst form and emerge in the stool in the dormant form. Cysts can survive outside of the body for more than a month and remain infective during that entire time. Trophozoites that pass out of the body when diarrhea is present die in a few minutes unless they can encyst before perishing. Figures 7.7 and 7.8 depict the structure of the trophozoites of *Giardia lamblia.*

Epidemiology   Giardiasis is transmitted by fecal contamination of food and water. Worldwide incidence is approximately 10%, with some populations showing infection rates as low as 0.1% and others as high as 50%. One of the major differences in infection rate lies in the efficacy of the sewage-gathering system and that of the drinking water purification system. Because giardia can be transmitted by the direct fecal route, personal hygiene also is an important consideration in transmission.

A recent study showed that hand-feeding of infants by infected mothers resulted in infection of almost 100% of babies tested. From this, it follows that the incidence of infection is higher among children than it is among

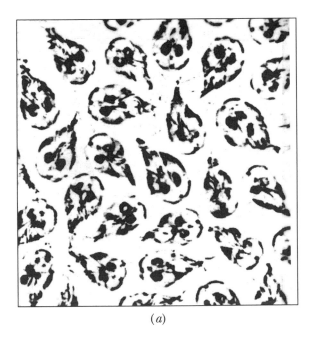 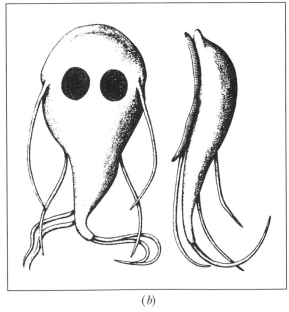

(*a*)　　　　　　　　　　　(*b*)

adults and higher in the lower socioeconomic levels than in the upper levels. It also is more likely to exist in endemic form in warmer climates than in colder ones, albeit it is present in all parts of the world.

Pathogenesis　Most infections of giardiasis are asymptomatic. Symptoms are more likely to appear in individuals that are medically compromised by existing malnutrition, predisposing disease, or other infections. The pathogenic mechanisms of this protozoan have been described only recently. Pathogenesis begins with the attachment of trophozoites to the cells of the intestinal wall with a specialized sucker plate (see Figure 7.8b), which allows the organism to obtain blood and body fluids from the host. Although a few microscopic parasites will do little damage to the intestinal tract of a human being, many millions cause irritation, ulceration, and, in extreme cases, even perforation.

In most cases, clinical symptoms are indefinite and vague but may include stomachache, abdominal tenderness, anorexia, and mild diarrhea. These symptoms are of little diagnostic value. Diagnosis of giardiasis is possible only by laboratory examination. The finding of giardia cysts, often together with trophozoites, in the stool for 3 days in succession constitutes definitive diagnosis. Treatment consists of repeated doses of quinacrine hydrochloride, which is normally 100% effective in eliminating the parasite.

Other mastigophorans, commensals with human beings, must be definitively identified to justify the identity of *G. lamblia.* An experienced microscopist can identify the various mastigophorans in the stools of human beings

Figure 7.8
Giardia lamblia trophozoites from upper intestinal tract of infected animal. Reprinted from the work of Kabnick and Peattie by permission of American Scientist, *journal of Sigma Xi, Scientific Research Society.*

Figure 7.9

Nonpathogenic flagellates normally found in intestinal tract of human beings.

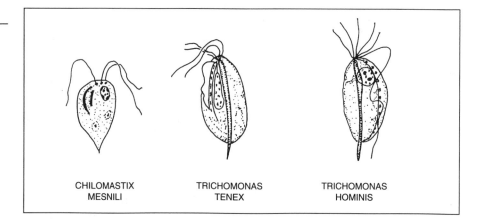

CHILOMASTIX
MESNILI

TRICHOMONAS
TENEX

TRICHOMONAS
HOMINIS

with little or no difficulty and will do so in a brief time. On the other hand, the inexperienced analyst can easily confuse one flagellated microorganism with another in fresh preparations of human stool.

There is no evidence that any of the other mastigophorans cause enteric disorders in human beings. *Trichomonas hominis, T. tenex,* and *Chilomastix mesnili,* shown in Figure 7.9, are normal inhabitants of the intestinal tract of human beings and other animals. These are considered commensals with human beings, and no pathogenic properties are attributable to them. It is assumed that the nonpathogenic protozoa are transmitted from one person to another by the feces-to-food route.

CILIATA

Balantidium coli

The only pathogenic protozoan of the subphylum Ciliata is *Balantidium coli,* a large (50 µm to 75 µm) ciliated organism found in the intestinal tracts of human beings, sheep, hogs, dogs, and other animals. These organisms are acquired by ingesting food or water contaminated with the cyst of the organism and may cause epidemics. These protozoa are found everywhere on Earth. They inhabit the intestinal tracts of healthy and diarrheic human beings, and in the latter case are said to be the etiologic agent of the diarrhea when no other pathogenic organism is found.

Most *B. coli* infections are asymptomatic, and the presence of the protozoa in the intestinal tract of human beings can be ascertained only by finding cysts and trophozoites by microscopic examination of fecal samples. When symptoms appear, they include nausea, vomiting, abdominal tenderness, and mild diarrhea. These symptoms possibly appear as a result of abrasions of the intestinal lining by movement of the protozoa.

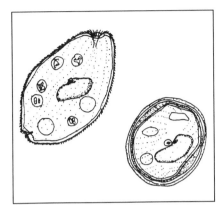

Figure 7.10

Trophozoite and cyst of Balantidium coli.

Autopsy materials show that the trophozoites burrow into the deeper layers of the intestinal lining, causing microscopic ulcerations and areas of irritation. The cilia on the outer surface of *B. coli* probably serve as the abrading and burrowing instrument activated by the spinning action of the protozoan as it moves. Heavy infestations show ulceration with exfoliation, but no penetration of the intestinal wall is evident. Balantidium infections are self-limiting and run their course in approximately 3 to 4 weeks. *Balantidium coli* are illustrated in Figure 7.10.

SPOROZOA

Isospora hominis

Isospora hominis, a protozoan of the subphylum Sporozoa, causes infections in human beings. Primarily a pathogen of animals, including pet dogs and other members of the dog family, isospora also may cause human infections, albeit at a low incidence. Human beings seem to be only incidental hosts, and human infections are the exception rather than the rule.

Infections in human beings are the result of ingestion of **sporocysts** in food and water or directly from hands and clothing contaminated by dog feces. Symptoms of the infection are vague and may include nausea, mild diarrhea, and abdominal pain, in that order of appearance. The only valid method of diagnosis depends on finding and identifying the sporocysts in the feces of the patient for 3 days in succession.

The incidence of human infection in areas where the infection is endemic in animals is of the order of 0.1% to 1.0%. It is not found in human populations in areas where the infection is not endemic in local animals. The major areas of infection are in Africa, South America, and the Philippines, and, to a lesser extent, in the countries surrounding the Mediterranean Ocean.

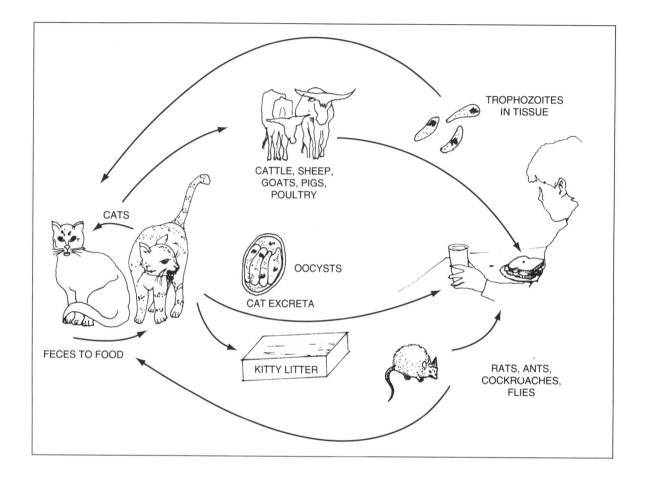

Figure 7.11

Transmission of Toxoplasma gondii.

Isospora is found in the United States at low frequency and generally in children of homes where pet animals have extensive direct contact with members of the household or where food, water, and other articles are exposed to contamination with their fecal matter.

Toxoplasma gondii

An organism normally found in many members of the cat family, *Toxoplasma gondii* also can cause infections in other animals including poultry, swine, sheep, cattle, and human beings. Epidemics have been observed although human infections are rare and the duration of the epidemic is brief. The incidence of infection in human beings is a function of the proximity and size of the cat population in contact with human populations. In some parts of the United States, approximately 20% of individuals tested showed evidence of antibodies specific for *T. gondii*. In human populations living in close contact with large populations of cats 100% of the individuals tested carried specific antibodies.

Toxoplasma gondii forms a resting stage called the **oocyst**, but oocysts are only in the intestinal tract of some members of the cat family. The oocysts mature in the expelled feces and remain viable in the soil in excess of a month. If animals or human beings ingest viable oocysts, the emerging **vegetative forms** will establish themselves in the intestinal tract and begin to reproduce there. In many animals, including human beings, the parasites invade the tissues of the body and form cysts. Ingestion of mature cysts by flesh-eating animals or of mature oocysts in soil and water will result in toxoplasmosis. Often epidemics develop, but the only evidence of this is in the levels of antibodies observed in the serum of human beings. Transmission of *Toxoplasma gondii* is shown in Figure 7.11.

Control The best method of controlling isospora and toxoplasma infections is to avoid intimate contact with animals, especially household cats, unless it is established definitively that they are free of toxoplasma and isospora. Not consuming improperly cooked meat, avoiding soil and water containing cat feces, and maintaining good habits of personal hygiene serve to reduce the number of human cases dramatically.

QUESTIONS TO ANSWER

1. Who discovered protozoa?
2. What is known of the pathogenic processes of protozoa?
3. What is antigenic mimicry?
4. What is the antigenic blocking reaction?
5. What are five chemotherapeutic agents that are effective against protozoa?
6. Who were F. Losch, and Walker and Sellards?
7. What is the infective form of *Entamoeba histolytica*?
8. What is the third syndrome of *E. histolytica* infection in human beings?
9. What is *Dientamoeba fragilis*?
10. What is *Endolimax nana*?
11. What relationship exists between one of the amebas found in human beings and *Enterobius vermicularis*?
12. How does *Giardia lamblia* damage the human intestine?
13. How large are the trophozoites of *Balantidium coli*?
14. Why can't the cysts of *Balantidium coli* be seen with the unaided eye?
15. What is toxoplasmosis?
16. Is toxoplasmosis dangerous for human beings?
17. What is isosporosis?
18. How is isosporosis transmitted to human beings in the United States?

19. What is *Chilomastix mesnili?*
20. What is *Trichomonas hominis?*

FURTHER READING

Baron, S. and L. M. Alperin. eds. 1982. MEDICAL MICROBIOLOGY. Addison-Wesley Publishing Company, Menlo Park, California.

Committee on Infectious Diseases, American Academy of Pediatrics. 1986. REPORT OF THE COMMITTEE ON INFECTIOUS DISEASES, 20th ed. American Academy of Pediatrics, Elk Grove Village, Illinois.

Farmer, J. N. 1980. THE PROTOZOA: INTRODUCTION TO PROTOZO-OLOGY. C. V. Mosby, St. Louis, MO.

Hill, M. J. 1986. MICROBIAL METABOLISM IN THE DIGESTIVE TRACT. CRC Press, Boca Raton, Florida.

Jones, A. W. 1976. INTRODUCTION TO PARASITOLOGY. Addison-Wesley Publishing Company, Reading, Massachusetts.

Reed, C. P. 1972. ANIMAL PARASITISM. Prentice-Hall, Englewood Cliffs, New Jersey.

Riemann, H. and F. L. Brown. 1979. FOOD-BORNE INFECTIONS AND INTOXICATIONS. 2nd ed. Academic Press, New York.

Enteric Diseases Caused by Viruses

8

Then turning to my love I said,
The dead are dancing with the dead,
The dust is whirling with the dust.

Wilde

Viral invasion of the body is accomplished by a strategy completely different from that used by the bacteria, the fungi, and the protozoa. Whereas the latter enter the cells and tissues of the body in efforts to obtain nutrients and an environment where they can reproduce and increase their numbers, the viruses have no such motive. Because viruses are in effect dormant, biologically inactive complex chemical **macromolecules**, rather than living entities in the same pattern as bacteria and fungi, they are incapable of reacting to their environment, of assimilating food, of growing, or even of reproducing. They possess no active enzymes, toxins, flagella, or capsules and have no means of attacking cells. Viruses have no moving parts and cannot carry out chemical reactions as do other living things. As a consequence, their strategy of life is different. As an extension of this, their mechanisms of pathogenesis also are different from those of other parasites. The structure of a common virus is depicted in Figure 8.1.

INFECTION

While different viruses employ many different means of attacking cells, the one described here may be seen as one that includes common elements from

Figure 8.1

Schematic diagram of a common virus.

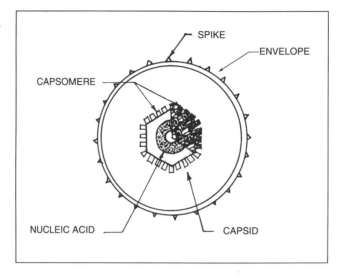

several known mechanisms. Before viral invasion can begin, viruses must be taken into the cell by a series of complex, energy-requiring actions that the victim cell itself must perform. The first step requires matching **chemical determinants** on the cell surface with **complementary determinants** on the viral particle. If the match is established, the viral particle adheres to the surface of the cell. Once the viral particle is attached, the cell surface **invaginates** and enlarges at the site of attachment to engulf the particle completely.

After engulfing the particle, the cell membrane vesicle that contains it migrates inward and moves about in the cell cytoplasm, dissolving as it does so. The viral particle eventually is stripped of its protective protein coat; that is, it is **uncoated**, and the naked nucleic acid moves to the outside surface of the host-cell nucleus. There it is acted upon by various nuclear enzymes present in the cell. The host's nucleic acid also is acted upon by the cell enzymes, and when the two are ready, they react with each other. The result of the reaction is that the viral DNA becomes a part of the host DNA. The **genes** on the viral nucleic acid now are genes of the host cell, and the viral genes are capable of acting as if they were host genes.

In some viral infections, although not in food-borne **viremias**, the viral genes may remain dormant within the host's cells and even be reproduced together with the cell as if it were a part of the host's substance. The period of dormancy may last from a few days to many years or even the entire life of the host. On the other hand, the viral genes may take control of the cell and direct it to synthesize viral substances for the purpose of producing viral particles identical to the one that initiated the invasion of the host. Viral genes affect host cells in many other ways, and some of these are so damaging to the cell, and consequently to the entire host, that disease results.

VIRAL REPLICATION

The effects of some viruses on the host result from a complex and lengthy series of chemical reactions that take place in the **target cells** of the body. These begin when the **viral nucleic acid** is acted upon by some of the cell's enzymes, resulting in the production of **messenger ribonucleic acid (mRNA)**. The messenger bears the genetic information present on the viral DNA. This step is called **transcription**, and it is followed by a step called **translation**.

Translation describes a large number of reactions that result in production of proteins made according to the genetic information contained in the viral genes. In the RNA viruses, viral RNA combines directly with the host-cell ribosomes, and protein synthesis proceeds according to the genetic information borne on the RNA molecule. These **viral proteins**, many of which are enzymes, carry out the replication of new viral nucleic acids, new proteins, and the other components required to build new viral particles. The components are assembled into the final form and released from the cell.

In some cases, the host cell is killed and all the viral particles in the cell are liberated at once. This method of replication is quite common among the viruses, but many others exist. Even though each group has its own characteristic processes, they all result in the same end: an increase in number of the infectious viruses.

Production of viral particles has many different effects on the host's cells and, by extension, on the host itself. The process described here, and illustrated in Figure 8.2, is typical, but in others the cells are not killed, only **transformed**. As such, these host cells often lose the ability to perform the function for which they were intended in the host's body, and disease results. An example of the former is poliomyelitis and of the latter, hepatitis.

VIRAL DISEASES

The major diseases of human beings caused by viruses that may enter the body with food and water are poliomyelitis, hepatitis, coxsackievirus, and viral gastroenteritis. The latter include several diseases caused by rotavirus, calcivirus, astrovirus, and echovirus infections. Viral gastroenteritis is the major cause of diarrheal disease in many parts of the world and is responsible for some 10 million deaths per year. In the United States, approximately 45% of diarrheal disease is viral in origin and transmitted by the feces-to-food route.

The feces-to-food route is well established, but evidence suggests that poliomyelitis, hepatitis, and coxsackievirus infection may be also transmitted by other routes such as saliva droplets and direct contact. Because epidemics are caused by the feces-to-food route, this obviously is the major means of

Figure 8.2

Events in the host cell subsequent to viral attachment.

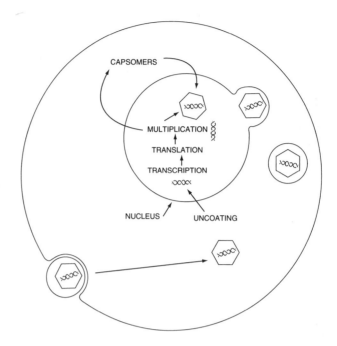

transmission during epidemics but the other methods of transmission play important roles. Probably many other enteric pathogens are found among the viruses, but only these are well documented at this time.

Even though the etiology of protozoan and bacterial enteric disease is well established, enteritis of viral origin remains much of a mystery. As many as 70% of the cases of diarrheal disease are of unknown etiology. The assumption is that many of these cases are of viral origin. Norwalk virus has been identified in the laboratory and may turn out to be a major etiologic agent of intestinal disease in the future. Table 8.1 summarizes some of mechanisms by which viruses cause disease in human beings, along with an example of each.

Table 8.1   Some of the Mechanisms by Which
Viruses Cause Disease in Human Beings

Mechanism	Example
Viral genes affect host cell functions; loss of natural functions results in damage, and illness, to host organism	Hepatitis
Viral particles affect host cell structure, thereby impairing normal function of host organism	Rabies
Viral genes destroy host cells, depriving organism of essential functions	Poliomyelitis

Table 8.2   Physical Characteristics of Poliovirus

Characteristic	Defining Feature
Size	28 to 30 nm
Nucleic acid	RNA
Capsid geometry	**cubic**
Number of **capsomeres**	32
Proteins	15
Viral coat	none; virus is naked

POLIOMYELITIS

The virus that causes poliomyelitis in human beings belongs to the group called the PICORNAVIRIDAE (pico = small + rna = nucleic acid is RNA + virus) and is named the **poliovirus**. Only one kind of virus causes poliomyelitis, but several types of the virus have been identified. The major types are designated Type 1, Type 2, and Type 3. These are characterized by the kind of disease they produce in human beings and also by immunological determinants. At one time, poliomyelitis was a major cause of illness and death in the United States, but a successful vaccination program in the late 1950s essentially eliminated the disease. The distinguishing physicial characteristics of poliomyelitis are presented in Table 8.2.

Etiology

The virus that causes poliomyelitis is one of the smallest of the viruses and one of the simplest in structure as well. Simplicity of structure and chemical composition make it relatively hardy when outside of the human body. It survives the sewage treatment process, including chlorination. It also is one of the most resistant viruses to chemical agents such as soap, alcohol, and ether, and also to physical agents such as desiccation, radiation, and high temperatures. These characteristics allow it to survive in foods and beverages for extended periods of time.

Symptomatology

The incubation period of poliovirus infection is from 3 to 7 days after ingesting the virus in contaminated food or beverage. The vast majority of poliovirus infections are asymptomatic, and infection can be ascertained only by laboratory diagnosis. To show that infection has taken place, the virus must be isolated from throat washings and fecal samples for 3 days in succession. Ten days after infection, and for an indefinite time thereafter, the virus may be found in fecal samples but not in the throats of those infected.

The infection persists several weeks and eventually disappears, leaving no residual effects except an elevated antibody **titer**. In approximately 10% of cases, clinical symptoms consisting of headache, sore throat, and slight fever appear, but these are often indistinguishable from the signs of a light summer cold. These two syndromes constitute approximately 99% of all poliovirus infections in susceptible individuals. In the other 1% of cases, the virus enters the bloodstream, producing a mild viremia. The symptoms that result from viremia are generalized low-grade fever and malaise. Only rarely are the signs of poliovirus infection sufficiently severe to require medical attention. During the viremia the infectious agent may be found in saliva, blood, and feces.

Approximately 1% of clinical poliomyelitis infections (1 in 10,000 infections) result in **central nervous system** involvement. The symptoms may be limited to slight fever and mild muscle stiffness, or they may appear as a severe spinal meningitis. Regardless of the severity of symptoms, death is rare and recovery is complete and uneventful in 2 to 3 weeks. During this phase of the disease, poliovirus is shed and respiratory cell viruses may give rise to droplet-borne epidemics.

Only 1% of cases with central nervous system involvement (one in 1,000,000 infections) progress to the paralytic phase of the disease. **Paralytic poliomyelitis** results from the destruction of **motor neurons** infected with the virus. Clinical symptoms appear only after neurons have been destroyed in sufficient numbers to affect nerve function. The first symptoms include pain in the neck muscles, stiffness and pain in shoulder and trunk muscles, and gradual paralysis of different parts of the body. If only the nerves of the trunk and limbs are involved and one or more limbs become paralyzed and atrophied, the disease is called paralytic poliomyelitis.

If the cervical nerves are attacked and destroyed by the virus, the result may be paralysis of the palate, tongue, and the respiratory apparatus. If this condition is sufficiently severe, respiratory collapse and death may result. In the period from 1927 to 1960, victims of respiratory collapse were maintained in the "iron lung" or **respirator** invented by Phillip Drinker and Louis A. Shaw of Harvard University in 1927. This **iron lung** housed the polio victim and, mechanically, performed the breathing function he/she could not perform. A mobile respirator that permitted the patient to walk and move about was invented by **Bo Sahlin** in 1937. If nerve damage is extensive enough to paralyze the heart and other organs, death is inevitable and occurs rapidly.

Epidemiology

Poliovirus is found only in the higher primates and produces disease only in human beings. It exists in all parts of the world where human beings dwell and can be recovered from fecal matter whether clinical poliomyelitis is or is not present in the population. The incidence of poliovirus antibodies may be as high as 100% in populations with low standards of sewage management and water purification. In some communities 100% of children under age 10 harbor the virus in the intestinal tract whereas only 50% of those older than 15 years of age show the same incidence of infection. Also, populations in which the incidence of infection is high during childhood are reported to have few cases of paralytic poliomyelitis, whereas communities in which the incidence of childhood infection is low have the most cases of paralytic poliomyelitis. Infection early in life, some authorities have proposed, may confer immunity to paralytic polio but probably not to poliomyelitis infection. Prior to the inception of polio vaccination, the United States had approximately 20,000 cases of paralytic poliomyelitis while many Third World countries had none. Immunization began in 1950 when Jonas E. Salk prepared a killed virus vaccine but its effectiveness was limited. In 1961, Albert A. Sabin prepared a vaccine using live, **attenuated** viruses which were then transmitted from one individual to another. The result of this is an "epidemic" of immunizations which resulted in the drastic reduction of poliomyelitis. In 1951 there were 5 cases of paralytic poliomyelitis per 100,000 inhabitants but only 2 per 100,000,000 from 1974 to the present.

Human beings are the only natural reservoir of poliovirus. Although the origin of almost all infections of poliomyelitis is the fecal matter of individuals who harbor the virus, transmission by the droplet route and by direct contact is an important factor. Because the virus grows in the cells of the pharyngeal region, coughing, sneezing, kissing, and common use of eating utensils all are vehicles of transmission and often are responsible for the spread of epidemic polio.

Control

Because it is transmitted by different routes, poliomyelitis cannot be completely controlled by conventional measures that are effective against other enteric diseases. This means that effective sewage disposal and an adequate water purification system are not sufficient to prevent the spread of poliovirus. If the water purification processes are adequate, however, the frequency of childhood poliomyelitis will be lower, leaving the population vulnerable to paralytic polio when it reaches adulthood. Be that as it may, contaminated water cannot be prescribed as a control measure for poliomyelitis in view of all the other enteric diseases that are certain to be present in that water.

HEPATITIS

Hepatitis, an inflammation of the liver, has been recognized as a serious **clinical entity** ever since the fourth century B.C. Even though many infections

Figure 8.3

Structure of hepatitis A virus.

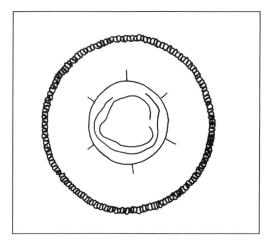

and intoxications can lead to hepatitis, the one caused by the hepatitis virus was identified and distinguished from all the rest just recently. It is called infectious or epidemic hepatitis, and the disease is characterized by its brief duration and relative mildness. Three forms of the virus were found to cause hepatitis in different manners. The one of interest in food microbiology is called hepatitis virus type A or hepatitis A virus (HAV) because it is transmitted by food and water. It is shown in Figure 8.3. The other is called hepatitis virus type B, or hepatitis B virus (HBV). It is transmitted primarily by blood and serum, and the disease produced is called serum, transfussion, or syringe hepatitis. Two other types, HCV and HEV, do not fit into either of the two categories described and cause different forms of hepatitis. These were formerly called non-A-non-B hepatitis but are now recognized as distinct clinical entities. The HEV virus is transmitted by food and water in the same manner as HAV, but at a much lower frequency. Recent reports indicate the transmission of HBV, the serum hepatitis virus, by food and water.

Etiology

The etiologic agent of type A hepatitis is a small, 28 to 30 nm, RNA virus of the enteric group. The hepatitis A virus is placed tentatively in the class *Picornaviridae*. Only one serotype is known at this time, but studies in progress indicate that some of the C and E hepatitis viruses may be antigenic variants of HAV.

Symptomatology

After ingestion of food or drink contaminated by the fecal matter of individuals with an active viral infection, an incubation period of 2 to 4 weeks ensues. During the incubation period no symptoms of disease appear but the virus is shed in the feces, indicating an active infection.

In young children infection by HAV is generally subclinical and the illness runs its course uneventfully. Hepatitis infection in adults is more serious and generally is associated with serious clinical manifestations. The most common of these are nausea, anorexia, malaise, abdominal pain, jaundice, and enlargement of the liver. Jaundice results from the death of virus-infected liver cells. Extensive cell necrosis leads to **biliary duct** occlusion and consequent inflammation. The infection may become chronic, but in cases where predisposing disease exists, the symptoms become acute and may even lead to death. Fulminating hepatitis may occur in the presence or the absence of predisposing conditions, but the incidence of this form of hepatitis is low. This syndrome leads to complete liver failure and death.

The vast majority of cases of hepatitis are either asymptomatic or self-limiting as a result of **circulating and tissue antibodies**. The death rate is approximately 1% of untreated cases but much lower in cases treated with **convalescent serum**. The normal course of the average infection is 8 to 12 weeks, and recovery is rapid. In some cases extensive residual liver insufficiency may require many months or even years to correct.

Epidemiology

Although hepatitis has been a well-known clinical entity for more than 2,000 years, the viral etiology of the disease was not established until the late 1940s and the first studies on dissemination were not performed until World War II. These studies showed that hepatitis was an enteric disease, although epidemiologic studies were confused with the epidemiology of type B hepatitis. In the early 1950s the disease was recognized as an important enteric illness that leads to severe liver impairment. The etiologic agent, HAV, was not characterized until 1982 and the first microscopic images were not obtained until 1983.

The full extent of the spread of hepatitis in human beings was not appreciated until it was learned that, of every 1,000 American soldiers in Korea during the period 1951–1952, 35 acquired clinical hepatitis. This is an incidence of 3,500 per 100,000 servicemen, one of the highest infection rates of all known diseases.

Studies of circulating antibodies in large numbers of Americans have shown that approximately 40% of the population of the United States has been exposed to the hepatitis virus. With a death rate of approximately 1% and more than 6,000 deaths from hepatitis every year, this clearly is a major food-borne disease. The same is probably true in many other countries, although the studies performed have not been as thorough as the ones carried out in the United States.

Hepatitis is especially associated with shellfish that are cooked improperly, particularly shrimp, oysters, clams, and mussels. Epidemics occur during the summer, and the number of infections is greater in areas where water quality is compromised and where sanitary conditions are less than adequate.

Control

Controlling the spread of hepatitis by measures that are effective in curbing other enteric diseases has been impossible because, like poliomyelitis, some aspects of the epidemiology of the disease are not yet understood. As only one serological type of HAV exists, finding an effective vaccine out of the several now being tested in selected populations may be possible.

COXSACKIEVIRUS

Coxsackievirus was first isolated from fecal samples of very young children in Coxsackie, New York. The agent was thought at first to be a variant of poliovirus but was shown later to be quite different, although the clinical symptoms observed in children were similar to those of poliomyelitis. A second type of the same virus was found later, and the two then recognized as a new genus quite apart from the poliovirus.

Subsequent studies revealed two variants, which were named Coxsackievirus A and Coxsackievirus B. The differences between poliovirus and Coxsackievirus were observed first in newborn mice, but further observations showed that clinical manifestations also were different.

Etiology

The viruses that cause Coxsackie disease are closely related to the poliovirus. They are members of the virus group *Picornaviridae*. The physical and chemical characteristics of Coxsackievirus and poliovirus are indistinguishable, but antigenic differences can be used to separate the two. In addition, the RNAs of Coxsackievirus and poliovirus are largely different. On the other hand, the numerical data given in Table 8.2 to describe the structure and chemical characteristics of poliovirus can be used to describe Coxsackievirus as well.

Symptomatology

Most Coxsackievirus infections are asymptomatic. The evidence of infection is often limited to the presence of specific antibodies in the blood of those exposed to the virus. A small number of infections show mild clinical symptoms, and in these cases the infection is thought to be confined to the alimentary canal and the respiratory tract.

On the other hand, when the infection progresses to the bloodstream, many body organs are affected, producing a variety of syndromes. The intestinal syndrome includes nausea, vomiting, diarrhea, and continued intestinal pain. The viremic syndrome represents a broad spectrum of symptoms that include **coryza**, fever, **exanthema**, meningitis, myocarditis, pericarditis, **myasthenia**, encephalitis, and paralysis. The syndromes that affect the heart and the nervous system often lead to death.

Epidemiology

The natural habitat of Coxsackievirus is the intestinal tract of human beings, but it can be found in other tissues of the body as well. The virus is spread by the feces-to-food route and is more prevalent in communities where the water supply is not purified adequately. It is more common in tropical and semitropical areas but can be found wherever the separation of sewage from food and water is not pursued scrupulously.

The greatest number of infections probably occurs in children younger than 5 years of age. This was found when it was proven that young children have antibodies for Coxsackievirus that are of the same serological type as the viruses found in the community, whereas adults carry antibodies against viruses not present in the community. In the latter case, the infection that gave rise to the antibodies found in adults obviously was acquired from other virus infections in other communities. These findings explain why clinical Coxsackievirus infection is rare in young children and more frequent in older children and adults. When Coxsackievirus infection is limited to the respiratory tract tissues, the virus, it is assumed, also may be transmitted by the saliva droplet route and perhaps by direct contact as well.

Coxsackievirus infections appear in the population in epidemic proportions in the summer and early fall, as do all other enteric infections. If the number of cases is low, the serotype of the virus isolated from fecal samples will be the same as the serotype of the antibodies found in the blood of individuals within the population. To the contrary, if the number of cases is large, the viral serotypes will not match the antibodies found in the population. This means that, in the latter case, several epidemics are in progress at the same time. The sources of these epidemics, undoubtedly, are infected individuals moving into the population independently, and probably also at different times. Such epidemics will last until all susceptible individuals are infected and develop antibodies that confer immunity.

Control

Control measures applicable to the elimination of Coxsackievirus infection are the same as those for the control of poliomyelitis. No vaccines have been developed that affect the rate of viral transmission or the incidence of Coxsackievirus disease, and none are likely to be effective because of the large number of serotypes. Because of the lack of crossreactivity among the different serotypes, at least 28 different vaccines would have to be used to attempt protective immunization.

OTHER VIRUSES

Other viruses suspected of causing enteric disease in human beings after being transmitted by the feces-to-food route are being studied at this time. Although none of these viruses has been proven to be important in human medicine as of this time, many indicate strongly that they may be of some, perhaps secondary, importance. As the techniques and instrumentation for the study of viruses improves, establishing cause and effect relationships in enteric virology becomes easier.

Echovirus

The echoviruses form a closely related group of enteric viruses classified as one species with 31 different antigenic variants. They are members of the *Picornaviridae* and, consequently, closely related to the polio- and Coxsackievirus. The physical and chemical characteristics of the echoviruses are similar to those of the other two but antigenically and genetically different from either one.

Echoviruses infect children before the age of 5, but generally infection is not accompanied by clinical manifestations. In a small number of cases, infection will develop into a viremia and may produce a variety of symptoms, each determined by the organs or tissues affected. The most common clinical syndrome is a nonspecific febrile disease. Another common syndrome includes symptoms similar to those of the common cold, but these may develop into pneumonia and, in untreated cases, lead to death.

Other symptoms include meningitis, exanthema, encephalitis, and, rarely, localized or generalized paralysis. Without intense medical intervention, the symptoms listed in the third category may result in death. The only method available for establishing the unambiguous diagnosis of echovirus infection resides in a complex set of observations.

1. Febrile illness
2. Isolation of echovirus from feces
3. Failure to isolate poliovirus or *Coxsackievirus* from feces
4. Increasing antibody titer specific for *Echovirus*.

The echovirus isolated from fecal samples and the antibodies rising in titer must be of the same antigenic type if the diagnosis is to be unambiguous.

Echovirus infections are spread by the feces-to-food route and are associated with poor sanitary conditions and lower socioeconomic status. The virus grows in the tissues of the pharyngeal region and also in the intestinal epithelium during the early part of the infection. At this time the virus can be isolated from either fecal matter or from saliva and nasopharyngeal washings. Spread of the infection during this phase of the disease may be effected by the respiratory route as well as by the fecal route.

In the latter stages of the disease, massive quantities of viral particles are shed in the feces but no viruses are detected in the respiratory tract. Fecal virus excretion continues for extended periods after all symptoms disappear, and many obviously healthy individuals with no clinical sign of infection excrete large numbers of viral particles. No effective control measures exist, and vaccines are thought to be impractical at this time because of the large number of antigenic variants found in these viruses. Fortunately, epidemics of food-borne echoviruses are rare in the United States.

Viral gastroenteritis is an inflammation of the tissues that line the intestinal tract. The viruses involved attack the cells of these tissues and affect their ability to regulate water passage resulting in diarrhea or the excretion of large quantities of water from the body. Some, e.g., rotavirus, affect the enzymes these tissues produce and make digestion of sugar such as maltose, sucrose, and lactose impossible. Nursing infants are particularly affected since lactose is the primary sugar of milk. Malnutrition and death are often the result.

Rotavirus

An agent found in many animals including birds and human beings. It is a double stranded RNA virus with a two-shelled capsid that gives this virus the appearance of a wheel (car tire) under the electron microscope.

Viral gastroenteritis caused by rotavirus is one of the major causes of diarrhea in the world. Authorities report that more than 500 million cases of diarrhea are caused by rotavirus and that it is a major cause of death in Third World countries. During epidemics, the morbidity of rotavirus infection approaches 100% in children less than 7 years of age. The major route of transmission is feces-to-food and the rate of transmission is directly related to the separation of sewage and water in the community. Rotavirus diarrhea is common in orphanages, prisons, camps, and retirement homes. Its effects on the aged and medically compromised are severe.

Calcivirus

Calcivirus causes a gastroenteritis in children under 4 years of age. The disease does not cause widespread epidemics but is endemic in all populations. It is transmitted by the feces-to-food route and the disease is characterized by diarrhea as severe as that caused by rotavirus. The symptoms of calcivirus infection are nausea, vomiting, diarrhea, and dehydration.

Viral diarrhea caused by calcivirus is not considered a serious disease in terms of extensive epidemics but a serious one in terms of small "hot spots" of disease in most parts of the world. At this time, no vaccine is available and development of a suitable one takes second place to development of a rotavirus vaccine.

Astrovirus

As the name implies, these viruses appear to be star shaped under the electron microscope. They are found in large number in the feces of children suffering from viral gastroenteritis. The virus is a single-stranded RNA virus and is normally transmitted by the feces-to-food route.

The incubation period for astrovirus gastroenteritis is 1 to 2 days after ingestion of contaminated food and most infections are subclinical. The symptoms of infection are nausea, vomiting, diarrhea, and high temperature. Unlike rotavirus and calcivirus infection, dehydration is not a result of astrovirus infection and, therefore, the need for hospitalization is not mandatory. No vaccine is available and, in general, no treatment is indicated.

QUESTIONS TO ANSWER

1. What are chemical determinants on viral particles?
2. What is the viral envelope?
3. How many genes do viruses have?
4. How many genes does *Escherichia coli* have?
5. How many genes do human beings have?
6. What are viral capsomeres?
7. What is poliomyelitis?
8. What are the three forms of poliomyelitis infection?
9. What is the Salk vaccine?
10. What is the Sabin vaccine?
11. Which vaccine was responsible for bringing poliomyelitis under control in the United States?
12. What is HAV?
13. What is HBV?
14. What is HCV?
15. What is rotavirus?
16. What is calcivirus?
17. What is astrovirus?
18. What is echovirus disease?

19. How is echovirus infection diagnosed?
20. What is Norwalk virus disease?

FURTHER READING

Baron, S. ed. 1982. MEDICAL MICROBIOLOGY. Addison-Wesley Publishing Company, Menlo Park, California.

Fenner, F., B. R. McAuslan, C. A. Mimms, J. Sambrook, and D. O. White. 1974. THE BIOLOGY OF ANIMAL VIRUSES, 2nd. ed. Academic Press, New York.

Finegold, S. M. and W. J. Martin. 1982. BAILEY AND SCOTT'S DIAGNOSTIC MICROBIOLOGY, 6th ed. The C. V. Mosby Company, St. Louis.

Milgrom, F. and T. D. Flanagan, eds. 1982. MEDICAL MICROBIOLOGY. Churchill Livingstone, New York.

Mims, C. A. 1982. THE PATHOGENESIS OF INFECTIOUS DISEASE, 2nd ed. Academic Press, New York.

Volk, W. A., D. C. Benjamin, R. J. Kadner, and J. T. Parsons. 1986. ESSENTIALS OF MEDICAL MICROBIOLOGY, 3rd ed. J. B. Lippincott Company, Philadelphia.

Food and Water-Borne Diseases Caused by Metazoans

9

The ardor of red flames is thine,
And thine the steely soul of ice:
Thou poisonest the fair design
Of nature, with unfair device.

Lionel Johnson

The material presented in this chapter briefly describes those diseases transmitted by food and water in which the causative agent is a member of the Kingdom **Animalia**. The word **infestation** is used to describe diseases caused by the presence of animal parasites in or on the body of the host. Although the definition of this word is well established, many authorities use it synonymously with the word **infection**. The terms will be synonymous in this text because distinctions of this nature are often subtle and maintaining objectivity can be difficult.

The word infestation is derived from Latin and means to "assault and molest." In modern science it describes the growth and habitation of animal parasites such as fleas, ticks, mites, and lice. It now has been extended to include the presence of any animal parasite. The distinction between parasites

Figure 9.1

Fragment of ancient medical papyrus describing intestinal worms in human beings.

that are found on the skin or outside of the body and those that invade the tissues is made on the basis of the names **ectoparasite** and **endoparasite**, respectively.

Hundreds of thousands, perhaps millions, of animals are classified as invertebrates because they have no vertebral column. Although they are, evolutionarily ancient and successful animals in terms of ability to survive, they nevertheless are said to be primitive organisms. The majority are small, but some, like the giant squid, may be many meters in length. They vary in structure, and many have complicated life cycles.

Invertebrates are, for the greatest part, free-living animals, but many of them have adopted the parasitic form of life. As such, they have **coevolved** with their hosts in elaborate systems of coexistence. These evolutionary developments have been both in form and in function. The most significant aspect of coevolution is seen in mutual tolerance of both parasite and host. Some invertebrate parasites have one or more metamorphoses and one or more hosts with intervening life phases as free **larvae** in nature. They are found in every **ecological niche** in every part of the world.

Research into the ancient practices of medicine shows that animal parasites of humans and animals were known as long ago as the 12th century B.C. Writings of the Egyptians, Hindus, Jews, Chinese, Greeks, and Romans in the pre-Christian era describe both ectoparasitism and endoparasitism. Many ancient texts give descriptions of worm infestations and of the parasites themselves from which valid diagnosis could be made today. The earliest medical records of the pre-Columbian Peruvian and Mexican civilizations give precise accounts of many parasitic diseases caused by intestinal worms and also the means by which these may be eliminated from the body. Figure 9.1 shows a fragment of an ancient papyrus describing intestinal worms in human beings.

MEDICAL USE

Parasitic animals have been used in the past, and still are in some societies, for medicinal purposes. The medicinal leech, an ectoparasite, is still used for removing blood from edematous areas of the body and to cure diseases such as high blood pressure and fever. Dried and ground hookworms were prescribed to prevent worm infestation and, in the Orient, to cure eye infections. Fly larvae, maggots, were used even in recent times to remove dead tissue from abraded skin in efforts to prevent gangrene.

PARASITISM

The many degrees of parasitism range from opportunistic to obligate parasitism. Opportunistic parasites are free-living animals that have the ability to live as parasites and will do so when the opportunity presents itself. Others live normally as parasites but may survive for extended periods in nature. Obligate parasites depend entirely on the host for all their needs and perish when the host dies.

Parasites may obtain their entire nutrient requirements from the host or simply use the host to supplement nutritional needs. Only a few of those that cause food- and water-borne diseases in human beings are described here. These include flatworms and flukes classified in the animal phylum *Platyhelminthes* and roundworms of the phylum *Nemathelminthes*.

PLATYHELMINTHES

Platyhelminthes is a word composed of the Greek words *platy* = flat and *helminthes* = worm to describe a large group of worms found in both soil and water and as parasites in the intestinal tracts of human beings and other animals. Most parasitic flatworms (see Figure 9.2) are large animals visible with the unaided eye, but a few are microscopic. The average size is a few centimeters, but some may reach lengths of 10 to 15 meters.

Two major groups of the platyhelminthes contain organisms that cause human infestations when their eggs or larvae are ingested with food or water. These are the *Cestoidea*, or tapeworms, and the *Trematoda*, or flukes. At any given time an estimated 100,000,000 and 300,000,000 human beings are infected by cestodes and trematodes, respectively, throughout the world.

Cestoidea

The major body part of all tapeworms is the **scolex**, a part somewhat equivalent to the head in other animals. The head and upper neck of the worm are

Figure 9.2

Anatomy of a typical tapeworm.

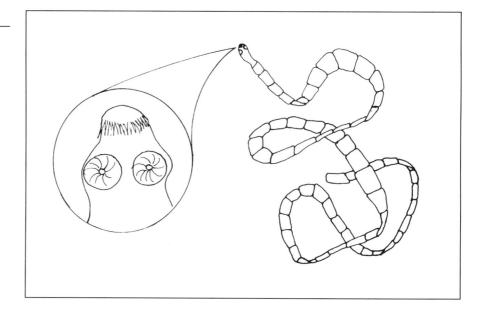

the only parts that persist throughout the life of the individual organism. The scolex has a specialized structure called a **rostellum**, which may be equipped with several suckers and many hooks and spines. The number of each and the arrangement vary in different species and sometimes are used to identify the animal. The scolex is the organ of attachment of the worm; with this it effects a fairly permanent hold to the intestinal wall of the host. Attachment to the intestinal wall allows it to suck blood and other body fluids for nourishment.

The scolex is relatively small—approximately 1 millimeter in diameter in the adult pork tapeworm, in comparison to the 2- to 3-meter length of the rest of the worm. Following the scolex is the lower neck region, and after this, the body proper of the worm. The neck appears to be a single structural unit, but closer scrutiny reveals that it is made of newly forming, individual segments. These segments begin where the lower neck region ends and grow downward as new segments appear.

The body is made of many identical segments called **proglottids**. They begin at the neck and mature as the distance from the scolex increases. As new segments appear, the segments immediately behind are pushed posteriorly and the ones at the very end may fall off. The entire chain of proglottids is called the **strobilium**. In a mature animal the strobilium may be composed of 2,000 to 3,000 proglittids. Each segment is an independent structure with little or no connection to the scolex.

As the segments mature, they begin to produce eggs. Each segment or proglottid is a **hermaphroditic reproductive structure** containing germinal

Table 9.1 Some Tapeworms That Infest Human Beings

Common Name	Scientific Name	Definitive Host	Intermediate Host
Pork tapeworm	*Taenia solium*	human beings	pig
Beef tapeworm	*Taeniarhynchus saginatus*	human beings	bovines
Fish tapeworm	*Dibothriocephalus latus*	fish-eating mammals	copepods and fish
Rat tapeworm	*Hymenolepis nana*	rodents and human beings	rat fleas, grain beetles, and cockroaches
Rat tapeworm	*Hymenolepis diminuta*	rodents and human beings	rat fleas and grain beetles

organs but few other body parts. In many tapeworms, nutrition, excretion, and growth take place by the diffusion of substances from the environment in much the same manner that more primitive organisms, such as the bacteria, do. Because both male and female reproductive glands are present in the same segment, fertilization of every egg produced is assured.

Proglottids that contain mature eggs are said to be **gravid**. Some tapeworms excrete eggs from special openings on each segment, enabling many proglottids to produce eggs at the same time. Other proglottids produce eggs until all the space is filled, and then they burst, releasing all the fertile eggs at the same time. In still others the gravid proglottids break off from the strobilium and pass out of the host's body, where the proglottid dies and becomes a protective case to aid in survival of the eggs in nature.

Five major tapeworm infestations are transmitted by food and water. Of these, three are large animals and two are small. Human beings are the definitive hosts of some of these and the intermediate hosts of others, as indicated in Table 9.1.

Pork Tapeworm   Human beings are the only definitive hosts of *Taenia solium*. The scolex of *T. solium* is approximately 1 mm in diameter followed by 800 to 1,000 proglottids to give it a total length of 2 to 3 meters. Human infections, taenaiasis, generally are acquired by ingesting **cysticerci**, the larval form of the worm, present in improperly cooked pork flesh. A coat which covers the cysticercus dissolves in the **jejunum**, allowing the immature scolex to emerge and attach to the intestinal wall. It begins feeding immediately and starts to grow into the adult worm. The ingestion cycle is shown in Figure 9.3. Cysticerci are sack-like structures, 5 to 10 mm in diameter which contain the larval worm. Pork meat with numerous cysticerci appears grainy or measley. The latter is an old term which is still in use.

In cases where *T. solium* eggs are ingested with food or water, they release larvae, onchospheres that burrow through the intestinal wall and invade the

Figure 9.3

Infestation cycle of Taenia solium.

body tissues. There they mature into cysticerci and dwell for extended periods. Cysticerci in muscle and other tissues, generally, do not cause serious disease. On the other hand the cysticerci may lodge in the liver, spleen, muscle, heart, eyes, or even the brain, causing severe disease and death. This disease is called cysticercosis. It has been recognized as a disease entity from antiquity and was often blamed on the "evil eye" and witchcraft. When the number of cysticerci is very large or when the worm larvae die and give rise to allergic reactions, cysticercosis may be life-threatening.

Normally the larval scolex grows into the adult form and begins to produce proglottids until the strobilium reaches its full length of 2 to 3 meters. Egg production begins, and the eggs, together with **gravid proglottids**, pass out of the body with the host's fecal matter. If the fecal material is available to pigs, they consume the eggs and the cycle is repeated.

The time from infection to acute clinical symptoms in human beings may vary from several months to 10 years, depending on the tissues involved and the parasite population. The disease is common in Africa, India, Russia, and America but rare in the United States, although the incidence of infestation in hogs is relatively high in some areas. *Taenia solium* infestations in human beings generally are treated with **niclosamide**, but cysticercosis is not affected, even with strong chemotherapeutic agents. These can be removed only by surgical procedures when the numbers are small or the cysticerci localized in specific tissues.

Figure 9.4

Infestation cycle of Taeniarhynchus saginatus.

EGGS
IN
NATURE

IMPROPERLY COOKED MEAT FROM
INFECTED ANIMAL

Beef Tapeworm   *Taeniarhynchus saginatus* and *Taenia saginata* are the names by which the beef tapeworm is known. Human disease results from eating improperly cooked beef or the meat of other animals such as buffalo, yak, llama, deer, and antelope. Humans are the definitive host of this tapeworm, and it causes the largest number of tapeworm infections in humans.

In areas of the world where human feces is used as an agricultural fertilizer, the incidence of infection may be greater than 75% in the human population and 100% in the bovine population. Cattle become infected by ingesting eggs from soil and water contaminated by human feces containing the eggs of *T. saginatus*. Eggs hatch in the intestinal tract of bovines, releasing larvae, which burrow through the intestinal wall and are carried by the blood and lymph to the tissues of the body including the muscles. The larvae mature to the cysticercus stage and remain alive in the tissues for 6 to 8 months. If not eaten by human beings in that time, they die in the tissues of the bovine host. As in pork tapeworm, the cysticerci were called "measles" and the meat is said to be "measley."

When human beings ingest live cysticerci, the larvae attach to the intestinal wall and begin to grow. When these mature into adult flatworms and egg production begins, gravid proglottids and eggs pass out in the feces and the cycle starts anew. The infestation cycle of *Taeniarhynchus saginatus* is depicted in Figure 9.4.

The adult worm lives many years and reaches a length of 4 to 20 meters, obtaining its nutrition from the tissues of its human host as well as from the digesta in the lumen of the small intestine. The person usually is asymptomatic, however, or may have abdominal tenderness, nausea, and weight loss. The infestation may go unnoticed until the proglottids appear in the stool, and definitive diagnosis requires microscopic identification of proglottids and eggs in the feces. Control depends on sanitary disposal of human wastes and avoidance of undercooked beef.

Fish Tapeworm   *Diphyllobothrium latum* and *Dibothriocephalus latus* are the old and new names, respectively, for the fish tapeworm. Fish tapeworms infest many mammals, including human beings. The highest incidence of infestation is found in the Great Lakes region of the United States and Canada, around the Baltic Sea, Scandinavia, and Russia. In these communities the incidence of infestation varies from 10% to 100% of human beings and 50% to 100% of fresh-water fish of certain species.

Human infection results from eating raw or improperly cooked fish that contain a larval form of *Diphyllobothrium* called the **plerocercoid** larvae. When ingested, the larval scolex emerges and attaches to the wall of the small intestine. The attached larvae begin to feed and mature into adult worms, which eventually grow to a length of 10 to 12 meters. Adult worms produce gravid proglottids and eggs, which pass out of the host's body with the feces. The eggs must reach water, where they mature in 10 to 15 days.

When the eggs hatch, they release a free-swimming larval form called a **coracidium**, which lives in the water approximately 24 hours. If the coracidia are not eaten by a specific **copepod** (a small freshwater crustacean) within a day, they die. Those that are eaten enter the copepod's intestinal tract and change into the oncosphere stage. The oncosphere bores through the intestinal wall and lives in the tissues of the copepod for 2 to 3 weeks and changes into another larval form, called a **procercoid**.

If the copepod is eaten by one of several kinds of freshwater fish, such as salmon, pike, or trout, the procercoid burrows through the fish's intestinal wall and enters the body muscles, where it changes into a plerocercoid. Mammals that feed on infected fish acquire the plerocercoid larvae, and the cycle begins again. Components of the life cycle are illustrated in Figure 9.5.

Definitive hosts of *D. latus* include bears, foxes, mink, seals, dogs, cats, and human beings. In human beings the infection generally is asymptomatic, but when symptoms occur, they include malaise, nausea, abdominal pain, and, in rare cases, anemia. In severe cases of infestation, anemia results because the tapeworm takes most of the vitamin B_{12} from the host, causing severe **avitaminosis** and attendant anemia.

Control resides in avoiding contaminated fish and the utensils with which these are prepared. Perhaps the best control is to avoid improperly cooked fish. Workers who clean and prepare fish run a high risk of infection as they

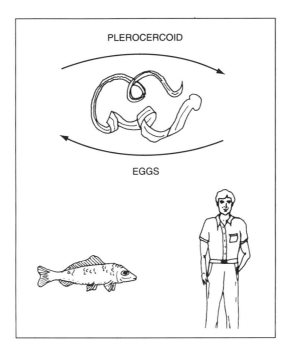

Figure 9.5

Elements in the life cycle of Diphyllobothrium latum.

are in contact with large numbers of the larvae. Accidentally ingesting these larvae or contaminating food and water by soiled hands are effective means of acquiring the infection.

Dwarf Tapeworms *Hymenolepis nana* and *Hymenolepis diminuta*, the rat tapeworms, are called dwarf tapeworms because they are much smaller than the other worms that parasitize human beings. The definitive hosts of these tapeworms are mice, rats, and various wild rodents. Although these tapeworms are minor parasites of human beings, the rates of infection can be quite high when sanitary conditions are unfavorable. The infection rate among rodents, and consequent transmission to human beings, may be high where the disease is endemic.

Tapeworms in the intestinal tracts of mice and rats excrete eggs in the feces of the host rodent. In the abbreviated cycle, *Hymenolepis nana* and *H. diminuta* eggs ingested in food contaminated by rodent feces develop in the intestinal tract of human beings and penetrate the tissues to reach the interior of the body. There they change into the cysticercoid form and find their way back into the intestinal tract, where they undergo metamorphosis and emerge in the form of the adult worms.

In the complete cycle both *H. nana* and *H. diminuta* infect various grain insects of the genera *Tribolium* and *Tenebrio*, as well as *Pyralis farinalis* and the larval form of the common rat flea, *Xenopsylla cheopis*. When these insects consume eggs in the feces of infested rats, the eggs develop into cysticercoid larvae in the intestinal tract and become established in the body tissues.

If human beings consume the grain containing infested grain beetles, the larvae from the insects invade the tissues of the human host and undergo metamorphosis into adult worms. The adult worms become established in the gut of the definitive host and begin the cycle again.

Adult *H. nana* and *H. diminuta* range in size from 1 to 5 centimeters and 50 to 100 centimeters, respectively. Human infestation generally consists of hundreds of adult worms. The total number of worms is larger in children than in adults.

Dwarf tapeworm infestations normally are asymptomatic or of very low virulence. After a long infestation, the worms may be eliminated while the host remains oblivious to the entire episode. The symptoms of *hymenolepiasis* are more severe in very young children than in adults, and may include nausea, vomiting, mild diarrhea, malaise, pain in the abdominal area, and anorexia. The dwarf tapeworms are found everywhere on Earth but most often are associated with human populations of the lowest socioeconomic ranks, where rodent populations tend to be larger and more firmly established.

Control consists of ridding living quarters and food-storage areas of rodent pests and insects such as the grain beetles. Foods suspected of harboring such insects may be rendered safe by thorough cooking as heat destroys all stages of the worms.

Formerly it was assumed that tapeworms had a life span exactly as long as that of the host animal it preyed on, since some studies showed that the parasites died when the host expired. Other studies in which marked tapeworms were repeatedly transferred from moribund hosts to young ones showed that *H. diminuta* could live at least 15 years. This is a much longer life span than that of the host mouse or rat. The strobili live only long enough for the eggs to mature, but the scolex survives and continues to produce new strobili as it ages.

Control can be attained by disrupting the feces-to-food route of transmission and by properly cooking all animal flesh or cereal flours. Control of tapeworm infestation is more than 99% effective if meat is cooked to the point that the center of the cut is not "pink." Some tapeworms have life cycles that include more than one animal host. In some of these, human beings are the definitive host and in others the secondary host. When an alternate host is involved, spread of the worm can be controlled by eliminating the animal host.

Adult tapeworms are killed with **niclosamide**, **atabrine**, and **paramomycin**, and the larvae are killed by **mebendazole** and **praziquantel**. Eggs of three tapeworms that infest human beings are shown in Figure 9.6.

Other Tapeworms   The other tapeworms that afflict human beings do so because humans are the intermediate hosts. These include *Echinococcus granulo-*

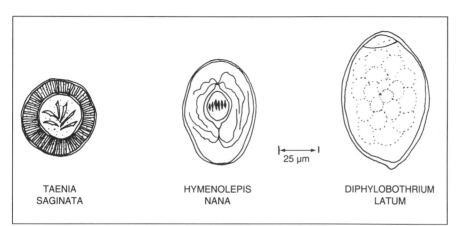

Figure 9.6

Eggs of three tapeworms that infest human beings.

TAENIA
SAGINATA

HYMENOLEPIS
NANA

25 µm

DIPHYLOBOTHRIUM
LATUM

sus, E. multilocularis, and *Multiceps multiceps*. The incidence of infestation is low for these tapeworms, and they usually are associated with primitive housing conditions and low socioeconomic level.

Trematoda

Like the tapeworms, flukes are members of the phylum *Platyhelminthes* and belong to the family *Trematoda*. Flukes are hermaphroditic animals that live in specific tissues of the host, where they produce sufficient physical damage to cause serious disease. They are fairly small, ranging in length from 1 to several millimeters in length and are quite complex in body structure. Although many of the flukes are parasitic and cause diseases in many animals, few cause diseases in human beings, and fewer still cause diseases transmitted by food and water. Of the diseases transmitted by food and water, the most significant are caused by the organisms named *Fasciola hepatica*, the liver fluke; *Clonorchis sinensis*, the Chinese liver fluke; *Fasciolopsis buski*, the intestinal fluke; and *Paragonimus westermanni*, the lung fluke.

The complicated life cycle of flukes begins when the eggs are released by the adult worm and fall into fresh water. There the eggs mature and hatch into a larval stage called the **miracidium**. The miracidia then must find snails of a specific species and invade their soft tissues. Failure to do so results in the death of the larval fluke. Once inside the body, miracidia begin to grow and eventually change into a sporocyst. By repeated division of germ cells, the sporocyst develops many reproductive structures.

Eventually the sporocyst undergoes metamorphosis and becomes a **redia**. Some of the reproductive structures in the redia develop into new rediae, and the rest develop into **cercariae**. In a last round of multiplication, the rediae change again, this time emerging as new cercariae. The grown cercariae crawl or swim on the bottom of the river or lake until the second intermediate host, a freshwater crawfish, crab, or other shellfish, is found. The cercariae invade the body of this second host, and there change to **metacercariae**. Metacercariae are inactive cyst forms that do little harm to the shellfish.

If the shellfish is eaten raw, or in an undercooked condition, by a definitive host such as a human being, the metacercariae become active in the small intestine and penetrate the wall. They enter the tissues of the body, where they migrate from part to part for long periods, eventually settling in the tissues for which they have a predilection. The liver flukes, lung flukes, and intestinal flukes are so designated because of the tissues they prefer.

The cercariae of some flukes have the ability to attach themselves to aquatic vegetation and change to the metacercarian form there. They then remain on the vegetation in the dormant form until ingested by a definitive host. Although this general plan has many variations, it is basically the same for all flukes. The redia, cercaria, and metacercaria forms of the fluke are illustrated in Figure 9.7.

Liver Fluke   The adult liver fluke is approximately 2 to 3 centimeters long and normally found in cattle, horses, sheep, goats, and primates including human beings. Individuals who work with such animals are said to be at risk. *Fasciola hepatica* metacercariae are also found on water plants such as watercress, water chestnuts, and others commonly eaten in many parts of the world. Human infestation begins when metacercariae are consumed and the larvae excyst, penetrate the intestinal wall, and invade the tissues of the body. The larvae eventually concentrate in the tissues of the liver and finally settle in the bile duct. As they wander around eating blood and liver cells, they irritate the tissues of the liver and bile duct. Figure 9.8 depicts the life cycle of the liver fluke.

When the populations of adult flukes become large, they obstruct the flow of bile. Obstruction of the liver passages eventually leads to **hepatomegaly** and liver diseases of various clinical descriptions. In its acute form, liver enlargement, hepatitis, jaundice, cirrhosis, and generalized liver failure may lead to death of the host.

In the bile duct of the definitive host, liver flukes release fertile eggs (see Figure 9.9) that pass into the intestinal tract and thence to the exterior in the feces of the host. Human infection is diagnosed by finding the eggs in the stool of those suspected of having the disease.

Fasciola hepatica is found everywhere on Earth but is more prominent where human and animal fecal matter contaminates natural waters. The highest incidences of infestation, according to a 1988 report of a survey conducted by the World Health Organization, are in France, Algeria, and Peru.

The most effective control of fasciola consists of sanitary disposal of human and animal wastes, protection of food plants from contaminated waters, avoiding uncooked plants in areas where the disease is endemic, and

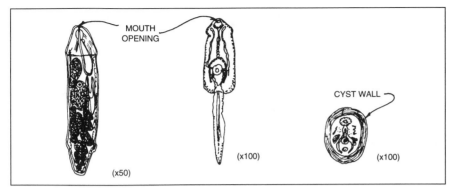

Figure 9.7

Redia, cercaria, and meta-cercaria.

MOUTH OPENING

CYST WALL

(×100)

(×100)

(×50)

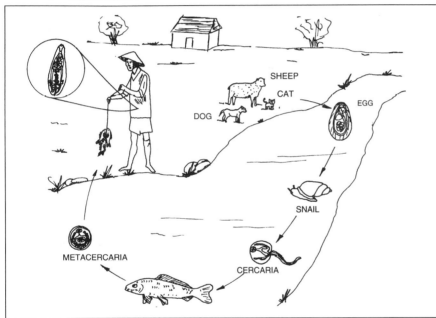

Figure 9.8

Life cycle of typical liver fluke of human beings.

SHEEP

CAT

DOG

EGG

SNAIL

METACERCARIA

CERCARIA

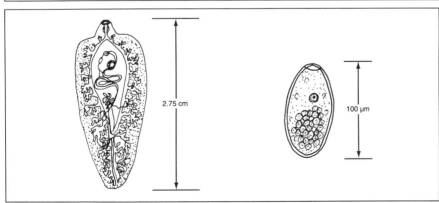

Figure 9.9

Fasciola hepatica.

2.75 cm

100 μm

avoiding contact with animals that are ill. Removing intermediate hosts is also an effective method of control. In the United States the most common source of infection is the consumption of watercress or handling the flesh of infected animals. The most important aspect of fascioliasis is economic in that the livers of cattle, sheep, and goats that are infested cannot be sold and also in that horses, pigs, rabbits, and other valuable animals lose their value when infected, because of illness resulting from liver impairment.

Chinese Liver Fluke   The Chinese liver fluke, illustrated in Figure 9.10, has an older name, *Clonorchis sinensis*, and also a newer name, *Opisthorchis sinensis*. It is a typical fluke in every respect and undergoes the same type of complex metamorphosis seen in fasciola. It passes through two or more intermediate hosts, finally establishing itself in the tissues of certain freshwater fish.

Eating the raw or improperly cooked flesh of these fish leads to invasion of the tissues, as described before, with eventual growth of adult flukes in the liver. *Clonorchis sinensis* causes liver abscesses, severe liver dysfunction, hardening of the liver, and even death. Because interference with the normal flow of bile affects proper digestion, diarrhea and weight loss occur in the later stages of infestation. The liver of a single patient may contain one to several hundred flukes. In severe cases the number may reach several thousand individual flukes, but these cases are generally seen at autopsy. Heavy infestation results in severe hepatic ulceration and functional insufficiency. *Clonorchis sinensis* damages liver tissues because it not only feeds on blood cells, as does *F. hepatica*, but also directly on the liver tissue itself.

Although **clonorchiaiasis** is found worldwide, Vietnam, Cambodia, Laos, Burma, Thailand, Malaysia, and the Philippines have disproportionate concentrations. In that area of the world, rice culture is common, and the intermediate hosts of the fluke, snails and fish, live in large numbers in the waters

Figure 9.10

Clonorchis sinensis

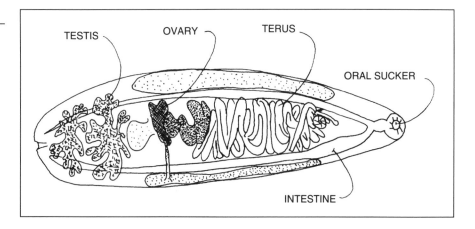

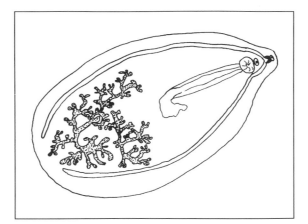

Figure 9.11

Fasciolopsis buski.

that make rice paddies. In addition, people who inhabit these areas of the world prefer a diet that includes raw and partly cooked fish. The incidence of infection in some populations from this part of the world may reach 100%.

Intestinal Fluke   The intestinal fluke, *Fasciolopsis buski*, illustrated in Figure 9.11, is commonly found attached to water vegetation in the same manner as *F. hepatica*. Infestation results when animals and humans eat the metacercariae along with water plants. These flukes also are acquired by using the teeth to peel the husks from water chestnuts and other aquatic plants where the metacercariae attach.

The metacercariae develop in the duodenal region of the small intestine, attach to the lining by hooks on the anterior part, and mature there. They begin to release eggs, which pass out in the feces, and the cycle is repeated.

Each adult fluke releases approximately 20,000 fertilized eggs per day for the 6 to 8 months of its life. The reproductive potential of this animal shows that one egg may give rise to approximately 4,000,000 new eggs per fluke per generation.

The symptoms in human beings are not as severe as those of liver flukes, but in heavy infestations the disease is serious enough to result in death. Symptoms of infestation result from irritation of the lining of the duodenum by the burrowing action of the flukes and consequent chronic diarrhea. In advanced cases, diarrhea may be followed by obstruction of the flow of digesta through the duodenum. In addition, an allergic reaction is produced by the presence of the flukes in the intestinal tissues. In cases where the number of flukes is large, a toxic reaction occurs that may be even more severe than the symptoms of the infestation. As in other fluke infestations, control depends on avoidance of raw water plants in the diet, control of snail populations, and sanitary disposal of human and animal waste.

Figure 9.12

Paragonimus wester-
manni.

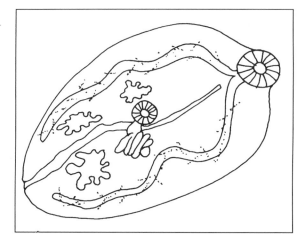

Lung Fluke　*Paragonimus westermanni* is found in almost all parts of the world, but it causes serious problems only in areas where crabs, crayfish, and other shellfish are eaten raw. This includes many parts of China, Japan, Korea, Burma, Thailand, Laos, Vietnam, Cambodia, the Philippines, and some parts of Africa. In these areas of the world, the incidence of infection may average about 7% of the population, but in certain localities it may be as high as 50%. Much lower incidences are found in Mexico, Peru, and Ecuador, but these countries still have higher average infection incidences than the world average.

Illustrated in Figure 9.12, *Paragonimus westermanni* is approximately 10 millimeters in length and covered with tiny **spicules** that probably help in its movement over body surfaces. The flukes are hermaphroditic and produce several thousand eggs each day throughout their entire life. Many mammals, including dogs, cats, mink, pigs, tigers, and human beings, serve as the definitive host.

The life cycle of this fluke begins when the adult fluke, present in the host's body, reaches the lung and releases eggs into the alveoli. The eggs ascend the bronchi, are swallowed, and pass out of the body with the feces. If they reach a body of water within a few days, they will survive. In water the eggs mature, releasing miracidia, which swim in the water until they find snails of the genera *Semisulcospira* or *Thiara*. Cercaria produced in the host eventually infest a crab or other shellfish and dwell there until this intermediate host is consumed by a human or other animal.

The metacercariae excyst in the duodenum of the human being, penetrate the intestinal wall, and move through the body for prolonged periods, finally emerging in the alveoli. In lung tissue they mature and begin to produce eggs, and the cycle is repeated.

The natural reservoir of *paragonimus* consists of animals that feed on crabs and other shellfish. The infestation is endemic among these animals and among human beings who eat infected fish. The major source of human infection is crab meat that is marinated and eaten raw. Although the marinated crab meat takes on an appearance that suggests it has been cooked or steamed, it is raw, and the metacercariae, because they are in a dormant stage, survive the treatment successfully. In other preparations, the fluid pressed out of infected crabs and crayfish is used as a salad dressing, and this constitutes an excellent means of transmitting the metacercariae.

Symptoms of infection include coughing, throat irritation, copious expectoration with flecks of blood, and chest pain. The body reacts to the presence of the parasites by forming fibrous exudates in the lungs, to entrap the moving flukes. As the flukes move away to avoid the fibrous exudate, they disrupt and irritate other tissues, complicating the clinical picture. Alveolar tissue involvement as a result of fluid exudation, accumulation of fibrous material, and increasing numbers of fluke cysts results in considerable lung impairment. In addition, if migrating flukes reach the central nervous system tissues and the brain, the prognosis of the disease changes from poor to critical. Diagnosis is made by identifying the animals and their eggs in the feces and also in the sputum of human hosts.

NEMATODA

Parasitic intestinal roundworms found in large numbers of human beings in all areas of the world belong to the phylum *Nemathelminthes*. The incidence of infestation may be as high as 85% in many populations of Africa, Asia, and South America. Individuals often harbor more than one type of parasitic roundworm, which means that the cases of nematode infestation may be more numerous than the total population of a given community. Roundworm infections are most common in the lower socioeconomic strata of Third World countries. Table 9.2 reports the numbers for some roundworm infestations.

Table 9.2   World Incidence of Roundworm Infestation*

Roundworms	Infestations	Deaths
Ascaris	1,000,000,000	20,000
Necator/Ancyclostoma	500,000,000	60,000
Trichuris	500,000,000	
Strongyloides	35,000,000	

*WHO Report, 1990.

Coevolution

Parasitic roundworms and their hosts, including human beings, have evolved simultaneously so that today a certain amount of mutual toleration exists between host and parasite. Little or no harm may be evident in the human host, even from the presence of large numbers of worms in the intestinal tract for prolonged periods. On the other hand, if the number of worms becomes excessive, damage to the linings of the intestinal canal may result, and the ensuing irritation may lead to serious consequences for both parasite and host.

In the extreme case, the host will die, and so will all the parasitic roundworms that inhabited that host. In cases where nutrition is marginal, the parasites may sequester sufficient nutrients to precipitate conditions of vitamin deficiency or insufficient caloric intake. In these cases, secondary damage to the host results from aggressive competition for nutrients on the parasite's part.

The larvae of some roundworms may enter the body tissues to complete their life cycles, and in doing so may attack vital organs of the host. When this happens, severe disease and even death may result. Although human hosts that harbor large populations of nematodes produce antibodies specific to the parasite, immunity, if it exists at all, probably has more to do with keeping the infestation under control than providing a defense for the body, as is the case with microorganisms. This, too, may be a result of coevolution and adaptation on the part of both host and parasite.

Enteric nematodes lay eggs in the intestinal tract of the host, and these pass out with the fecal matter. Eggs survive for extended periods in warm, moist soil and can be transported from there to any other place by insects, animals, or natural agents such as water and wind. The eggs are resistant to moderate cold and heat and survive the sewage treatment process, including chlorination.

Morphology   Roundworms are small, bilaterally symmetrical organisms with well developed alimentary, muscle, and reproductive systems but a rudimentary nervous system. The organs seem to float freely in a viscous fluid-filled body cavity. They are covered by an outer integument which has openings for mouth, anus, and genital structures. The female is larger than the male and there is generally a larger number of females than males in an established population.

All roundworms pass through several stages beginning with the egg and ending in the adult form. These are depicted in generalized form in Figure 9.13.

Although more than thirty roundworms parasitize human beings, only eight normally are transmitted in food and water. Of these, only the ones found in the United States in major numbers will be discussed in this

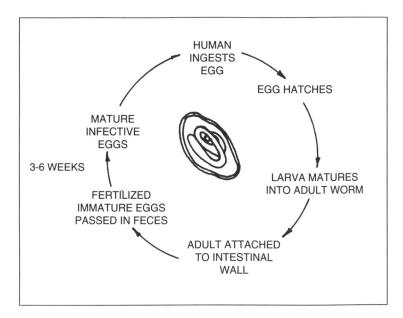

chapter. Common roundworms transmitted by food and beverage are summarized in Table 9.3.

Hookworms

Hookworms are the most prevalent enteric parasites of human beings. *Necator americanus* is called the "new world hookworm" and *Ancylostoma duodenale* the "old world hookworm" because they are most common in these areas,

Table 9.3   Common Roundworms Transmitted by Food and Beverage

Scientific Name	Common Name	Length of Female	Distribution
Necator americanus	New world hookworm	1 cm	worldwide
Ancylostoma duodenale	Old World hookworm	1 cm	worldwide
Trichuris trichura	Whipworm	5 cm	worldwide
Ascaris lumbricoides	Roundworm	35 cm	worldwide
Trichinella spiralis	Trichinella	5 mm	Europe, North America, Arctic
Trychostrongylous orientalis	Strongyloides	2 mm	worldwide

respectively. The natural habitat of *A. duodenale* is the European soil and the gastrointestinal tract of Europeans. On the other hand, *N. americanus* was brought from Africa to the new world, but it now is so well established in the Americas that it is regarded as native here. The parasites are similar to each other, and only experts with many years of experience can distinguish one from the other with certainty. Their eggs also are indistinguishable to all but experts.

Female hookworms are approximately 1 centimeter in length, and the males are approximately three quarters that length. The male and female of *Necator americanus* have a set of **tooth-plates** in a large mouth, with which they cut and burrow into the intestinal tissues of the host and by which they attach for prolonged periods. *Ancylostoma duodenale* have long, sharp teeth that accomplish the same function.

The most common place of attachment is that part of the small intestine called the jejunum. There, males and females copulate and produce large numbers of eggs, which pass out of the host's body with the feces. *Necator americanus* females produce as many as 10,000 eggs per day, and *A. duodenale* may produce as many as 20,000 eggs per day during the largest part of their 2 to 5 years of life.

The eggs develop and mature in the soil, yielding **rhabditiform larvae**, which feed on bacteria and other microorganisms in the soil until they molt and become **filariform larvae**. The filariform larvae are the infectious form of the hookworm and may enter the new host's body by penetrating the skin of the feet or ankles. They may also be swallowed in food and water contaminated with the soil where the larvae exist. If the filariform larvae fail to find a suitable host in 1 or 2 weeks, they will die in the soil because they do not feed during this stage of development. They also will die at temperatures lower than 2°C or higher than 30°C.

Infection by Skin Penetration   The first sign of hookworm infestation is an allergic rash that appears at the site of skin penetration. The filariform larvae may wander about just under the skin, leaving a trail of skin irritation and rash, called cutaneous **larval migrans**, before going into the deeper tissues. They are picked up in the lymphatic and blood circulation and carried through the heart into the lung. There they leave the capillary vessels and penetrate into the exterior of the alveoli. From the alveoli the larvae are swept up into the trachea and then the esophagus. From the pharynx they are either expectorated out of the body or swallowed into the stomach.

Hookworm infection results in damage to the lungs as the larvae break out into the alveoli, causing bleeding and serum exudation. This may manifest itself by throat irritation, coughing, bloody sputum, and malaise. The larvae that are swallowed attach to the lining of the jejunum and develop into the adult worm. They begin to lay eggs, and the cycle begins again. Figure 9.14 shows this cycle.

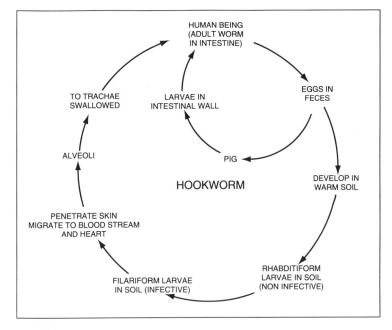

Figure 9.14

Life cycles of hookworm.

If the number of intestinal worms is large, when they begin to take nourishment from the jejunal wall of the host, signs of blood loss and intestinal damage appear. The most significant of these signs are nausea, vomiting, diarrhea, anemia, and weight loss. Severity of the symptoms depends on the "worm load," or number of worms present. Other factors that affect the prognosis include the host's size and age, state of health, level of nourishment, and the presence of complicating bacterial infections.

Infection by Feces-to-Food Route   The symptoms and the effects produced by larvae that enter the body through the skin are much more serious than the symptoms produced by eggs or larvae that enter by way of the mouth. The latter reach their final habitat, the intestinal wall, develop into the adult form, and begin to lay eggs.

Chronic Symptoms   Chronic hookworm disease is a debilitating illness in which the victims suffer lassitude, malaise, and, often, malnutrition. Under these circumstances normal life may be robbed of its natural vigor, making victims less productive and more subject to frustration and melancholia. Although the data are not entirely clear, hookworm disease also may be an important factor in the shorter life span of otherwise healthy individuals. Whatever the known and unknown physiological effects of hookworm infection are, the quality of life is lessened seriously for the 500 million human beings who are afflicted chronically with this infection.

Control   Control of hookworm infection is effectively accomplished by sanitary disposal of sewage, by wearing shoes, and by avoiding skin contact with contaminated soils. Hookworm disease is found most often in areas where human feces spill onto the soil and where the population is likely to go barefoot. This combination of circumstances is found in Central Africa, much of Asia, tropical America, and the Southeast region of the United States.

Diagnosis depends on identifying the eggs in fecal samples. Chemotherapy is most effective with **mebendazole**, **pyrantel**, and **embonate**. In the past, camphor, carbon tetrachloride, and tetrachloroethylene were used with little chance of killing worms and greater chance of killing the host or at least causing serious damage.

Whipworm

These roundworms are called whipworms because they resemble a whip. The microscopist can imagine the posterior end of the worm as a well-defined whip handle and the anterior end as the whip proper. Previously these were called hairworms, as the scientific name *Trichuris trichura* indicates, because they are long and thin. Adult female worms are approximately 5 centimeters long, and males are slightly smaller. The bodies of the female worm end in a straight or slightly curled round tail, whereas the bodies of the male end in a spiraled, abruptly pointed posterior. These features are illustrated in Figure 9.15.

Figure 9.15

Male, female, larva, and egg of Trichuris trichura.

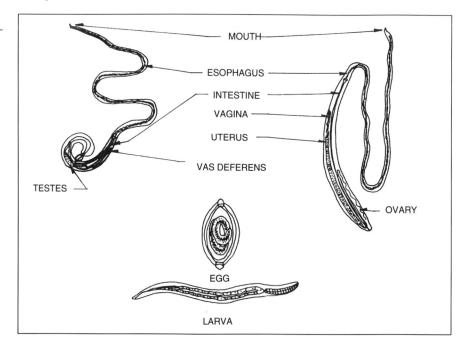

Infestation Whipworms generally enter the body in food and water contaminated with feces or soil containing eggs with mature embryos. The eggs develop in the intestinal tract, and the larvae simply attach to the walls of the small intestine. They grow and mature at the site of initial attachment and then migrate to the area of the large intestine called the **caecum**. In the caecum the anterior end of adult worms penetrates the intestinal wall and the posterior end remains in the lumen of the intestine. After burrowing through the intestinal wall, the worms feed on blood and juices from the tissues with which they come in contact.

Whipworms mature while attached to the intestinal wall and mate without having to release their attachment because males are always close by. Females live 7 to 8 years, during which they lay between 1,000 and 10,000 eggs each day. The eggs are expelled from the host's body with fecal matter and, if mixed with warm, moist soil, embryos will develop. Under proper conditions of temperature and humidity, full development of the embryo requires 4 to 6 weeks. Fertile eggs will survive for prolonged periods but are killed by desiccation, high temperature, freezing, and various chemicals.

Symptoms Clinical symptoms of whipworm infestation result from penetration and attachment of the worms to the intestinal wall. Severity of the symptoms depends in large part on the number of adult worms present in the intestinal wall. In heavy infestations hundreds or even thousands of worms are present, causing severe irritation and ulceration. Symptoms include abdominal tenderness, diarrhea, bleeding, weight loss, anemia, and ulceration attributable to worm damage. Ulceration often leads to bacterial infection and the more serious consequences of peritonitis. Also, a toxic reaction manifests itself in nervous irritation, anorexia, and lassitude.

Epidemiology These parasites are most common in tropical areas where the soil is moist and warm. In some countries the infestation rate is as high as 85%, and in others, such as the United States, the incidence may be as low as 2.5%. Approximately 500 million cases of trichuriasis are present in the world's population currently. The areas of highest infection rate are the tropical and semitropical regions that girdle the Earth.

Control The basis of effective control is sanitary removal of fecal matter to keep the soil from becoming contaminated. The most effective method in personal control resides in scrupulous removal of all traces of soil that may harbor eggs from all food and drink that is to be consumed without cooking. Washing the hands to prevent contamination of food and utensils also is a highly effective control measure. Heating destroys both larvae and eggs, so cooking food is an excellent control measure.

Figure 9.16

Adult, larva, and egg of Ascaris lumbricoides.

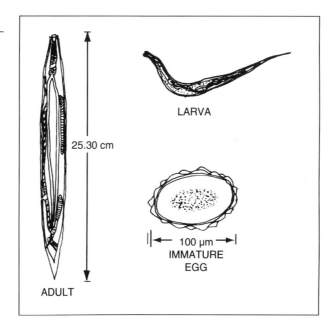

Diagnosis of whipworm infestation is accomplished by identifying the eggs in fecal samples. Whipworms are susceptible to mebendazole, thiobendazole, and dithiazonine iodide.

Ascaris lumbricoides; the Human Roundworm

One view of the world asserts that this is a wormy world, and another view is that *Ascaris lumbricoides*, the human roundworm, is what makes it so. This intestinal worm probably causes more food- and water-borne infections than any other parasite. The estimated incidence of ascaris infections is 20% of the world's population making a total of more than one billion human infections at present.

Ascaris lumbricoides females are 25 to 30 centimeters in length and approximately 0.5 centimeter in diameter at the widest part of the worm (see Figure 9.16). The males are smaller and exist in close proximity to the females, as fertilization of the eggs is accomplished by copulation. Each adult female may carry as many as 25 million eggs at any time and release approximately 250,000 of these per day during an egg-laying period of approximately one year. Several hundred worms may exist in a single infected host, resulting in the release of vast numbers of eggs during the illness.

Among the many species of the genus *Ascaris* are worms that preferentially parasitize dogs, cats, horses, swine, and poultry. The human ascarid worm, *A. lumbricoides*, exists in two races: one that parasitizes human beings exclusively and the other, *A. lumbricoides suum*, that parasitizes pigs exclusively.

Life Cycle   Eggs pass out of the body in the feces of the infected human host and, if deposited in the soil, will develop to the embryonic state. The fertile, embryonated egg is the infectious form of the organism. These eggs may live in the soil for a year or more if conditions of temperature, moisture, and humidity are ideal. They will die rapidly if the soil dries or if the temperature becomes hotter than 38°C or colder than 0°C. Because the eggs are normally found in the soil, vegetables and fruits that grow in contact with the soil are the primary sources of the parasite eggs.

When the eggs are ingested as contaminants in food or water, the embryos finish development in the intestinal tract of the host and emerge as larvae when the eggs hatch. The larvae move to the intestinal wall and penetrate the linings, gaining entrance to the interior of the body. They are picked up in the lymph and blood circulation and carried to the walls of the alveoli in the lungs. They reside there for a brief period while growing to a length of approximately 1 millimeter. They then burrow through the alveolar walls and escape to the outside of the body. The normal breathing motion of the lungs and periodic coughing eventually sweep the larvae up to the pharyngeal region, where they lie for a short time. Then they are swallowed and enter the small intestine, where they attach to the intestinal wall. Here they begin to grow, mature, and eventually mate and begin laying eggs to complete the cycle that started in a different host.

Adult ascarids lie free in the lumen of the small intestine, feeding on digesta. They also may attach to the intestinal tissue and feed on blood and body juices. These worms can travel the entire length of the intestinal tract, returning to the jejunum at will.

Symptoms   Ascariasis is a disease of children, and particularly children of the lower socioeconomic population. The presence of a few worms in the intestinal tract of a healthy individual probably will not produce symptoms, but large numbers of larvae migrating from the intestinal wall to the lungs cause damage to the tissues through which they pass. They also become lodged in small spaces in the body, giving rise to abscesses and swollen membranes.

If a large number of larvae break out of the alveoli at one time, the amount of damage and consequent bleeding may produce serious respiratory effects. These include bloody sputum, excessive coughing, and pneumonia. A large number of adult worms in the intestinal tract may cause nausea, abdominal pain, diarrhea, and weight loss. In extreme cases, especially in children, the host may starve while receiving adequate and sufficient nutrition, as the result of the increase in number and size of the worm population in the intestinal tract. An additional complication arising from this may be intestinal obstruction and perhaps even complete blockage.

Complications   Attempts to kill ascarid worms in the intestinal tract with medicinal chemicals must be carried out carefully and effectively, because

Figure 9.17

Heavy infestation of ascarids in pig stomach.

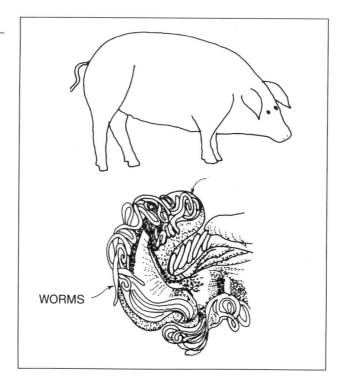

WORMS

failure to kill them rapidly may result in damage to the host. Alarmed and irritated worms attempt to avoid the harsh chemicals used as vermicides by leaving the intestinal habitat. In doing so, they may penetrate the intestinal wall and invade the viscera, causing severe damage as they move through the vital organs of the body. They certainly will terrorize the host as they emerge from mouth, nose, ear, vagina, or other body opening.

Ascarids also have a tendency to "explore" the ducts that open to the intestinal tract and then may become lodged in the visceral organs, causing serious medical complications. When the worm load becomes large, the tangled mass of worms (Figure 9.17), may fill the stomach. Often, only surgical removal will alleviate this condition.

The presence of larvae in the body tissues also gives rise to allergic reactions of varying symptoms.

Epidemiology   The tropical areas of Asia, Africa, and South America have the greatest incidence of ascaris infestation. In some areas the incidence may be as high as 100%, but 85% is the rule in many parts of the world. Although parts of the rural Southeastern United States have incidences of this magnitude, the national average is only 2.5%. The worldwide incidence of 20% infestation makes this considerably more than 1 billion cases of ascariasis at present. The incidence of ascariasis in a given population has been suggested as an inverse function of the cost of the sewage treatment plant in that community.

Control   Control of parasitic ascaris is simple in industrialized, urban societies served by modern sewage systems capable of effective removal of human and animal wastes from the living environment. To the contrary, effective removal of human and animal waste is almost an impossible task in emerging, rural settings where the sanitary facilities often consist of an area down a footpath behind the house. The incidence of ascaris is always highest in communities where human wastes, euphemistically called **night soil**, are used as fertilizer for food crops.

Control depends primarily on rigorously avoiding contaminated soil or contaminated food and water. Cooking is effective in killing both larvae and eggs, and this can be the first line of defense for those who live in endemic areas. The eggs of ascaris resist many deleterious agents and can survive passage through the sewage treatment plant, but they are killed by desiccation and by moderately high temperatures.

Trichinella spiralis

Trichinellas are small roundworms in which the female measures approximately 4 millimeters and the male 2 to 3 millimeters in length. These worms have a life span of 6 to 8 weeks and produce progeny most of this time. They are unusual among the nematodes in that they are **viviparous** rather than **oviparous** as are most of the other roundworms. These nematodes are primary parasites of animals and infect many flesh-eating mammals including humans. They have been found in animals as diverse as human beings, rats, walruses, and lions, and tigers, and bears.

Life Cycle   Trichinella enter the bodies of human beings in the larval stage when they are ingested in the uncooked or undercooked flesh of infested animals. The worm larvae are present in animal flesh in the encysted form and emerge during passage through the stomach of the new host. The larvae burrow deep into the tissues that make up the wall of the small intestine and here molt several times, eventually maturing into the adult worm. The males pass out of the host's body with the feces almost immediately after mating, but the gravid females burrow deeper into the lining of the intestinal walls. The adult female gives birth to live larvae within the tissues of the intestinal wall, and the young move into the interior of the body.

The newborn larvae eventually are picked up by lymphatic and blood circulation and dispersed to all parts of the body. In this stage they are extremely aggressive and can penetrate any tissue barrier in the human body. They can penetrate the placenta and infect the fetus of infected women. They

Figure 9.18

Trichinella spiralis *larvae in pork or beef muscle.*

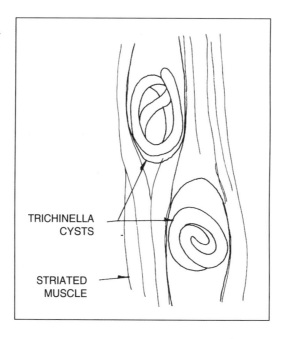

TRICHINELLA
CYSTS

STRIATED
MUSCLE

also may infect the mammary glands of lactating females and be transmitted to suckling offspring by breast milk. The infection also may be transmitted by handling the infected cadavers of game animals or in cutting and preparing the infected flesh.

As they move about the host's body, the larvae grow and mature into the final larval form. When the larvae are mature, they lodge in the various tissues of the body but preferentially in the striated muscle bundles (see Figure 9.18). Here they feed and grow from a size of 0.1 millimeter to approximately 1 millimeter and then enter the resting stage.

The host's body reacts to the presence of trichinella by forming a fibrous cyst structure around the worm. Part of the reason for doing this is to alleviate the chemical irritation the worm causes to the body. The worms may remain in this condition for many years and eventually die without emerging from their encysted form. The cycle can be repeated only if another susceptible host consumes the raw flesh.

Autoinfection   Autoinfection or self-infection is rare but can occur in untreated cases. It results when adult worms in the intestinal tract produce eggs that mature, releasing new larvae that enter the body and take residence in the muscle tissues. The entire cycle takes place in the individual, resulting in multiplication of the parasite population many-fold. The number of larvae in the muscles continues to increase until the host is overwhelmed by the parasites.

Symptoms Clinical symptoms of the early infection include diarrhea, abdominal discomfort, vomiting, fever, and facial edema. The host may have a strong allergic reaction to the presence of the larvae in the tissues, and the effects of this can vary. In the later stages of the infection, the symptoms include lassitude, muscle pain, and fever.

Another, indeterminate set of symptoms results from the effect the larvae may have on specific tissues and also from the total number of larvae present in the tissue. If the lungs are invaded heavily, hemorrhage and fluid infiltration may cause respiratory insufficiency or even collapse. Infection of the heart muscle may lead to myocarditis and invasion of the pancreas to chronic pancreatitis. Infection of the liver and other visceral organs may lead to organ dysfunction and atrophy, often accompanied by abdominal distension with great discomfort and pain. If the eye muscles are invaded, loss of motor control may lead to loss of effective vision.

Etiology *Trichinella spiralis* is found everywhere on Earth, but the greatest number of cases exists in Europe, Scandinavia, Canada, and the United States. It is a rare disease in Muslim countries and in communities where pork and beef are not abundant or where they are eaten only after thorough cooking. Although any flesh-eating mammal can acquire trichinosis, the major sources of infection for human beings are improperly cooked pork or pork sausage in which the larvae survive.

The spread of trichinosis can be controlled easily by avoiding the uncooked flesh of any carnivorous mammal and especially pork products. The best evidence of effective control of trichinosis is seen in Jewish and Islamic countries where religious dietary laws proscribe the eating of pork. The incidence of trichinella in these people has been lower than in others for more than a thousand years.

Trychostrongylus orientalis

Trychostrongylus orientalis is a parasite of herbivorous animals but also can infect human beings. Infection normally occurs when vegetables and fruits that grow on the ground are consumed raw and without thorough washing. The infective larvae are found in the soil and readily invade the intestinal wall of human hosts. They live for 5 or more years in the small intestine, and autoinfection can result in large populations.

Infections tend to be asymptomatic and self-limiting. The disease is more common in the orient, Asia, Africa, and certain areas of South America. New evidence indicates that **strongyloides** infections may be more common in the United States than was assumed previously. A recently published report (WHO, 1990) claimed that approximately 1% of the world's population may harbor these roundworm parasites.

QUESTIONS TO ANSWER

1. What is the name of the fish tapeworm?
2. How are human beings infected with the fish tapeworm?
3. What is the name of the beef tapeworm?
4. What is the life cycle of the beef tapeworm?
5. What is the name of the American (new world) roundworm?
6. What is the name of the European (old world) roundworm?
7. How are the American and the European roundworms identified?
8. What is the most common liver fluke in the United States?
9. What is the name of the lung fluke?
10. What are copepods, and why are they important in food microbiology?
11. What is *Hymenolepis nana*?
12. What is the life cycle of the dwarf tapeworm?
13. What is *Hymenolepis diminuta*?
14. How does the whipworm attach to the intestinal wall?
15. What is the infective stage of trichinosis?
16. What is the infective stage of trichostrongloides?
17. What does *Trichuris trichura* mean?
18. What are rhabditiform larvae?
19. Are human beings subject to *Trichostrongylus orientalis*?
20. What are filariform larvae?

FURTHER READING

Chen, T. C. 1973. GENERAL PARASITOLOGY. Academic Press, New York.

Cohen, M. R. and I. E. Drabkin. 1958. A SOURCE BOOK IN GREEK SCIENCE. Harvard University Press, Cambridge, Massachusetts.

Jones, A. W. 1967. INTRODUCTION TO PARASITOLOGY. Addison-Wesley Publishing Company, Reading, Massachusetts.

Katz, M., D. D. Despommier, and R. W. Gwadz. 1989. PARASITIC DISEASES. Springer-Verlag, New York.

Noble, E. R. and G. A. Noble. 1982. PARASITOLOGY: THE BIOLOGY OF ANIMAL PARASITES. Lea & Febiger, Philadelphia.

Read, C. P. 1972. ANIMAL PARASITISM. Prentice-Hall, Englewood Cliffs, New Jersey.

Villee, C. A., W. F. Walker, Jr., and R. D. Barnes. 1979. INTRODUCTION TO ANIMAL BIOLOGY. W. B. Saunders Company, Philadelphia.

Intoxications

10

*Chemical medicines are nearly all
deadly poisons but chemists prepare
them so that their poisonous nature
is removed, and that, besides their
derivation from poisons, permits them
to act as antidotes to all poisonous
ills, by their very nature.*

Basil Valentine, 1604

Bacterial toxins were discovered late in 1889 by Emile Roux (1853–1933) and Alexandre Yersin (1863– 1943) through their work on the mechanism of action of *Corynebacterium diphtheriae.* This discovery led Emile von Behring (1854–1917) and Shibasaburo Kitasato (1852–1931) to examine the immunity that resulted from injecting small amounts of aged toxin preparations into experimental animals. By 1890 they showed that the blood of animals injected with sublethal doses of toxin was able to neutralize the effect of the toxin when the two were mixed before administering the mixture to susceptible animals.

It subsequently was shown that the body produced **antitoxins** when toxins were injected into animals. This phenomenon was put on a quantitative basis by Paul Ehrlich (1854–1915) in 1897 so that today the amount of toxin and antitoxin in the animal body and culture fluids can be measured with great precision.

In 1898 Ehrlich noted that toxins stored for a long time lost their potency but not their immunizing power. He developed the theory and the methods by which toxins could be converted into **toxoids** (from the Greek *-oid* = like). The toxin molecule is changed into a form (the toxoid) that can be injected into the body in sufficient quantities to cause immunity without danger of intoxication.

Today, toxoids are used as the main line of defense of human populations against many microbial infections. Toxoids now are produced by mild treatment of toxins with formaldehyde or phenol for several days instead of by aging a year or more, but the effect is the same. As a result of this and other work, the German government decreed in 1911 that Ehrlich be honored by the title *Excellenz.*

THE NATURE OF TOXINS

One of the reasons why protozoa, fungi, and bacteria cause disease in human beings is that they produce a variety of chemical substances that when introduced into the body, elicit physiological reactions detrimental to the host's well-being. Organisms that produce toxins have been shown to possess the genetic ability to cause disease, and those that lack the genetic information to produce toxins fail to produce disease. Some organisms possess only one toxin, others have several. An example of the former is *Vibrio cholera*, and an example of the latter is *Staphylococcus aureus*. In some organisms such as *Yersinia pestis*, the degree of virulence is directly dependent on the number of toxins while in others, the presence of a single toxin is sufficient to endow the organism with maximal virulence. The latter is exemplified by *Clostridium botulinum*. There are many diseases produced by bacteria and fungi for which specific toxins have not been identified yet. On the other hand, some organisms possess well-known toxins and yet are incapable of producing disease because these factors are not produced in the animal body.

SPECIFICITY

In some instances the symptoms of a specific disease are produced by a specific toxin or other **virulence factor** such as capsules or enzymes. In others, symptoms result from the interaction of several factors and the combined effect these have on the host. The interaction between virulence factors and the host's body components yields a set of extremely complex actions that are difficult or impossible to define. An example of the former case (specific toxin) is botulinum intoxication, and one of the latter (interaction of several factors) is typhoid fever.

TERMINOLOGY AND NOMENCLATURE

One of the most important areas of food microbiology is devoted to the study of illnesses caused not by infections but, rather, by the ingestion of toxins

Table 10.1   Classes of Toxins According to Site of Action

Class of Toxin	Site of Action
Neurotoxins	Nerve tissues
Enterotoxins	Enteric system
Cytotoxins	Cells
Hemotoxins	Blood cells
Dermotoxins	Skin cells

produced by protozoa, fungi, and bacteria. The resulting illnesses are properly called **intoxications**, and those produced by microorganisms must be distinguished from venoms, poisons, and other toxic agents produced by plants and animals.

An example is represented by staphylococcal food poisoning in which the bacterium *Staphylococcus aureus* forms a toxin as it grows and divides. The toxin is excreted into the growth medium, where it accumulates in large amounts because it is stable and will not degrade spontaneously. If the growth medium is food, and if this food is consumed by human beings or animals, the toxin acts almost immediately to produce certain effects on the body.

In contrast, when the same organism enters the tissues of a susceptible host, a given amount of time is required for the bacteria to grow in the cells of the body and produce the symptoms of active infection. The difference between intoxications and infections, therefore, resides in whether the toxin is produced by microorganisms outside the body or inside the body as the microorganism grows.

As a result of the predilection of toxins for specific tissues, one of the most common methods used to classify toxins is to name them according to the tissue they affect. A systematic method such as that by which enzymes are named has not yet been proposed for the toxins. As a result, a mixture of designations includes disparate kinds of names such as **choleratoxin**, **staphylococcus food toxin**, and *Clostridium botulinum* type A toxin.

Establishing classes of toxins such as those shown in Table 10.1, though convenient, still has many shortcomings. For example, some enterotoxins also are cytotoxins. In general, the primary effect is used for classification, but many toxins had been given names before the primary function was known, and consequently they are classified according to effects of secondary functions.

BACTERIAL TOXINS

All bacterial toxins can be divided into two great classes: the **exotoxins** and the **endotoxins**. The toxins by which protozoan, fungal, and viral pathogenicity is manifested, with few exceptions, are not as well understood as are those of the

bacteria. Even so, many bacterial toxins have not been properly characterized yet and, undoubtedly, many are still to be discovered. The material given
herein is not an exhaustive compendium but rather a sampling to show something of the armamentarium that the microorganisms can bring to bear on
the body defenses of human beings.

Exotoxins

Exotoxins are products of bacterial metabolism that are elaborated and
excreted into the growth medium as the cells grow. They are called waste
products of metabolism because there are few indications that the exotoxins
play a role in the well-being or survival advantage of the organism that produces them. Exotoxins are synthesized as the result of genes in the chromosome of the cell, but others such as the staphylococcus food toxin are the
product of genes borne on plasmids that parasitize the bacteria. Other toxins
are produced by viruses, bacteriophages, that parasitize the bacterial cell.

Exotoxins have dramatic effects on specific parts or functions of the host's
body. They react with specific tissues on the host and elicit highly specialized
reactions. In general, toxins have no effect on the other parts of the host's
tissues or on all the other countless functions of the body. The specific part
of the body where the toxin acts is called the **target tissue**, and this reaction
often is used as a characteristic by which the toxin can be identified.

Exotoxins are made of two protein molecules combined into a single
entity. One bears chemical **complementarity** for the target tissue and ensures
attachment at the proper site; the other penetrates the cell membrane and
interacts with the cell constituents. Because these are specific proteins, the
body forms specific antibodies against exotoxins, thereby providing strong
and longlasting immunity against such toxins.

As a consequence of being unique protein molecules, exotoxins elicit
strong **immunological responses** in the host. This makes it possible to attain
effective, longlasting immunity to exotoxins such as those of *Clostridium botulinum*. Immunity against exotoxins depends on antibodies produced by the
animal host in response to the presence of the exotoxin, or the analogous toxoid, in the body. The antibodies are specific for the exotoxin and will react
with it, neutralizing its toxic properties when the two come in contact. This
neutralizing reaction also occurs in the body tissues and bloodstream of animal hosts, saving the host from the effects of the toxin.

On the basis of this reaction, mass immunizations can be carried out successfully and entire populations can be protected from illnesses caused by specific toxins. Such is the case with tetanus and diphtheria in the United States
today. As long ago as the 10th century A.D., Mexican animal herders knew
that a dog bitten by a poisonous snake would become immune to subsequent
bites if it survived the first one. These "snake dogs" were considered valuable
property by those who lived in danger of the bite of poisonous snakes.

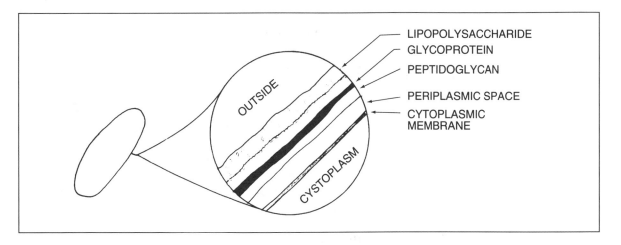

LIPOPOLYSACCHARIDE
GLYCOPROTEIN
PEPTIDOGLYCAN

PERIPLASMIC SPACE
CYTOPLASMIC
MEMBRANE

OUTSIDE

CYSTOPLASM

Today, dozens of **immunogens** produced from microbial cells are used to immunize entire populations against dozens of diseases. Although the words of Basil Valentine usually are mocked, little doubt exists that he was correct in his notion that clever chemists have the power to convert the deadly poisons into antidotes, i.e., toxoids or antibodies.

Finally, exotoxins are produced by only a small number of bacteria. Those that produce exotoxins are generally Gram-positive, but a few Gram-negative organisms produce them as well. *Pseudomonas aeruginosa*, a Gram-negative bacterium that produces several exotoxins, is important in the food and beverage industry because it has the ability to grow in almost any environment and is a major cause of food spoilage. Also, it is the major cause of "blue milk," an anomalous condition resulting in great monetary losses in the dairy industry. Most exotoxins produced by Gram-negative bacteria have not been fully characterized yet, although some, like the toxins of pseudomonas and *Yersinia pestis*, are quite important.

Figure 10.1

Lipopolysaccharide layer of the Gram-negative cell wall.

Endotoxins

Endotoxins are chemically more complex than exotoxins. The majority are found in Gram-negative bacteria and are made of three major chemical parts:

1. The **core polysaccharide**.
2. The **side chain sugars**.
3. The **lipid A** molecule.

Together, the entire molecule makes a complex entity classified chemically as a **lipopolysaccharide** (LPS). Figure 10.1 depicts the lipopolysaccharide layer of the Gram-negative cell wall. Partly as a result of their chemical nature, the endotoxins are not good antigens, and the antibodies produced against them by the host are not protective as are those of the exotoxins. Also, unlike exotoxins, endotoxins are not secreted into the growth medium, as they are one of the structural components of the bacterial cell wall. As a consequence, they are not active until the bacterial cell is disintegrated by action of the **phagocytic cells** of the body.

As the bacterial cell wall is degraded by the enzymes of phagocytes, the structural units of the wall are dispersed. Fragments of cell wall that contain the endotoxin then react with the **receptive site** of the target host cell and manifest the toxic effect of the endotoxin. The side chain sugars are strong antigenic determinants and give rise to a large variety of antigenic types among some of the Gram-negative bacteria. On the other hand, lipid A is fairly similar in all Gram-negative bacteria. Because much of the toxicity resides in lipid A, the toxicity of Gram-negative bacteria is quite similar from one organism to the next.

Effects   The effects of endotoxins on the human body are not as well understood as are those of the exotoxins. Endotoxins tend to involve many different tissues and also diverse physiological functions of the host. Some of the endotoxins of different Gram-negative bacteria have different effects on the same tissues of the same animal species, and others produce the same effect. Although the mechanisms of action of the endotoxins are not as well understood as those of the exotoxins, their effects on many different animal species have been well documented. The effects of endotoxins on one animal species or another in different dosages include:

- Disruption of water permeability of cells and tissues
- Increased susceptibility to stress
- Increased resistance to infection
- Altered mitochondrial metabolism
- Capillary vessel dilation
- Skin rashes and lessions
- Cardiac shock
- Fever

Many of the Gram-negative bacteria that fall into the group called the Enterobacteriaceae possess endotoxins, and much of their pathogenicity is attributable to these toxins. Other lipopolysaccharides produced by these bacteria are not endotoxins. As a general rule, all endotoxins are LPS, but not all LPS has toxic properties.

Variety   Although all bacterial toxins are placed in one or the other category (either exotoxin or endotoxin), some cannot be merged into this simple system. Some bacteria of the genus *Clostridium* produce toxins and secrete them into various parts of the cell, where they are stored and released only when the cell disintegrates. These toxins are not part of the cell wall as endotoxins are,

Table 10.2   Bacteria That Produce Toxins Associated with Foods and Beverages

Organism	Site of Action	Likely Source
Bacillus cereus	Cell membranes	Puddings, rice, cereals
Clostridium botulinum	Blocks nerve impulses	Undercooked canned foods and smoked meats
Clostridium perfringens	Water retention	Undercooked meats, honey, poultry
Vibrio cholera	Intestinal lining	Water, shellfish, vegetables
Escherichia coli (pathogenic)	Diarrhea	Water, milk, salads, undercooked meats
Salmonella typhi	Cell membranes	Water, poultry, milk, meat, vegetables
Salmonella paratyphi	Cell membranes	Water, poultry, milk, meat, vegetables
Shigella dysenteriae	Water imbalance	Water, milk, vegetables
Staphylococcus aureus	Vomiting reflex	Protein foods, mayonnaise, milk, meats

and they are not excreted into the growth medium as the cell grows, as are the exotoxins.

The large and complex families of toxins produced by fungi fall into different categories and are not classified by this convention. Although fungi produce hundreds of different toxins, the occurrence of these in food and beverage is rare in comparison to the bacterial toxins.

BACTERIAL INTOXICATIONS

Several of the most important illnesses to which human beings are subject are caused by the ingestion of preformed toxin. Food poisoning, or intoxication, has been known from the beginning of human history, but not until the 20th century was the etiology and nature of these illnesses understood.

The most important of the intoxications are caused by *Clostridium botulinum* and *Staphylococcus aureus*. The former are more rare but usually are fatal, and the latter are common but rarely fatal. Other bacteria that produce toxins of importance in the food and beverage service are *Bacillus cereus* and *Clostridium perfringens*. These and other bacteria are listed in Table 10.2 along with the foods and beverages with which they are associated and the body sites affected.

Botulism

Botulism is caused by ingestion of a toxin called **botulin** or, more commonly, botulinum toxin. This exotoxin is produced by a group of closely related bacteria placed in the genus and species *Clostridium botulinum*. They are large, motile, rod-shaped, Gram-positive, anaerobic bacteria commonly found in the soil.

Spores resistant to desiccation, high temperature, chemicals, and other environmental agents are produced in large numbers and released into the soil. The spores of *C. botulinum* accumulate in the soil and air in large numbers and are present constantly on all food and water in either small numbers or large numbers depending on conditions of cleanliness. The spores grow when conditions are adequate and the newly developed **vegetative cells** produce toxin readily. At least eight different strains of the organism exist and each one produces, predominantly, one kind of exotoxin. The toxins are named Botulinum Toxin:

Type A, Type B,   Type E, Type F,
                Type C1, Type C2,   Type D, Type G.

Of these, only A, B, E, and F affect human beings, whereas C, D, and G affect other animals. *Clostridium botulinum* also produces wound infections when large numbers of organisms are introduced into damaged tissues such as those resulting from massive skin abrasions and severe destructive trauma. In this case, however, the infection precedes toxin production, making this a completely different disease in this respect.

Etiology   Bacteria that produce botulinum toxin are ubiquitous in nature and are present in all food products. When food so contaminated is ingested, the bacteria are killed by the digestive processes of the stomach. If spores are ingested, they germinate in the intestinal tract, but under the conditions that prevail there, they are not able to grow extensively or to produce toxin.

During the last 100 years the most common form of botulism in the United States has been that associated with home preservation of foods. Most people in the United States lived in rural communities until World War II and were, by necessity, more self-dependent than we are now. Much of the family's food was produced on the family farm, and it was processed and prepared for the table there as well. A typical case from the archives of a rural hospital may read as follows:

> String beans were washed carefully and readied for preservation. They were placed in glass jars, the lids put in place, and the jars heated in a home pressure cooker for the time recommended in a popular cookbook. When heating was finished, the jars were allowed to cool and the lids tightened to make perfect hermetic seals. The jars were then stored in the basement of the house and left for several months. When one of the jars was opened for use the following winter,

some of the beans were added to a three-bean salad and the rest were cooked with ham and served as a vegetable dish.

Of six people at dinner, six ate beans with ham and four ate salad. Of the four who ate salad, three became ill and died before the next day and the fourth person became ill and lived five days in a coma before dying. The two who ate ham with beans but no salad showed no symptoms of intoxication and did not become ill.

If the situation had been analyzed in light of present-day knowledge, it would have been found that when the jars were sterilized, the temperature of the pressure cooker never attained its maximal point because the gasket was damaged. In effect, the highest temperature attained was 110°C for 25 minutes. The heating was insufficient to inactivate the spores of *C. botulinum* but sufficient to destroy the majority of the microorganisms present in the various ingredients. When the jar lids were sealed, only a small number of spores from the soil survived, including some *C. botulinum*. These were **activated** by the heat and germinated as soon as the jars cooled to room temperature.

The heating also served to drive off all the air in the liquid, creating anaerobic conditions, while the hermetic seal served to preserve anaerobiosis until the jars were opened. Because the temperature in the basement was less than optimal for the growth of *C. botulinum*, the cells grew slowly and the population never became large, but a small amount of exotoxin was produced nevertheless.

Other factors that impinge on this must also be considered. If the temperature of the basement had been higher, there would have been enough growth of clostridium to produce gas bubbles in the jars. When the jars were opened, a strong, pungent, unpleasant odor would have been obvious, and the string beans would have been discarded as spoiled. Under the conditions described, however, the amount of growth was so small that the beans appeared to be well preserved and no detectable odors were present.

The beans cooked with ham were safe because botulinum toxin is destroyed by exposure to a temperature of 80°C for 30 minutes or 100°C for 10 minutes. Ham and string beans have to be boiled approximately 60 minutes to cook properly. This time and temperature were more than sufficient to inactivate the toxin present.

On the other hand, the salad contained sufficient toxin to kill human beings. Although botulinum toxin is poorly absorbed by the intestinal tract of human beings, the **minimal lethal dose** of botulinum toxin is quite small and the toxin is effective at very low levels. It is thought that 0.000,000,004g (4×10^{-9}) gram of botulinum toxin are required to kill an average human male. If we assume that 5,000,000,000 people live on Earth, then:

$$4 \times 10^{-9} \times 5 \times 10^{9} = 20 \text{ grams}$$

This means that approximately 20g of toxin would kill all the human inhabitants of Earth. No other substance is as powerful as botulinum toxin.

This potential has not been ignored by those who plan on ways of killing their fellow human beings by the techniques of biological warfare.

History   The name *botulinum* is derived from the Latin (*botulus* = sausage + ism = of) because the disease was associated first with sausages made from pork intestines filled with blood and ground parts of the animal. These sausages were boiled and then smoked, a process that killed most bacteria but not those that produce spores such as *Clostridium*. If the sausages then were stored at a mild temperature, sufficient growth occurred to yield lethal quantities of botulinum toxin.

Reports as that described about the beans were not uncommon in the United States in the earlier part of this century and even until the 1940s. As the country changed from a predominently rural society to an urban one, and as households came to depend more and more on food prepared in canning plants rather than in the home, outbreaks of botulism intoxication became less and less frequent. In the period 1900 to 1967, only 640 deaths were reported to the public health authorities. The interim from 1967 to 1975 averaged some 30 cases per year with a death rate of approximately 10%. The rate from 1975 to now is approximately 5 cases per year with no fatalities. The peak period for deaths from botulinum intoxication in the United States was the time from 1930 to 1940, the depression years, when 384 deaths were reported.

The exact number of cases of botulism intoxication in the United States during the 1930s cannot be assessed accurately because reports to public health authorities were rather informal during that time and many cases went unreported. Failure to report was more likely in rural areas than in urban settings. Furthermore, in the absence of autopsy findings, death from botulism could be passed off easily as "the colic."

Control   Antiserum therapy has been the major weapon in the treatment of botulin intoxication. On the other hand, knowledge of the causative organisms and the toxins elaborated by them has been the major force in preventing intoxication. Although the organism *C. botulinum*, is still present in all soils and as a contaminant in all foods, the art of preservation of vegetables, meats, and fruits is so well understood that millions and millions of cans of preserved food can be prepared without a single can being contaminated with *C. botulinum*.

Toxin   Botulinum toxin is a protein with a molecular weight of 1×10^6, which is chemically stable and can be purified by crystallization. It is lethal against a great variety of animals, and 1 milligram can kill as many as 10 billion laboratory mice. The lethal dose for an adult human male weighing 80 kilograms may be as low as 4 nanograms, a ratio of 1 part toxin to 20,000,000,000,000 (2×10^{13}) parts of body tissue. Think of having one needle in 20,000 billion pieces of hay.

The toxin can be converted readily into toxoid and used for immunizations, but, as mentioned previously, immunization of human beings is not practical. It is practical, on the other hand, to use the toxoid for **hyperimmunization** of animals such as horses and goats. As a result, they produce high levels of antibodies in their serum. The antitoxin or botulinum antiserum then can be used to treat individuals who ingest toxin. If the antitoxin is administered before the toxin reaches its target tissues, effects of the toxin on the tissue can be avoided. The reaction between toxin and tissue is **irreversible**, so the antitoxin is ineffective if it is given after the toxin reaches the target tissue.

Mechanism of Action and Symptoms   When toxin is ingested with food, it is absorbed slowly into the cells of the intestinal tract and passed into the bloodstream. From the bloodstream it goes to the cells of the body, including the neurons. In the neurons the toxin travels to the distal end, where it prevents the release of **acetylcholine**. This results in an irreversible inhibition of muscle contraction. The overall effect is a **flaccid paralysis** with some of the following symptoms: dryness of the mouth, tightening of the throat, blurring of vision, **diplopia**, intense head and neck ache, weakness of the limbs, and progressive paralysis finally involving the respiratory apparatus. Death is generally caused by asphyxiation or cardiac arrest.

Infant Botulism

A special designation, infant botulism, is made for botulism of the newborn. In these cases the organism grows in the baby's stomach on the carbohydrate components (e.g., honey) used in the preparation of infant formula. Some evidence indicates that the organism becomes established as a member of the normal intestinal flora of the newborn. Evidence also indicates that some cases of sudden crib death may be caused by botulism intoxication.

Clostridium perfringens Food Poisoning

The production of toxins by *Clostridium perfringens* has been known from the early days of microbiology, but the production of food poisoning toxins was discovered only in 1945. Perfringens food poisoning was first recognized in Europe, but scientists and physicians in the United States did not consider it a serious problem until the 1950s. Generally, the symptoms are subclinical and limited to slight diarrhea, abdominal pain, headache, and nausea. Symptoms last from 12 to 24 hours; recovery is complete within 48 hours with no residual effects. Because of the fleeting nature of perfringens food poisoning, it normally goes unrecognized and, as a consequence, unreported.

Toxin formed during cell growth and spore formation is stored in the spore coat and released when the cells **autolyze** to free the spores. Perfringens intoxication is common when cooked meat dishes are allowed to stand at temperatures favorable to the growth of these bacteria (10°C to 40°C).

Figure 10.2

Release of toxin by Clostridium perfringens.

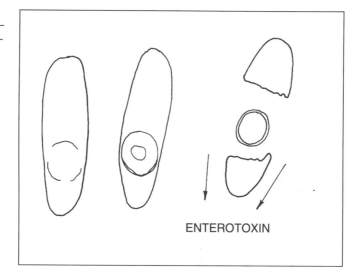

ENTEROTOXIN

Cooking drives off the air, creating anaerobic conditions, and meat with a low **oxidation-reduction potential** serves to maintain the anaerobic condition. This makes meat dishes ideal media for the growth and sporulation of *C. perfringens*. Studies show that the majority of the U.S. population has circulating specific antibodies indicating exposure to the toxin.

Because *Clostridium perfringens* toxin is heat-labile, individuals who eat reheated food are not in danger of being intoxicated. On the other hand, people who consume meat that sits without refrigeration and is not reheated run high risk of intoxication.

Another form of poisoning may occur when large numbers of organisms are ingested and undergo sporulation in the host's intestinal tract. In this case the toxin is released in the intestinal tract rather than in the food. Infection is not established, and recovery is as rapid as when the preformed toxin is ingested. The exotoxin, an enterotoxin (see Figure 10.2), acts by increasing the water permeability of the cells of the intestinal lining in much the same way that cholera toxin does. This is not a reportable disease, so no reliable data are available on the incidence of this intoxication, but it is thought to rank among the top five food intoxications in the United States.

Staphylococcal Food Poisoning

Staphylococcus aureus is a common pathogen of human beings and other animals. An inhabitant of the skin and the nasopharyngeal region of human beings, it has the ability to produce diseases ranging from simple skin rashes and boils to pneumonia and meningitis.

It produces approximately ten different toxins and other factors associated with virulence. Among these is an exotoxin that acts on the enteric system of human beings. Staphylococcal food poisoning toxin is the most common source of intoxication, responsible for thousands of cases of food poisoning in the United States each year.

Organism   *Staphylococcus aureus* is a Gram-positive, spherical bacterium that measures approximately 1 μm in diameter. These bacteria grow in clusters that, when stained, look like clusters of grapes. Staphylococci are facultative, heterotrophic organisms found everywhere on Earth. They are found in all foods that have been handled by human beings or that have been contaminated by animal matter. These bacteria grow well in all prepared foods including meat dishes, vegetable preparations, fruit dishes, pastries, and milk products. They are halophilic, growing well on nutrient agar containing 10% sodium chloride.

Staphylococcal Enterotoxin   Eating food in which *Staphylococcus aureus* has grown will produce symptoms of staphylococcus food intoxication including nausea, vomiting, diarrhea, abdominal distress, headache, blurred vision, and generalized body weakness. All or some of these symptoms may last as long as 12 hours and then disappear completely. Normally, recovery is uneventful and no residual effects remain even after severe intoxication.

Responses of individuals receiving the same amount of enterotoxin in laboratory experiments showed varying symptoms and varying degrees of severity. This indicates personal differences in susceptibility to the enterotoxin or perhaps different histories of previous exposure and consequent immunity. Within wide limits that accommodate the differences described here, the extent of intoxication of any individual depends on the amount of toxin consumed and the serological type of the toxin. Staphylococcal enterotoxin type A produces more severe symptoms than does type D. Normally, staphylococcal intoxication requires no medical attention. Death is rare, but consequences can be serious when medically compromising conditions exist.

The consequences of staphylococcus enterotoxin food poisoning are in direct contrast to botulinum food poisoning. The symptoms of both staphylococcal food poisoning and botulism are somewhat ambiguous, however, and can be confused easily. Because of the seriousness of botulism a distinction must be made as rapidly as possible lest the botulinum toxin have time to react with its target tissues. Sublethal doses of botulism result in crippling disease and even coma that may linger for many months or even years. Figure 10.3 gives the protocol for a differential diagnosis.

Although modern tests based on the identity of the toxin are available, many physicians and technicians prefer the animal test; a test in which some of the suspected food material is mixed with antitoxin and injected into mice. If mice that received the suspected food material die and those that received

The ability to distinguish between botulinum and staphylococcal toxins can mean the difference between life and death. This can be done in several ways, but the most common and most effective method is the animal test. The animal of choice is the chimpanzee, but in its absence, newborn kittens will do as well. Whereas botulin is heat-labile, staphylococcus toxin is heat-stable, and this difference is the key to the test.

The sample suspected of having toxin will be treated in the same manner whether it be food, beverage, vomitus, feces, or other. Mix it with a small amount of water and shake vigorously for a few minutes. Centrifuge briefly to obtain a clear supernatant, pour it off, and divide it into two equal parts in small test tubes. Place one portion in a boiling water bath for 10 minutes, and leave the other at room temperature. Cool the hot sample and bring it to room temperature quickly.

Obtain nine to twelve suckling kittens and divide into three groups of three to four each by random selection. Using a stomach tube and syringe, introduce approximately 5 milliliters of the heated sample to the kittens in one group, the same amount of unheated sample to another group, and leave the third group as control. Administer to the third group 5 ml of sterile water, in the same manner as that used for the two test groups.

Do all this with extreme care so the kittens are not alarmed, irritated, frightened, or in any other way disturbed lest they exhibit the effects of mishandling instead of intoxication. After administering the suspected samples, place the kittens in a quiet, dark, comfortable place, and observe them for several hours.

The observations may be interpreted as follows:

Observation	Interpretation
All kittens receiving unheated sample ill, disoriented vomiting; all others normal	Botulinum toxin
All kittens receiving sample sick, vomiting, disoriented; controls normal	Staphylococcus food poisoning enterotoxin
All kittens normal	No bacterial intoxication
Control kittens ill, disoriented, vomiting	Test invalid

the combination of food and specific antitoxin do not, the test is considered positive for botulinum toxin.

The most precise and convenient test is based on the enzyme-linked immunosorbant assay (ELISA) in which the antibody for botulism toxin is attached to a molecule of some dye or other indicator system in a manner such that the antigen (toxin) –antibody reaction releases the indicator system, which can be easily assayed. If these tests are negative for botulism, the symptoms are assumed to be the result of staphylococcal food intoxication.

Two types of staphylococcus enterotoxin have been identified and studied from the point of view of the genetic origin of the toxin molecules. Strains of the organism that produce the enterotoxin designated type A were found to do so only when they carry specific **tox gene-bearing plasmids**. Nontoxigenic strains of *S. aureus* can be converted to toxigenic ones by infection with specific plasmids, and toxigenic ones can be "cured" by removing the plasmids. On the other hand, no indication that strains that produce enterotoxin type B have to be infected with plasmids to produce the toxin. In these strains of *S. aureus*, toxin production is **genomic**, i.e., it is a genetic characteristic of the organism.

By either mechanism, the enterotoxin of these bacteria seems to be synthesized when the organism grows, and no special conditions other than those conducive to growth are required. Within wide limits, environmental factors that inhibit growth have been shown also to inhibit toxin synthesis, and those that enhance growth to enhance toxin synthesis.

An exception to this rule may be the case of sodium chloride. *Staphylococcus aureus* grows well in laboratory media with only a trace of sodium chloride and produces large amounts of enterotoxin but fails to produce toxin even after long periods of incubation in the presence of 10% sodium chloride. The reason for this phenomenon is not yet understood, and neither is its practical value if such exists.

Epidemiology Staphylococcal enterotoxin is a small protein molecule that is not destroyed by stomach acids or by heating to 100°C for 10 minutes. It also is chemically stable, lasting for many weeks in refrigerated or dehydrated foods. These properties make it possible to have food which contains large amounts of toxin but no live cells of *S. aureus* since the bacteria can be killed by cooking or reheating the food while the toxin will not be affected.

Staphylococcal food poisoning is quite common in the United States and all other countries of the world. Outbreaks are numerous during all seasons of the year, with a noticeable increase during the summer months. A common scenario of staphylococcus food intoxication may be as follows.

The microbiologists from Goof Laboratory at Humbug University decide to have a picnic "with all the trimmings" on the first warm day of May. All being chauvinists, the guys get the beer and the gals the potato salad. One of the women has a small, almost invisible "zit" on her hand, which, as everyone knows, is infected with *S. aureus*. This is a "must" before one can have a picnic of the genre envisioned by these people. They put the beer on ice and the salad under a tree, where the warm May sun will shine on it with all its splendor. While they play ball and other childish games, explore the river, get poison ivy, and take their shoes off, the day warms to a temperature of 38°C, and so does the potato salad.

Finally, after giving the staph sufficient time to grow abundantly, the future microbiologists and their august professors decide to eat. Within 2 hours most of those who ate potato salad begin to experience blurred vision, nausea, and vomiting. They all begin to shout and yell that everyone will die of botulism, and they all rush to the hospital. There they are told that it is most likely staph and that they all will live. After writhing in pain, vomiting, crying, and making a spectacle of themselves in general, they slink out of the hospital one by one and head for home.

Monday morning, each one will tell everyone else, in private, how he or she knew all along that it was staphylococcal food poisoning and not botulism. The ones who did not eat potato salad will tell everyone, in a loud voice, that they knew the potato salad had staph toxin. The one point of truth in this scenario is that they all forgot that the same thing had happened last year.

Medical treatment is not prescribed for staphylococcal food poisoning unless there is excessive loss of electrolytes from vomiting and diarrhea. In this case, electrolyte replacement and treatment of symptoms is the only measure indicated.

Control   Control depends entirely on preventing the growth of *S. aureus* in foods. This is accomplished easily by adequate refrigeration because the bacteria do not grow at temperatures below 5°C. In addition, individuals with septic sore throat or boils and other similar lesions on their skin should not have contact with foods in preparation or in serving. The milk from cows with mastitis should be avoided as *S. aureus* often is the causative agent of this disease in cattle.

Bacillus cereus Food Intoxication

Bacillus cereus is a large, spore-forming, Gram-positive, aerobic, rod-shaped organism, ubiquitous in nature. It was thought to be a nonpathogenic soil organism until quite recently. In the 1950s European microbiologists established that *B. cereus* was associated with epidemic gastroenteritis, and soon thereafter the pathogenicity of the organism was confirmed in the United States. The gastroenteritis this organism causes is mild, with no recorded deaths of healthy adults resulting from intoxication. Medical attention is required rarely, and most intoxications probably go unnoticed.

A large degree of individual immunity or resistance is apparent in that human beings who eat the same contaminated food may show great variation in the symptoms that accompany intoxication. In most outbreaks a few individuals will suffer the symptoms of frank diarrhea and vomiting. The latter syndrome has an incubation period of approximately 10 hours, and the episode of diarrhea and vomiting lasts approximately 24 hours. This form of food intoxication has no serious side effects and no discernable residual effects after the primary symptoms disappear.

Growth   *Bacillus cereus* grows well at room temperature in almost any kind of food material. If the food is cooked, most of the other soil and water organisms will be killed, but spore formers such as *B. cereus* will survive. Heating activates the spores, and if the food then is allowed to sit at room temperature, the spores germinate and give rise to large populations of vegetative cells.

If the food is reheated or allowed to stand for extended periods, the vegetative cells will autolyse, releasing enterotoxin. If, on the other hand, the food is consumed while the cells are still alive, they will be lysed by stomach acids, releasing the enterotoxin in the stomach. In either case intoxication results if the number of cells in the food is large enough to release critical quantities of enterotoxin.

Toxin   In the laboratory this bacterium produces two enterotoxins. One is a heat-labile toxin that induces diarrhea, and the other is a heat-stable toxin that induces vomiting. The former resembles the toxin of *Vibrio cholera*, and the latter is a yet-to-be classified entity that affects both human beings and apes in the same manner.

Bacillus cereus can colonize the intestinal tracts of human beings and laboratory apes with resultant chronic diarrhea and excretion of toxin in the feces. Upon treatment with antibiotics that eliminate the bacteria, diarrhea ceases and so does production of enterotoxin.

Control   Proper refrigeration of food and generalized sanitary conditions in food preparation and serving areas will prevent outbreaks. Because these bacteria are present in all soils and waters, other methods of control are impractical. Of the methods recommended, control is effected best by proper refrigeration or by reheating of foods before serving them a second time.

MYCOTOXINS

Although few bacterial toxins are of importance to human beings, at least 165 species of the more than 250 pathogenic fungi produce toxins active against human beings. Many more are known, but these have not been properly described yet. Although fungal poisons were recognized as early as the 5th century B.C., most of the toxigenic fungi were not identified until the last 25 years. The physiological effects produced by fungal toxins affect various parts of the body in different manners. The effects vary from simple intestinal disorders to mental aberration and even death. Some fungal toxins are **oncogenic**, giving rise to both benign and malignant tumors of various kinds. Other **mycotoxins** are **teratogenic** when consumed by animals or human beings. Teratogenic or **carcinogenic toxins** serve as unique tools in the study of the mechanisms by which cancers and congenital malformations are induced in human beings.

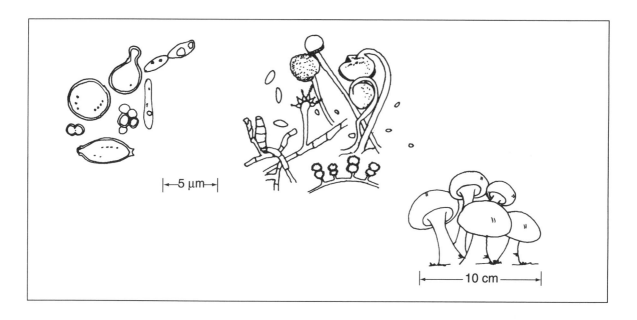

Figure 10.4

Yeasts, microfungi, and mushrooms.

For many purposes it is possible to divide the fungi into two major classes.

1. **Microfungi.** These are the yeasts and other fungi that form aerial or aquatic mycelia which are made of simple hyphae with associated fruiting structures. In general, intoxication results when these organisms are incorporated into foods and beverages as contaminants.
2. **Fleshy fungi** are those in which the fungal colony assumes a defined and recognizable structure. These are the mushrooms. Intoxication by mushroom toxins is the result of consuming the fungus as an item of food when the poisonous organism is confused with one that is edible.

Toxins produced by microfungi are called mycotoxins, and those produced by the fleshy fungi are called mushroom toxins.

Many of the mycotoxins belong in a class of chemicals called alkaloids; others are placed in different classes of chemical substances. Some toxins are small and simple molecules; others are large and complex. The genera of fungi that produce most toxins active in human disease are *Aspergillus* and *Penicillium*. These also are the genera most common in producing valuable fermentation products such as antibiotics, alcohols, and acids. The entire spectrum of fungal intoxication is far too broad for the scope of this text, so only a few examples are described in the following material.

Table 10.3 gives the names of a few of the toxins that have been identified and characterized. These were taken from a list of more than 165 different substances known to be toxic to human beings and other animals.

Table 10.3 Toxins of Microfungi and Their Source in Nature

Toxin	Organism Genus	Source
Akakabi-byo	*Fusarium*	Rice, barley
Luteoskyrin	*Penicillium*	Rice flower
Ochratoxin	*Aspergillus*	Barley and corn
Patulin	*Penicillium*	Fruits
Roquefortine	*Penicillium*	Cheese
Rubratoxin	*Penicillium*	Corn, grains
Stachybotryotoxin	*Stochybotrys*	Grasses, hay
Sterigmatocystin	*Aspergillus*	Oats and wheat

Toxin Hunters

During the 1940s countless major studies designed to find new antibiotics for combatting microbial infections were carried out in the United States and many other countries. These yielded a cornucopia of antibiotics and also a large number of new mycotoxins. Most of the toxins have not been studied or named, and it is not known whether they are or are not produced in nature. When the results of these studies were first contemplated, the number of fungal products that exerted drastic physiological effects on animals and man was astonishing. The extent of this phenomenon still is not known, and its significance is not understood. A few of the mycotoxins that are well known are described briefly in the following pages.

Ergotism

The fungus *Claviceps purpurea* is a common soil organism that produces a powerful toxin called ergot (see Figure 10.5 for the appearance of the ergot and

ERGOT

Figure 10.5

Ergots of Claviceps purpurea *on rice stalks.*

Figure 10.6

D-Lysergic acid diethylamide (LSD).

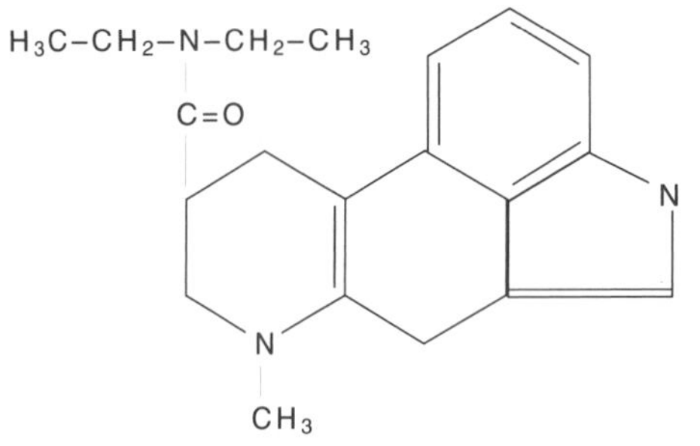

$H_3C-CH_2-N-CH_2-CH_3$

$C=O$

N

CH_3

D-LYSERGIC ACID
DIETHYLAMIDE (LSD)

Figure 10.6 for the chemical structure of one form of the toxin). In nature this mold grows on cereal plants such as wheat, rye, barley, oats, and many others. Normally the amount of growth is not sufficient to yield enough toxin to cause clinical symptoms in human beings, but under special conditions of temperature and humidity, large amounts of fungal growth occur with the consequent accumulation of large amounts of toxin.

In times before the nature of *C. purpurea* was known, epidemics of ergotism were common and large numbers of deaths often resulted. During the Middle Ages in Europe, this intoxication was called "St. Anthony's fire," and the affliction was seen as a punishment for disobedience to authority. The most common sequence of events leading to epidemic ergotism previous to the 20th century started in the fields as cereal grain crops began to mature. Because *C. purpurea* forms spores that are easily carried in the air, all cereal plants are inoculated readily by aerial spores. As the grain matures, the fungi grow to produce purple-brown, hard ergots in the cereal stalk. Ergot is the common name for *sclerotia*, which are accumulations of fungal spores protected by a hard membrane. If the grain cereal is harvested and stored while still moist, the fungi will grow abundantly, producing large numbers of sclerotia.

Fungal growth and consequent toxin production also will take place if the grain is stored in poorly ventilated buildings or other places where the humidity is high. In either case, if the cereal with ergots or abundant growth of fungi is ground into flour or prepared in some other form, the toxin survives extended storage, cooking, and even fermentation, as in the making of beer.

Toxin Ergot is a stable molecule that still will be present at full strength in foods prepared from contaminated grains. Cereal meal, soups, gruel, and beer made from infected cereals will produce the intoxication when these are consumed. *Claviceps purpurea* and other closely related fungi produce a large variety of toxins that are closely related to ergot and collectively are called ergot alkaloids. Other forms of the toxin are called **ergotamine** and **ergonovine**. **Lysergic acid diethylamide (LSD)**, a derivative of ergotamine, is a potent hallucinogen that has been a commonly abused, illegal drug. It became "cultish" during the 1960s in the United States and was favored by the "drop acid, drop out" crowd. The chemical structure of lysergic acid diethylamide is shown in Figure 10.6 in standard chemical notation. Because of their powerful physiological effects on the human body, many ergot alkaloids are used in modern medicine. These uses are as **vasoconstrictors**, central nervous system affecters, muscle contractors, and as mind probes in psychiatry.

Illness The symptoms of ergotism include nausea, vomiting, malaise, mental confusion, hallucinations, epileptic seizures, limb paralysis, gangrene, abortion in pregnant women, and death. Different species of *C. purpurea* produce different toxins, and two syndromes can be identified in human beings. One is the **gangrene syndrome**, and the other is the **epileptic syndrome**. The former is characterized by a predominance of blood vessel involvement and the latter with central nervous tissue involvement.

As with bacterial toxins, the extent of intoxication depends on the amount of toxin ingested, the presence of predisposing disease, and the person's general state of health. The alkaloids that make up ergot have different effects on various tissues and cells of the body, and symptoms depend on the tissues affected. The major target tissues include those of the circulatory system, the **endocrine system**, nerve and brain tissues, and muscle tissues. The symptoms resulting from involvement of such diverse parts of the body are complex, as they are caused by combinations of several effects rather than by a single action caused by one toxin, as is the case in some of the bacterial intoxications.

Epidemiology Epidemic ergotism was at one time a major cause of death in Europe and probably other parts of the world as well. Medical records of the past indicate that epidemics of ergotism in Europe at the end of the 10th century took more than 50,000 lives in France alone and probably similar numbers in surrounding countries. The last great epidemics with much loss of life occurred in France in the 1770s, Russia in the 1940s, and France in 1951.

The last epidemic of ergotism in the United States took place in 1824–1825. Medical historians who have studied the pertinent facts believe that some of the children said to have been the victims of witchcraft in Salem, Massachusetts, in 1692 may have been the victims of epileptic ergot poisoning. The scores of women and men who were tortured, strangled, drowned, or hanged under convictions of witchcraft may well have been the innocent victims of false accusations by those suffering from ergot intoxication.

Table 10.4 Substances Used to Inhibit Mold Growth in Cereal Products

Inhibitor	Concentration; % (w/w)
Ammonia	2*
Propionic acid	1
Chlorine	(vapor)
Calcium propionate	0.1 to 0.3
Sodium diacetate	0.1 to 0.3
Sorbic acid	0.1

*w/w (2 g ammonia per 100 g of ammonia-water mixture sprayed on cereal)

Most of the wheat seed planted in the United States and other advanced countries comes from plants resistant to *C. purpurea* infection, and ergotism consequently is now a rare disease worldwide. The only method of control required is inspection of all cereal crops to assure that the fungus *C. purpurea* is not present in large quantities at harvest and that grain cereals are stored in dry, well-ventilated, clean areas.

The picture is somewhat different now because many other fungi are known to grow on damp cereals and produce other, equally devastating toxins for human beings and animals alike. Chemicals also are employed in controlling *C. purpurea* and the other organisms that contaminate cereal foods. The most commonly used additives for the control of fungi are shown in Table 10.4.

Aflatoxins

In 1960 an epidemic called **Turkey - X poisoning** killed more than 100,000 turkeys in several poultry farms in England. Shortly thereafter, 14,000 ducklings died, and numerous epidemics followed in the calf populations of England and other European countries. Investigations revealed that these animals had been fed a cereal meal with peanuts as one of the ingredients. Upon examination, the peanuts in the meal were found to be contaminated with the fungus *Aspergillus flavus*. This is a common organism once considered a nonpathogenic soil and water saprophyte but now known to produce the toxin designated **aflatoxin** (a = aspergillus + fla = flavus + toxin) as well as several other toxic substances. This organism and other closely related species such as *A. parasiticus* and *A. columnaris* produce a variety of some 20 different forms of aflatoxin when grown on the following food crop plants:

barley	cowpeas	pecans	sesame
cassava	millet	rice	sorghum
corn	oats	yams	soybeans
cottonseed	peanuts	rye	wheat

Subsequently it was found that human beings are susceptible to intoxication and that many previously misdiagnosed or undiagnosed cases of human illness were probably caused by aflatoxin intoxication. Further research showed that aflatoxin was carcinogenic and teratogenic as well. Aflatoxins can induce cancers and other genetic lesions on animals ranging from fish to human beings.

Symptomatology **Aflatoxicosis** is characterized by malaise, nausea, vomiting, central nervous system involvement, convulsions, paralysis, coma, and death. When the dose of toxin is at sublethal levels, the effect is **encephalopathy** with **progressive fatty degeneration** of the viscera. This syndrome, called EFDV disease has been known for many years.

Doses of purified toxin required to kill laboratory animals are in the order of a few milligrams per kilogram of body weight. The LD_{50} (L = lethal + D = dose + 50 = needed to kill 50% of animals injected) for ducklings is 0.36mg per kilogram of body weight; 0.072mg of aflatoxin is required to kill 50% of ducklings weighing 200g. Smaller quantities of toxin cause an increase in the number of liver cancers and congenital malformations. The amount of toxin required to produce cancer and other genetic abnormalities in ducklings that weigh approximately 200g each is of the order of 0.0002 to 0.004mg (0.2 to 4 micrograms) of toxin. Figure 10.7 gives the calculation of LD_{50} for ducklings.

Figure 10.7

Graphic calculation of LD_{50} of aflatoxins for ducklings.

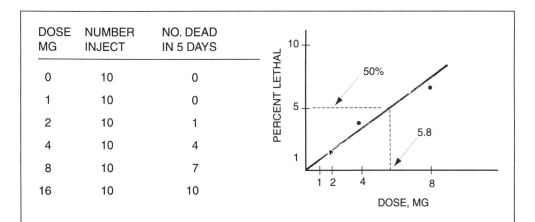

DOSE MG	NUMBER INJECT	NO. DEAD IN 5 DAYS
0	10	0
1	10	0
2	10	1
4	10	4
8	10	7
16	10	10

1. The LD_{50} is 5.8 mg.

2. If average weight of ducklings was 500 g, then results are reported as 5.8 mg per 500 g duckling or 11.6 mg toxin per 1,000 g; LD_{50} = 11.6 mg/kg BW.

Carcinogenesis   The incidence of liver cancers in human beings is higher in communities where corn, rice, peanuts, and other cereal foods form a large part of the daily diet. This is specially so in communities where grains are harvested and stored under less than ideal conditions. Often this also is the case in areas where warfare places added pressures of time and space on the preparation of crop harvests. Situations conducive to aflatoxin production are harvesting during the rainy season, allowing cereal stalks to fall to the ground, picking grain before it is mature, and using grain from crops with heavy fungal growth from the field.

Epidemiology   A crop of peanuts that contains a given quantity of aflatoxin can be divided into two parts; one part will be made up of a small number of peanuts containing very large quantities of aflatoxin, and the other part, a large number of peanuts containing no trace of the toxin. The toxin-containing peanuts can be recognized because the shells are broken, bruised, or have fungal growth on them. If the damaged or moldy peanuts are removed and the crop is stored in dry, well-ventilated, clean areas, no toxin or only traces of toxin at most are present.

If the toxin-bearing nuts are not removed, the quantity of aflatoxin for the entire crop will increase even if the conditions for storage are adequate. This phenomenon results from growth and spread of the fungus from one nut to the next during storage. If contaminated crops are used in preparing foods, the foods may be contaminated, too. Beer made from corn carrying aflatoxin will contain the toxin, and peanut butter or peanut oil made from contaminated peanuts also will have aflatoxin.

Toxin   The toxin of *A. flavus* is a small, stable molecule easily extractable from the plants in which the fungus grows because it is only slightly soluble in water but soluble in lipid solvents. A mixture of 97 parts of chloroform and 3 parts of methanol can extract all of the aflatoxin from almost any kind of food product, whether it be natural or prepared. Chemical analyses of the extracts reveal at least 20 closely related chemical substances classified as aflatoxins. Of these, the most common in the United States, and the most toxic, is one designated B_1, shown schematically in Figure 10.8. Its effects on various animals are listed in Table 10.5.

Toxin Distribution   Many different methods have been devised for separating toxin-bearing peanuts from those that contain no toxin. The easiest involve air blowing and **density gradient separation**, as the **specific gravity** of peanuts infected with *A. flavus* is less than that of healthy nuts. By contrast, the most effective method is individual inspection and hand-picking, but this is by far the most expensive. In the United States all suspected food products are inspected for aflatoxin contamination.

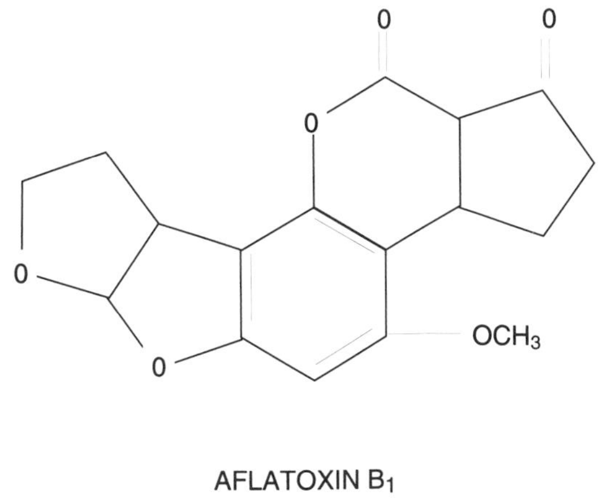

Figure 10.8

Aflatoxin B$_1$.

AFLATOXIN B$_1$

If the quantity detected by chemical analysis is greater than 20 micrograms of toxin in each kilogram of food material, the product is acceptable for human consumption. As determined previously, consumption of foods containing aflatoxin at this level will not result in cancers or teratogenesis in human beings. This standard was established in 1963 by the Food and Drug Administration (FDA) for peanut products in the United States. In 1973 an international standard of 30 micrograms per kilogram of food was set by the World Health Organization (WHO). This standard was based on feedings to monkeys and has a safety factor of 50 times the dose. It is used in all international food inspection procedures and has proven to be adequate protection.

Milk products often contain aflatoxins that enter the milk from pasture. Pastures contaminated with *A. flavus* often contain aflatoxin B$_1$ in quantities

Table 10.5 Effects of Aflatoxin B$_1$ on Various Animals

Lethal to	Carcinogenic to	Teratogenic to
Cats	Ducklings	Ducklings
Chickens	Hamsters	Sheep
Ducklings	Mice	Trout
Guinea pigs	Rats	Turkeys
Hamsters	Sheep	
Human beings	Turkeys	
Mink		
Monkeys		
Rabbits		

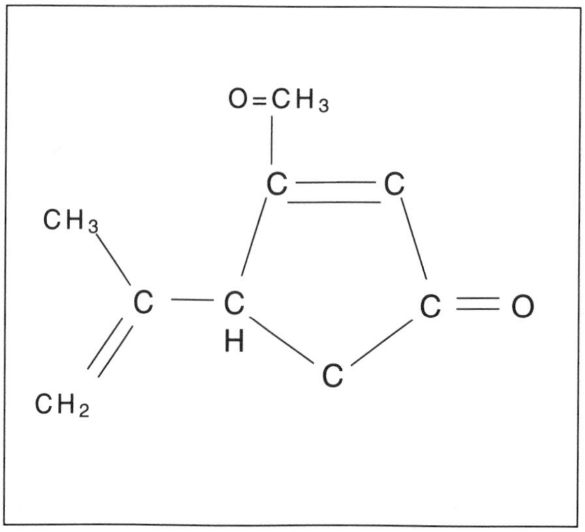

Figure 10.9

Chemical structure of penicillic acid.

below toxic level. Aflatoxin B_1 is converted to aflatoxin M_1 and other forms in the cow and excreted into the milk as M_1. A survey conducted in 1975 showed that approximately 1% of milk samples in the United States contained **trace amounts** of M_1 toxin. In addition, the general lack of concern for the consumer during the 1980s resulted in the marketing of products with levels of aflatoxins higher than those permitted by law.

Penicillic Acid   Penicillic acid was first extracted from moldy corn in 1896 and found to be toxic for swine, cattle, and other animals including laboratory mice and rats. Since then, it has been proven that penicillic acid is carcinogenic. Many documented cases show that sublethal human poisonings by penicillic acid are not rare but, rather, that they often go unrecognized.

It first was determined that penicillic acid was produced by *Penicillium puberulum,* but later it was shown to be produced by at least twenty different species in the genus *Penicillium* as well as by many fungi in other genera. Penicillic acid is produced when fungi grow on various foods including oats, wheat, barley, rice, corn, and several varieties of beans. Its chemical structure is shown in Figure 10.9.

Intoxication   Penicillic poisoning in human beings is seen mostly under unusual circumstances because many of the fungi that infect corn or bean crops cause discolorations and disfigurement of the plant leaves, giving a frank warning of fungal infection. Intoxications in animals are more common but generally occur as a result of false economy. Farmers who find that their bean crop cannot be sent to market often use it for animal feed in an effort to avoid a complete loss.

The final result may be much more costly than the mere loss of the crop, as the value of sick and dead livestock now must be added to the account. Furthermore, leaving the crop in the field without burning will result in the accumulation of vast numbers of toxigenic fungi in the soil and an almost certain repetition of the infection in subsequent crops.

Toxin Toxin potency was studied using laboratory mice and found to be on the order of 100 milligrams per kilogram of body weight for the LD_{50} dose. The carcinogenic character of this toxin is evident in doses of 1 milligram per kilogram of body weight in mice and rats. The toxin is effective against many **invertebrates** and some microorganisms as well. Because it has the ability to kill intestinal worms, it was investigated thoroughly as an antihelminthic drug but found to be too toxic for use in human beings. It also was studied as a possible antiviral agent but was discarded for the same reason.

Tricothecenes

Some 30 different but closely related chemical substances collectively called trichothecenes are produced by different genera of fungi including:

Trichothecium	*Myrothecium*
Fusarium	*Stachybotrys*
Cephalosporium	*Trichoderma*
Gibberella	

Distribution Fungi that produce trichothecenes are ubiquitous in nature and produce toxins when growing on wheat, corn, sorghum, rice, cranberries, coffee, and several species of grass. The intoxication that results from ingesting foods contaminated with trichothecenes is best known as **aleukia** or **alimentary toxic aleukia** (ATA). The illness is characterized by multiple symptoms involving both the skin and the alimentary tract. It has long been recognized as a disease of cattle and other livestock that can be transmitted to human beings.

Intoxication In the older Russian medical literature, aleukia is described as an epidemic alimentary tract intoxication associated with excess rainfall. Today the epidemic intoxication is known as **red mold poisoning** in the Orient and as **moldy corn poisoning** in the United States. It is regarded as an inevitable result of corn storage, as the corn brought from the field always bears one or more of the organisms that produce trichothecenes. These chemicals are considered a major problem in livestock feeds since animal illness or death from trichothecene intoxication is an ever present threat. Trichothecenes also are a hazard to human beings who work with infected plants or grain crops, since they may absorb sufficient toxin through the skin to develop serious skin intoxication and damage.

The symptoms observed include edema, irritation, inflammation, and necrosis of the skin, in addition to intestinal tract bleeding, **leukopenia**, and death. An aspect of universal toxicity about trichothecenes is evidenced by the variety of animals that are susceptible to the toxin. These include human beings and other mammals, many kinds of fish, shrimp, insects, plants, and bacteria. The effects may bring about illness, death, or carcinogenesis, and these effects vary from one kind of victim to another.

Toxin   The most common of the trichothecene toxins are **vomitoxin** and **T-2 toxin**. These are stable, small molecules that resist microbial degradation and remain active for extended periods. They are produced at low temperatures, and storing infected materials in the refrigerator will not inhibit mycotoxin synthesis. Laboratory studies have shown that plant materials stored at 4°C develop sufficient toxin to cause toxicosis in experimental animals.

MUSHROOM INTOXICATIONS

The organisms commonly called mushrooms or fleshy fungi belong to the classes of fungi named *Basidiomycetes* and *Ascomycetes*. Many of these produce chemical substances that can be characterized by their strong physiological effects on many animals including human beings. The effects produced in human beings may not be the same as those experienced by Alice when she traveled in Wonderland, but they nevertheless are strange enough. Eating certain mushrooms results in physiological changes that vary from simple and fleeting nausea to mental aberration, and even to rapid death.

Human beings have used mushrooms for their hallucinogenic and poisonous properties from earliest times. The hallucinogenic varieties often are associated with religious rites, while poisonous mushrooms were used equally on spear points as in wine glasses. Notwithstanding this, mushrooms also have been considered a dietary staple by all human beings. The differences between toxic and edible varieties were fully appreciated even by early human beings who made the identification of mushrooms an important task of all food gatherers.

Mushrooms as Food

Little has to be said about mushrooms as a source of food today. In the United States, mushrooms have become a common item in the diets of most individuals, although previous to World War II, mushrooms were considered more a food for the gourmand and epicure than for the average person. Even so, consumption in large quantities is limited to one or two species that can be grown economically on an industrial scale.

Figure 10.10

Toxic or edible?

In Europe and other places, mushrooms are gathered from wood and field by individuals who are skilled in differentiating the edible and the toxic varieties. To those who dine on wild mushrooms, the major pleasure lies in savoring the different tastes and textures characteristic of the different species and strains. The mushrooms that can be found in the field in any part of the United States offer a rich gourmet reward to anyone who is capable of distinguishing successfully between the edible and the inedible varieties.

On the other hand, individuals who venture into the field with basket in hand and mushroom identification guide under arm are more likely to rue their decision than to enjoy their mushroom meal. No one who feels the need for help with the scientific identification of mushrooms should attempt to gather mushrooms from the field. Everyone who thinks that toxic mushrooms can be distinguished from edible ones by a rule of thumb should obtain mushrooms only from a trusted grocer. All who identify mushrooms for their supper gamble their life on their ability to tell the difference between edible and poisonous species and to do so unerringly.

Variety   Because more than 40,000 species of mushrooms are in existence, a great deal of knowledge is required to find those few that make all the trouble worthwhile. Wild species of mushrooms can be divided into four categories of food: delicacy, edible, safe but unappetizing, and poisonous. Those that are safe but not desirable can be divided further into species that have no desirable qualities and those that are not toxic but are endowed with odors or tastes sufficiently unpleasant to make them a nuisance.

Poisonous Mushrooms

For the purpose of this discussion, all the mushrooms that cause illness or distress when eaten by human beings are regarded as poisonous. Although fewer than 200 species of mushrooms contain toxins sufficiently strong to cause serious illness in human beings, they present a serious problem since they are scattered among the edible mushrooms. The degree of intoxication that can result is variable and depends on many factors. Since mushroom intoxications result from the meal itself rather than from contaminants, as is the case with bacterial and microfungus poisoning, intoxication must be viewed from a different perspective, including the understanding that some mushrooms are toxic only under certain conditions and that others are toxic but may be eaten in total safety if prepared properly before eating.

Another factor of great importance in considering the poisonous mushrooms must be the understanding that the vast majority of poisonings occur as the result of misidentification of the mushrooms. Typical candidates for intoxication are the amateurs who do not understand that they are betting their life on their ability to recognize the difference between edible and poisonous mushrooms.

Mushroom Murder   Mushroom toxins have long been the choice of professional, and probably also amateur, poisoners. The dual qualities of easy availability and extreme potency probably have been common knowledge to human beings since the beginning of human history on Earth.

One of the most famous of the mushroom murders in history may be the poisoning of Claudius, Emperor of Rome, by his wife, Agrippina. The lady prepared the Emperor's favorite mushroom dish, and she gave it a little extra snap by adding a few pieces of poisonous mushroom. The Emperor's favorite mushroom now bears the name *Amanita caesarea* and is considered a favorite of most gourmets. Agrippina probably spiked the Emperor's supper with *Amanita muscaria*, a toxic mushroom that can be confused easily with the edible *A. caesarea*.

Identification   The number of deaths due to consumption of poisonous mushrooms is not known because determining whether the mushrooms were the cause of death or simply part of the last meal can be difficult. Since mushrooms are identified by their form and structure, one often cannot see much of either trait in mushrooms that have been partially digested.

Another factor that makes compilation of mushroom intoxication statistics difficult is that most field-picked mushrooms are eaten in rural areas and obtaining the expert medical diagnosis necessary in such instances is not always possible. One authoritative source states that approximately 1,000 deaths worldwide result from mushroom intoxication each year. Most experts agree, however, that the data available are not reliable.

Figure 10.11

Variety of mushrooms.

Commercial Mushrooms   Commercial producers depend on the advice of experts and take all precautions to avoid even the hint of confusion in the identification of the products they sell. Although identifying mushrooms produced for sale, generally poses no difficulty, the danger always exists that wild, poisonous species may enter into the population by accident. All precautions are taken to prevent this, but it must be borne in mind that mushrooms, like all fungi, propagate by microscopic spores that are blown by the wind. Figure 10.11 depicts the manner in which commercial mushrooms may be displayed for sale.

Amanitas

The mushrooms most commonly associated with fatal intoxications belong to the genus of fleshy fungi named *Amanita*. Although only six or seven species in the genus produce toxins strong enough to kill human beings, many other members of the same genus are among the most desirable of the edible mushrooms. Probably the best known of the poison mushrooms are the ones commonly called **destroying angel** and **fly amanita**. Other toxic mushrooms with common names are **blushing amanita** and **death cap**.

The scientific names of some of the poisonous mushrooms are given in Table 10.6. The value of scientific names lies in the fact that such names are universal and unambiguous, whereas common names change with time and are valid only in small areas with small populations. The confusion that results from common names includes using different names for the same mushroom and also using the same name in one place as the designation for an edible species and in another for a toxic one.

Table 10.6 Species of *Amanita* found in United States, With Characteristics

Species	Characteristics
Amanita phalloides	very poisonous; causes most deaths
Amanita silvicola	beautiful; may be poisonous
Amanita porphyria	hallucinogenic and poisonous
Amanita ocreata	matures in deep winter; poisonous
Amanita pantherina	hallucinogenic and poisonous*
Amanita muscaria	hallucinogenic and poisonous*
Amanita caesarea	edible; confused with *A. muscaria*
Amanita calyptroderma	edible; a delicacy

*smoked and eaten in religious rites in small doses

All the toxic species of Amanita are found in the Americas, including the United States. Several are commonly called "destroying angel mushrooms," although this common name has some variations even within the U.S. population. All of these species now are common to North and South America, although they all are native to Central Europe and were brought to America by Central European settlers. They probably were imported in the spore or mycelial stage unknown to those who transported them, because the Europeans knew the nature of these mushrooms quite well and would not have brought them on purpose. *Amanita phalloides* (see Figure 10.12) is now well established in Canada, the United States, and Mexico and, in limited populations, even in South America.

Figure 10.12

Amanita phalloides.

The majority of mushroom intoxications in the United States occur in California, and most of these result from eating *Amanita phalloides.* Most poisonings are found among those who collect wild mushrooms for home consumption and who, by mistake, add one or two poisonous stems to a basket of edible mushrooms.

Other Toxic Mushrooms

Mushrooms of other genera and species also produce toxins that are lethal to human beings. Among these are *Lepiota josserandii, Galerina autumnalis, G. marginata,* and *G. venenata.* These fungi produce toxins chemically quite similar to those of the amanitas. Mushrooms of the genera *Lepiota* and *Galerina* are small, unattractive, and not likely to be selected by mushroom hunters. The danger lies in the fact that they are small and inconspicuous and thus may get into collections of field mushrooms by accident.

These mushrooms grow in secluded areas, tree stumps, and wood litter, making them hard to see. In effect, they are rarely eaten, and only a few cases of human intoxication are on record, although this is sufficient to establish the fact that they are poisonous. Chemical analyses of the stomach contents of individuals who have consumed lepiota or galerina and subsequently shown the symptoms of mushroom intoxication have indicated the presence of **amanitoxins**. In addition, the symptoms of intoxication are much the same as those produced by the amanitas.

MUSHROOM TOXINS

Amanitoxins and Phallotoxins

Several toxins are produced by the poisonous mushrooms of the genus *Amanita.* These are divided into two classes. The most toxic are called **amanitoxins** and those of lesser, but still lethal, toxicity are called **phallotoxins** (Table 10.7). Other toxins of less potency are produced by various species of amanita, but they are of less significance to the food preparation services than are the amanitoxins or the phallotoxins (Table 10.7 and Figures 10.13 and 10.14).

All amanitoxins are not lethal. Some, like **muscarine**, cause serious effects when eaten by human beings and animals, but fatalities are rare and generally are limited to the very young, very old, and the medically compromised. Muscarine is found in the mushrooms of the genera *Amanita, Clitocybe, Omphalotus,* and *Inocybe.*

Symptoms of Intoxication   Symptoms of intoxication develop 10 to 15 hours after ingestion of mushrooms that contain amanitoxins. It takes

Table 10.7   Some Toxins and Their LD$_{50}$*

Amanitoxins	LD$_{50}$*
Amanitoxins	
alpha-amanitin	0.39
beta-amanitin, thiophenyl ester	0.60
beta-amanitin, methyl ester	0.80
beta-amanitin	0.97
beta-amanitin, anilide	4.45
gamma-amanitin	0.15
epsilon-amanitin	0.5
Phallotoxins	
Phallocidin	1.5
Phallacin	1.6
Phalloin	1.8
Phalloidin	2.0
Phallisin	2.5
Phallisacin	4.4

*milligrams purified toxin per kilogram of body weight; injected into adult mice intraperitoneally.

that long for the toxin to pass from the intestinal tract to the bloodstream. At any time during the first 2 to 3 hours after ingestion, the mushroom tissue and its toxin can be extracted by emptying the stomach. After the toxin passes from the stomach into the intestinal tract, removal is much more difficult, but with medical intervention the effects still can be ameliorated. The appearance of symptoms is an indication that lethal quantities of toxin are in the bloodstream already and that death is imminent.

Vomiting Reaction   Many mushrooms contain substances that induce **emesis**. When these mushrooms contain toxins also, poisoning is averted because the toxin is vomited before it enters the intestinal tract or the bloodstream. Unfortunately, the amanitoxins do not activate the vomiting reaction, and ingesting the toxin usually leads to intoxication as no sign of poisoning appears until the toxin is in the bloodstream. Undoubtedly, this results in a high proportion of fatalities among those who ingest the amanitas.

Prognosis   The first symptoms to appear are headache, neck pain, abdominal distress, nausea, vomiting, and diarrhea. These symptoms are short-lived and disappear completely, giving the impression that the episode of food poisoning is at an end. Some 36 to 48 hours after the first symptoms disappear, a second, and final, syndrome begins. The new symptoms include colitis, liver impairment, kidney failure, prostration, and death. Autopsy findings show extensive intestinal, liver, and kidney damage, and massive hemorrhage dam-

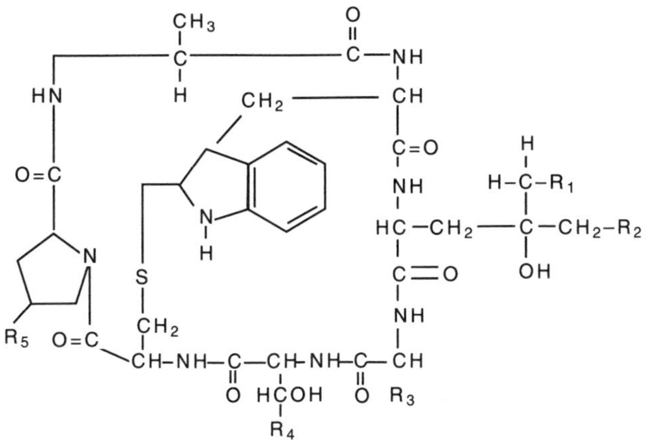

Figure 10.13

Chemical structures of some amatoxins.

	R_1	R_2	R_3	R_4	R_5
Phallocidin	–OH	–H	–CH(CH$_3$)$_2$	–COOH	–OH
Phallacin	–H	–H	–CH(CH$_3$)$_2$	–COOH	–OH
Phalloin	–H	–H	–CH$_3$	–CH$_3$	–OH
Phalloidin	–OH	–H	–CH$_3$	–CH$_3$	–OH
Phalloisin	–OH	–OH	–CH$_3$	–CH$_3$	–OH
Phallisacin	–OH	–OH	–CH(CH$_3$)$_2$	–COOH	–OH

age to many tissues in addition to intestine, liver, and kidney. Also, almost all the visceral organs show extensive cell and tissue destruction.

Mechanism Microscopic examination at autopsy reveals that the nuclei of damaged cells in many tissues of the body have been destroyed selectively. Many investigations have shown that amanitoxins react with **RNA polymerase II** and thereby inhibit protein synthesis. Polymerase II is an enzyme in the nucleus of cells that is involved in the synthesis of mRNA, a molecule indispensable for the synthesis of proteins according to the genetic information carried in the cell's DNA. Failure to synthesize RNA polymerase II, and subsequently proteins, leads to disorganization of the nucleus and death of the cell. The effects of amanitoxin are not reversible.

Cortinarius

In 1950, Polish physicians studied a large number of unresolved mushroom poisonings and found that 100 cases and 11 deaths resulted from consumption of mushrooms of the genus *Cortinarius*. These studies included chemical analyses, circumstantial evidence, autopsy findings, and animal feedings. Prior to this time, it was not known that cortinarius was a poisonous mushroom, although it had been suspected of being so for many years and normally was

Figure 10.14

Chemical structures of phallotoxins.

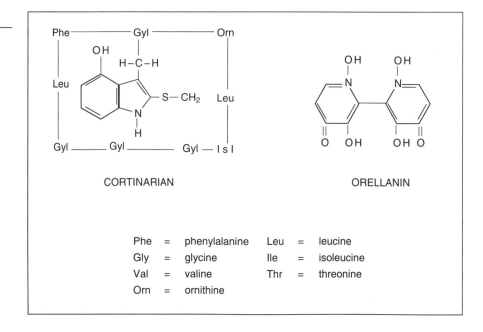

not recommended as an edible mushroom. Some of the chemical analyses showed the presence of amanitoxins in those who were ill or dead after eating cortinarius, but the amounts found at autopsy were too small for reasonable statements regarding cause of death.

In the interim from 1952 to 1957, 135 cases of intoxication and 19 deaths were recorded for human beings who ate mushrooms of the genus *Cortinarius.* These data were derived from an extended investigation in a population of individuals who ate mushrooms they picked from the field. Because individuals such as these pick a wide variety of genera and species of mushrooms, the investigation required keeping exact records of all the mushrooms a given population consumed. By the method of comparative matching, the researchers were able to conclude that individuals who consumed mushrooms of the genus *Cortinarius* became ill and that some died. It also was noted that symptoms of intoxication often did not appear until 2 or 3 weeks after eating cortinarius.

The major toxins of *Cortinarius orellanus* and *C. speciosissimus* are shown in Figure 10.14. The toxins produced by cortinarius are less toxic than the amanitoxins but are still capable of causing death in many animals and human beings as well.

Ingesting small amounts of cortinarians (Figure 10.15) will give no sign of intoxication in normal, healthy human beings. If the amount of toxin ingested is large but still sublethal, symptoms of intoxication begin some 8 to 10 days after ingestion. The first symptoms include nausea, vomiting, abdom-

Figure 10.15

Cortinarius orellanus
and C. speciosissimus.

inal pain, and diarrhea, among others. Sometime later the signs of kidney and liver damage appear. The extent of this damage is roughly proportional to the amount of toxin ingested. Damage to the kidneys and liver may become chronic and linger for 2 to 3 years after the intoxication.

HALLUCINOGENIC MUSHROOMS

Many mushrooms, such as *Psylocybe cubensis, Amanita panterine,* and *A. muscarine,* produce substances that act on different parts of the human brain to cause mental aberrations not experienced with any other chemical or physical agent. Although the active ingredients in many mushrooms have been isolated and identified, their mechanisms of action are, for the most part, still unknown. The degree and extent of hallucination can be determined in some cases by the amount of mushroom consumed.

Many native American populations refer to these as "magic mushrooms" and consider them essential in many traditional religious rites. The mushrooms are prepared in different ways and consumed for the purpose of reaching "higher states of consciousness" and also for inducing visions. Quite often these are secret rituals, and preparation of the mushrooms is left to the high priests of the religion. In other communities the role of the mushroom material is simply to produce altered states of mind in the communicants. In some religious rites the state of mental alteration may be maintained for many

Figure 10.16

Gyromitra.

days by continuous administration of measured quantities of mushroom.

Fungi that produce hallucinogens are not easily consumed accidentally because they are small and unattractive and appear unappetizing when growing in the field. Gathering these mushrooms in the wild requires great skill both in locating them and in identifying those that bear the desired hallucinogens.

Monomethylhydrazine

Gyromitra esculenta is an edible mushroom enjoyed in all parts of the world and appreciated as a delicacy by many. Mushrooms of the genus *Gyromitra* (Figure 10.16) are commonly called brain mushrooms or **lorchels**. Traditionally, these mushrooms are prepared for eating by soaking them in cold water and then changing the water before boiling for an hour or more. Because the mushrooms give off several substances that turn the water a dirty, dark color, the boil water usually is discarded and the mushrooms washed well before cooking. **Gyromitrans** have formed part of the food of human beings since time immemorial, yet three disturbing observations have occurred frequently and have been confirmed by scientific observation. These observations can be stated simply:

1. Chefs, cooks, and homemakers who boil mushrooms sometimes become ill, and some have died as a result of inhaling the fumes given off during cooking. Symptoms of their illness are the same as those of individuals who become intoxicated by eating the mushroom.

2. According to the medical literature, of several people eating the same gyromitran meal, one of the group may become ill or die and the others might show no ill effect. The person or persons who became ill, it was noted, generally were those who had eaten the gyromitran mushrooms the previous day.

3. Symptoms of intoxication appear at maximum effect with no evidence of low-level intoxication as is seen with other mushroom poisons. This "all or nothing" phenomenon has been known from antiquity. It also has been known for many years that the full effect of intoxication can be manifested whether a small amount or a large amount of mushroom material is consumed.

These three observations were made long ago, but those who addressed the problem tendered no explanation, nor did they succeed in understanding the nature of this unusual set of circumstances. The explanations for these peculiar effects came suddenly and very recently.

Mechanism of Intoxication

The mechanism of action of gyromitran intoxication did not begin to emerge until very recent times, and its origins were found in an unlikely place. Workers in ordnance plants where solid rocket propellants are made often become ill, and their symptoms are confused routinely with those of *Gyromitra esculenta* intoxication. When deaths occurred, autopsy findings were identical to those of gyromitran intoxication. It was rapidly shown that rocket fuel workers who became ill or died from exposure to rocket fuel contained high concentrations of **monomethylhydrazine**, one of the constituents of the fuel, in their blood and body tissues.

Subsequent study showed that mushrooms of the genus *Gyromitra*, and especially *G. esculenta*, produce several chemical substances with a basic chemical structure called **N-formyl-N-methylhydrazone**. A series of experiments designed specifically to test the toxicity of N-formyl-N-methylhydrazine showed that this and closely related substances are not toxic to animals or to human beings even after ingesting large amounts in various forms. Previous chemical studies had shown that the monomethylhydrazine formed part of the molecule of N-formyl-N-methylhydrazine.

Finally it was proven that when human beings consume mushrooms such as *G. esculenta*, stomach acids hydrolyze N-formyl-N-methylhydrazine to yield **monomethylhydrazine** (MMH).

When all the evidence described above was brought together, the following picture emerged:

1. *Gyromitra esculenta* contains powerful toxins that take several days to eliminate from the body.
2. The toxins are water-soluble and can be removed completely by washing and boiling in several changes of water. This makes the mushroom safe to eat under normal circumstances.
3. The fumes from boiling mushrooms contain some monomethylhydrazine released from the N-formyl-N-methylhydrazine by the action of boiling water. These fumes can cause death when concentrated in small, poorly ventilated kitchens. If the mushrooms are boiled in an open space with freely flowing air, insufficient toxin will exist to cause ill effects.
4. If the mushrooms are not washed thoroughly, some toxin will remain, but not enough to cause death or produce symptoms of intoxication in human beings. If the mushrooms are eaten on 2 consecutive days, sufficient toxin may accumulate to constitute a lethal dose. This explains why one person in a group is affected—why an individual who consumed gyromitra on the previous day and again on the subsequent day will accumulate sufficient toxin in his or her body to form a lethal dose.

It now is known that monomethylhydrazine causes several serious effects in the human body. These are designated as the three toxicities of the gyromitran poison.

First Toxicity   Monomethylhydrazine passes rapidly from the intestinal tract to the bloodstream. The blood carries it to each cell of the body, where the monomethylhydrazine reacts with vitamin B_6. Vitamin B_6 exists in three different chemical forms, **pyridoxal** being one of these. The reaction between pyridoxal and monomethylhydrazine results in the formation of a compound called **pyridoxal monomethylhydrazone**, as diagrammed in Figure 10.17.

Pyridoxal monomethylhydrazone is released by the cells where it was formed and circulates throughout the body. In a way not yet completely understood, pyridoxal monomethylhydrazone and monomethylhydrazine work together to produce convulsions. This combination of chemical substances has the highest potency of any known material for producing convulsions in animals, including human beings.

Second Toxicity   In addition to the above, the binding of vitamin B_6 to form pyridoxal monomethylhydrazone creates a deficiency of this vitamin in the cells. As a result, many cells die, causing tissue disruption and the failure of many body organs.

Third Toxicity   Monomethylhydrazone reacts with hemoglobin converting the hemoglobin to **methemoglobin**. The change in the hemoglobin molecule interferes with its ability to bind and carry oxygen to the cells of the body.

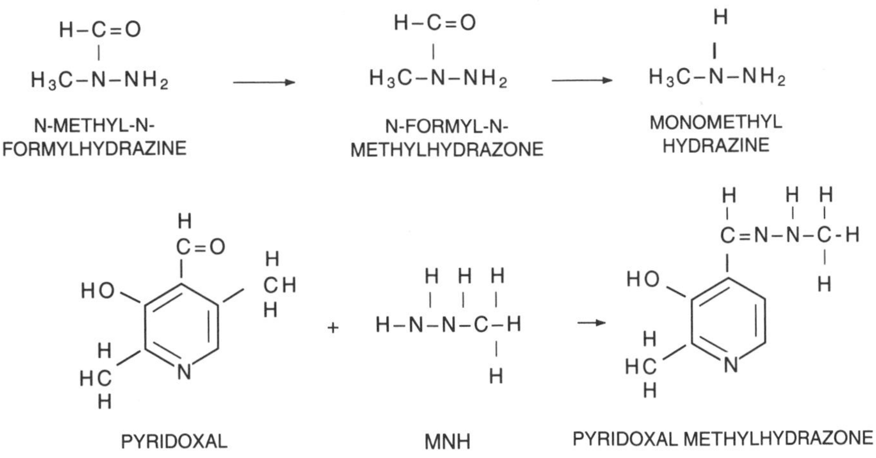

This results in the death of many tissue cells from oxygen deprivation.

Coprine

Coprinus atramentarius, commonly called **inky cap**, is an edible mushroom fancied by mushroom eaters all over the world. The same genus contains two of the most appetizing of the edible field mushrooms: *C. comatus* and *C. micaceus*.

History   It has long been known that drinking alcoholic beverages after eating *C. atramentarius* brings on a series of frightening physiological reactions that last for varying periods. The first sign of atramentarius intoxication starts a few minutes after consuming alcohol and lasts for a time roughly commensurate with the amounts of alcohol and mushrooms ingested.

Symptoms   The first sign of intoxication appears as a persistent tingling of the lips and the area around the mouth, followed by a throbbing sensation of the face, neck, and upper chest. Additional signs are an increase in blood pressure, heavy heartbeat, tightening of the chest area, disorientation, dizziness, and alarming mental confusion. These symptoms are accompanied by severe nausea, copious salivation, and vomiting.

Those who suffer this intoxication experience terror, panic, and terrible dread, as well as fright at the rapid progression of symptoms. In spite of the extreme physiological reaction the intoxication produces, permanent damage

Figure 10.17

Formation of pyridoxal monomethylhydrazone from pyridoxal and monomethylhydrazine.

Figure 10.18

Toxic reaction of coprine and alcohol.

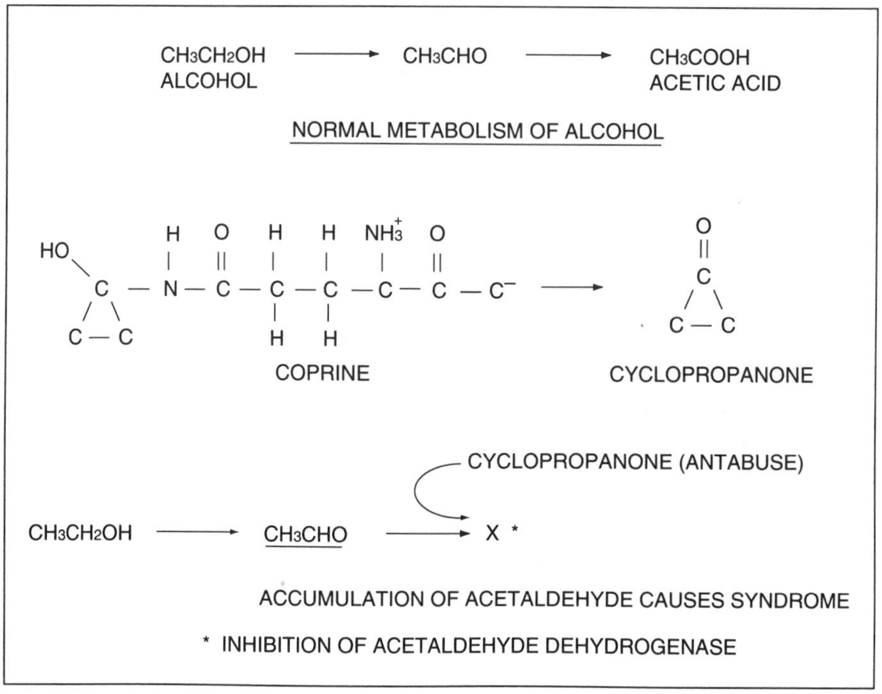

NORMAL METABOLISM OF ALCOHOL

CH_3CH_2OH ALCOHOL → CH_3CHO → CH_3COOH ACETIC ACID

COPRINE → CYCLOPROPANONE

CYCLOPROPANONE (ANTABUSE)

CH_3CH_2OH → CH_3CHO → X *

ACCUMULATION OF ACETALDEHYDE CAUSES SYNDROME

* INHIBITION OF ACETALDEHYDE DEHYDROGENASE

or death is extremely rare and all symptoms disappear within an hour or two.

Many studies have shown that a nontoxic, physiologically active substance named **coprine** is normally present in *C. atramentarius*. Also, when coprine passes through the stomach of human beings and other animals, it is converted to **cyclopropanone hydrate**. When this conversion takes place in the stomach, the cyclopropanone hydrate passes to the intestinal tract and then to the bloodstream. The blood carries it to all the cells of the body, and it is here where its effects are manifested.

Normally an enzyme in the cells of the body called **acetaldehyde dehydrogenase** converts alcohol to **acetaldehyde** in order to detoxify it. In the normal sequence of events, acetaldehyde reacts with the enzyme **aldehyde dehydrogenase**, and in so doing the acetaldehyde is converted to acetic acid. The cell then metabolizes acetic acid as if it were a carbohydrate. Cyclopropanone from *C. atramentarius*, however, reacts with aldehyde dehydrogenase causing the acetaldehyde to accumulate in the cells. This produces the symptoms of intoxication described above. This sequence of reactions is shown in Figure 10.18. Cyclopropanone is metabolized itself by the body cells and disappears from the cells in a short time, leaving no aftereffects. The intensity of the toxic symptoms depends on the quantities of coprine and alcohol that are in the bloodstream simultaneously.

Practical Use Cyclopropanone is the active ingredient in the medication called **Antabuse**, used in aversion therapy with individuals who wish to refrain from drinking alcohol. Those who take Antabuse (cyclopropanone hydrate) and then drink alcohol suffer the same symptoms as those who consume coprine mushrooms and alcohol at the same time. Medical authorities generally consider aversion treatment of alcohol abusers with Antabuse to be effective.

SPOILED MUSHROOMS

One of the most serious intoxications of human beings derives from the consumption of mushrooms that have spoiled during improper storage. Bacteria such as *Clostridium, Staphylococcus, Pseudomonas*, and many others, including the enteric pathogens, grow well on the mushroom tissues and may produce bacterial intoxications. Although the individuals would become ill from eating mushrooms, in this case the intoxication would be of bacterial origin. Many individuals think that mushrooms do not spoil and that they need no protection. In actuality, mushrooms spoil just like any other food and, when spoiled, can cause intoxication or even infection.

Improperly cleaned and stored mushrooms may become contaminated with pathogenic organisms. This is a particular problem with mushrooms grown on prepared animal dung, as are the coprines and many others. In commercial production of mushrooms, manure used for the culture beds is steamed to kill pathogenic bacteria. Although this is an excellent procedure, it is not without failures in that the material may be contaminated by vermin, workers, or soiled utensils during preparation.

OTHER MICROBIAL INTOXICATIONS: ALGAE

In many areas of the world, dinoflagellates such as *Gonyaulax catanella, Alexandrium tamarense*, and *Pseudonitzschia pungens* "bloom" at irregular times and for unknown reasons, produce extremely large populations. These microalgae contain red pigments, and if the number of organisms in the water is large enough, the ocean water takes on a deep red coloration. This phenomenon called the **red tide**, has been known from antiquity when it was regarded as "blood in the sea."

During such blooms, **filter-feeding animals** such as oysters, clams, mussels, and scallops feed primarily on red tide algae because of their easy availability. The dinoflagellates contain a toxin called **paralytic shellfish poison**, saxitoxin, which is concentrated in the tissues of the shellfish as the algae cells are digested.

Figure 10.19

*Red tides recorded in 1990
throughout the world.*

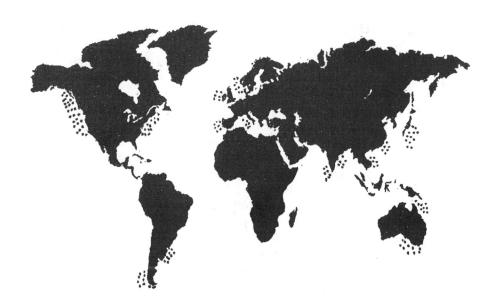

Animals that feed on the shellfish or other fish that eat algae also accumulate the toxin if they are resistant to its effects but sicken and die if they are not (Figure 10.19). Land animals that eat contaminated fish and survive also may accumulate the toxin, and those that feed on them may be affected by the paralytic toxin. Human beings are quite susceptible to the effects of this toxin and show symptoms of intoxication after ingesting small amounts of contaminated shellfish or of animals that feed on contaminated shellfish. A red tide on the East coast of the United States in 1987 resulted in 108 cases of intoxication and three deaths.

The symptoms of intoxication in human beings include a tingling, burning sensation of the face, neck, and upper chest. As the intoxication progresses, the tongue, lips, and area around the mouth become numb and then paralyzed, making swallowing difficult and later impossible. Vomiting, diarrhea, abdominal cramps, disorientation, and short-memory loss are also symptoms of intoxication. Breathing becomes more and more labored and asphyxia sets in. At about this stage in the intoxication, blood pressure plummets, cardiac arrest becomes imminent, consciousness fades, and death occurs.

The extent of intoxication is a function of the amount of toxin ingested. If the quantity is small, the syndrome of sublethal intoxication may last several months before fading. In some cases permanent damage takes the form of paralysis or neurological damage.

QUESTIONS TO ANSWER

1. Why do mushrooms produce toxins?
2. Why do viruses not produce toxins?
4. Why do most algae not produce toxins?
5. Why is botulinum toxin so powerful?
6. Why is staphylococcus toxin heat stable?
7. What is Antabuse ?
8. What is coprine?
9. What is acetaldehyde dehydrogenase?
10. What is cyclopropanone ?
11. What is monomethylhydrazine ?
12. What is methemoglobin ?
13. What are the three toxicities of *Gyromitra esculenta*?
14. What are amanitoxins?
15. What are phallotoxins?
16. Who discovered bacterial toxins ?
17. Who discovered antitoxins ?
18. What is the difference between endotoxins and exotoxins ?
19. What is the Red Tide and how does it affect human beings?
20. How does "antabuse" work?

FURTHER READING

Banwart, G. J. 1979. BASIC FOOD MICROBIOLOGY. AVI Publishing Company, Inc., Westport, Connecticut.

Beuchat, L. R. 1987. FOOD AND BEVERAGE MYCOLOGY, 2nd ed. Van Nostrand, New York.

Jay, J. M. 1979. MODERN FOOD MICROBIOLOGY, 2nd. ed. D. Van Nostrand Com-pany, New York.

Marteka, V. 1980. MUSHROOMS: WILD AND EDIBLE. W. W. Norton & Company, New York.

Riemann, H. and F. L. Bryan. 1979. FOOD BORNE INFECTIONS AND INTOXICATIONS, 2nd. ed. Academic Press, New York.

Wieland, T. 1986. PEPTIDES OF POISONOUS AMANITA MUSHROOMS. Springer-Verlag, New York.

Microbial Fermentations

11

I have drunk from wells I did not dig,
I have been warmed by fires I did not build.

Anonymous (Arabic)

Human beings learned very early to take advantage of natural processes, including microbial fermentations, that occur spontaneously. These processes include the production of cheese and the changing of grape juice first to wine and later to vinegar. The merits of fermented foods undoubtedly became apparent to early human beings quite by chance. The conversion of grape juice into wine, cabbage into sauerkraut, and inedible berries into olives developed into manageable technologies with the passage of time.

The most important aspect of this new technology was that fermentation and spoilage seemed to be two aspects of the same process. At least 10,000 years passed before the nature of these two natural processes was understood, but early human beings were able to benefit from them for many thousands of years even while ignorant of the forces at play.

THE VALUE OF FERMENTATIONS

The value of microbial fermentations to early human beings was not too different from the values that modern human beings derive from the same processes. These include:

1. conversion of food items from rapidly perishable to slowly perishable, as in the conversion of milk to cheese;
2. conversion of unpalatable foods to desirable ones, as is the case for the conversion of olives from the natural state to the fermented form;
3. conversion of bulky foods into an economical volume, as in the conversion of a ton of wheat or barley into a few barrels of distilled whiskey.

In the hands of skilled practitioners, the art of fermentation can be used for the creation of new foods. There is no natural beer, sauerkraut, cheese, or soy sauce. These exist only because human beings manufacture them by manipulating the natural processes of fermentation. Extension of these skills into modern times has led to creation of the solvent industry, the discovery and manufacture of antibiotics, the synthesis of hormones, and too many others to mention here.

Many hundreds, perhaps thousands, of fermentation procedures were worked out and perfected long before human beings understood the general principle of fermentation or the fact that all fermentations are brought about by the metabolism of microorganisms. The origins of leavened bread and beer are lost in antiquity, as are the origins of hundreds of other fermentation products. Complex and sophisticated formulas for the production of wine have been accurately dated to 4,000 B.C. in both the European and the Oriental civilizations. Above all this, all discoveries of fermentation processes were made by trial and error.

Obviously, only those procedures that yielded important products were preserved, while less successful ones were left by the wayside. The way is littered with the bodies of experimenters whose fermented food or drink resulted in toxic concoctions rather than valuable products. Although the road was treacherous and difficult, fermentation recipes laboriously and carefully worked out in caves in Northern Spain some 5 to 10,000 years ago form the basis for present-day, automatic, quality-controlled, and cost-effective processes now running on 24-hour-per-day schedules in all modern communities.

THE FERMENTATION INDUSTRY

Microbial fermentations are the starting point for a vast industrial complex that, in one way or another, touches almost every individual on Earth in the course of every-day life. Experts say that to live a single day in the modern world without recourse to one or more of the products of the fermentation industry would be difficult, if not impossible. And, the experts continue, if this could be done, that day would be uncomfortable, dull, and perhaps even dangerous.

Even in today's most simple and primitive societies, some of the more basic fermentations are always found at the center of communal life. For example, there are probably no communities on Earth where leavened bread and alcoholic fermentation are not developed to a high degree of sophistication. In the study of human beings, one point of view suggests that the production of alcohol by microbial fermentation may well be the first step toward the state of development generally regarded as civilized.

Only the highlights of some fermentations that are important in the production of foods and beverages are considered here. Many hundreds of other fermentation products play secondary but still important roles in the food and beverage industry. The preparation of additives, preservatives, conditioners, flavoring agents, dyeing agents, and packaging materials are all indispensable to food production and preparation. The examples described in brief detail in the ensuing material point out salient features of the methods and processes of fermentation. These are given to illustrate pertinent points of microbiology and are not intended as recipes or formulas to be used in preparing food products.

THE BEER FAMILY

Beer probably evolved from the practice of soaking cereal grains in water to make them sprout before eating them, or grinding them into meal for making bread. The first form of beer was the liquid resulting from soaking the grain for prolonged periods. Written directions for brewing large quantities of beer from barley or from millet date back to 7,500 B.C. in Sumerian history. Evidence indicates that as much as 40% of all cereal grains was devoted to producing beer, and an even larger share than this was devoted to the same purpose by the Mesopotamians of the same era.

Instructions with striking similarities for brewing beer have been found in the writings of the Babylonians, Egyptians, Chinese, Arabs, Greeks, and Indians, among many others. The similarities are astonishing because ample evidence suggests that the fermentation processes described in these writings were developed independently. Although the techniques described adhere to sound scientific principles, many include elements of magic, mysticism, and religion. The Egyptians considered beer a gift from the Goddess Osiris, and Imperial Romans denied beer to women on the basis that it was a beverage for men only.

Today, beer is made by methods based on a thorough understanding of the role of microorganisms in the conversion of sugar to alcohol. The following description for making beer includes only the basic elements of the process and is intended solely to describe the microbiology of fermentation. It is not a recipe for home brewing.

Malting

The first step in beer making takes place in the **malt house** and ends in production of the **wort**. Ripe cereal grain such as barley is rid of insects, sorted to remove weeds and other impurities, and washed thoroughly. It then is placed in a **steep vat** and allowed to **steep** at approximately 10°C for 2 to 3 days, during which time the barley grains imbibe water. During this time the grain is turned and mixed well, and then the temperature is raised to about 20°C.

    Germination of the barley takes place at this higher temperature and requires approximately 7 days before it is completed. After germination, the cereal grain sprouts and other solids are removed and the temperature is raised to evaporate much of the water, leaving a thick malt that contains many active enzymes, including the **amylases** and **proteinases** released by the developing embryos in the grain.

Mashing

A second malt, the **starchy malt**, is prepared by steeping grain at 10°C until it appears swollen. Then it is crushed in a roller mill to break the grain kernels and release the starch stored in them. The resulting steep liquid contains starch, proteins, and other constituents of the cereal grain.

    The two malts are mixed, and the temperature is raised to 50°C to allow **proteases** to degrade as much of the protein as possible. Next the temperature is raised to 60°C to activate the amylases. These are enzymes that hydrolyze starches and other complex carbohydrates in the starchy malt, liberating large amounts of simple carbohydrates. **Adjuncts** such as flavoring and coloring agents dissolved in hot water are added at this point, and the mixture, now called the **main malt**, is allowed to stand while amylases convert residual starches to simple sugars of various kinds and proteinases hydrolyze proteins liberating amino acids into the malt. Other enzymes from the grain embryo, break down other complex molecules yielding simpler substances which can then be metabolized by yeast cells. When this step is completed, the temperature is raised to 75°C for several hours to inactivate all enzymes in the preparation.

    This heating is called **kilning** and determines the kind of beer to be produced. Kilning at a lower temperature will result in pale yellow beer, whereas a higher temperature will yield a darker, heavier product. After kilning and cooling, the malt is transferred to a **lauter tub**, and the solids in the malt are allowed to settle. Grain solids, the **traub**, are removed from the malt house and used for animal feed. Prior to the 17th century human beings considered traub a dietary delicacy. After the solids are removed, the malt is called the **mash**.

Figure 11.1

Selection of hops and preparation of wort in 14th century monastery.

The mash is allowed to sit for several days at a temperature lower than 10°C or until **the fines**, a precipitate made of fine grain particles and **denatured proteins**, falls to the bottom of the holding vessel. This is called the **cold break**.

Wort

The mash with adjuncts is now clear and is called the **wort**. Part of the wort is moved to a separate container and heated. Hops or other flavoring agents are added and the mixture is brought to a boil to extract flavoring agents, acids, resins, tannins, and other water soluble substances. The extracted material contributes to the bouquet of the finished product and also serves to inhibit the growth of Gram-positive bacteria and other microorganisms as the beer is made. The **hoped wort** is filtered and returned to the main wort, and the hops fronds are discarded. The **finished wort** now is transferred to the **brew house**.

Hops, the flowering part of the plant *Humulus lupulus*, was first used in European monasteries during the 8th century. By the 14th century many of these monasteries were famous for their production of hops (see Figure 11.1). By the early 1500s hops constituted the predominant flavoring agent, having displaced all other flavorings. The flavor we associate with beer today is the flavor of hops. Modern commercial beer production depends on the use of

Figure 11.2

Modern fermentation vat; inset shows the yeast Saccharomyces ellipsoideus.

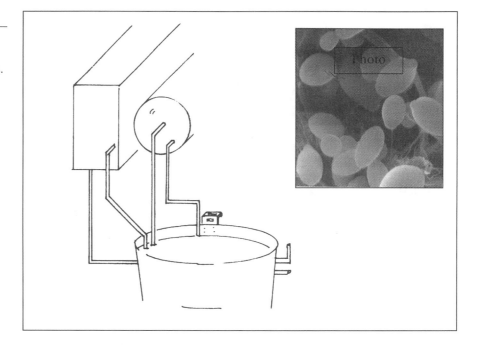

hops extract rather than hops fronds because of convenience of use, shipping, and storage.

Fermentation

In the brew house the wort is placed in a fermenter tank, brought to a temperature of 4°C to 10°C, and the **pitching yeast** is added (see Figure 11.2). The **pitching yeast**, or starter, is inoculum saved from a previous, satisfactory fermentation. Most American beers are made with *Saccharomyces cerevisiae*, or *S. ellipsoideus*, but many breweries use special yeasts selected for special qualities. *Saccharomyces carlsbergensis* used in the production of Carlsberg Beers is an example. Many commercial enterprises keep their production yeasts secret and under high security.

After the starter is added and fermentation begins, it is allowed to continue for several days, during which time the sugars in the wort are converted to alcohol and carbon dioxide. Because this gas is formed inside the yeast cells, they float and remain suspended in the beer, giving it a thick, heavy appearance. Escaping gas causes a boiling and foaming action, which continues as long as carbohydrates are available for fermentation. Proteins, amino acids, tannins, resins, and acids are also changed by yeast metabolism and in the end will provide much of the flavor and bouquet of the finished beer. When fermentation ceases, all yeast cells settle to the bottom of the fermenter

and the wort becomes still. It is allowed to remain still until there is a break in which all the suspended solids settle to the bottom of the fermenter, leaving a clear, crystalline liquid, the young beer.

Maturing and Aging

The young beer at this stage is called **green beer**. It is allowed to rest at about 4°C for several weeks (**lagering**) during which time proteins, resins, tannins, and all solids are allowed to settle. Acids, alcohols, esters, and other products of the fermentation are changed to milder substances. Acids react with some of the alcohol from the fermentation to form esters with pleasant tastes and aromas. Residual acids, esters, alcohols, aldehydes, and ketones contribute much to the final bouquet of the finished beer.

Carbonation

Traditionally, beer was consumed without carbonation, but the vast majority of modern beers are highly carbonated. Beer was carbonated by a "second fermentation" in which a small amount of sugar was added to the finished beer and the evolved carbon dioxide was caught in the final container. In modern breweries some of the carbon dioxide from the main fermentation is saved

Figure 11.3

This state of the art control panel enables the brewing process to proceed efficiently and uniformly, assuring a quality product. To the right is a brew kettle in which water, hops, malted barley and a corn adjunct are cooked. Photo courtesy of The Stroh Brewery Company, Detroit, Michigan

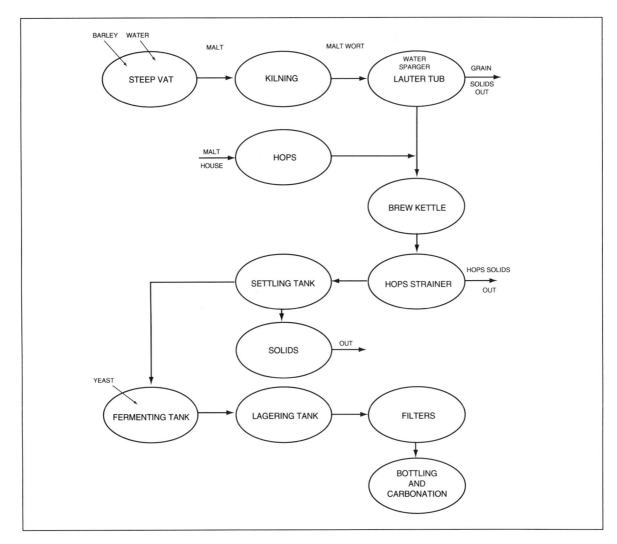

Figure 11.4

Flow diagram of modern brewery.

and added to the finished beer either as gas under pressure or as solid "dry ice." Figure 11.4 is a flow diagram of a modern brewery. Carbonation provides the zest and brilliance of the finished beer as well as sparkle to the final taste. The effervescence, bubbles or "perlen," can last for hours when the beer is undisturbed, or it can be discharged as an explosion if it is shaken vigorously just before opening.

The Finished Beer

The **finished beer** will have approximately 5% (vol/vol) ethyl alcohol, 0.5% (wt/wt) carbon dioxide, 3.5% (wt/vol) dextrins, and 0.3% (wt/vol) proteins.

It will contain vitamins, small quantities of lipids, sugars, and gums, as well as **trace** amounts of other alcohols such as **methanol, propanol, butanol**, and **pentanol**. It also will contain small amounts of formic, acetic, propionic, butyric, and pentanoic acids, among others. Beer also contains trace amounts of esters, aldehydes, ketones, ethers, phenolic acids, resins, and alkaloids. Many of these substances are toxic even in small amounts, but the quantities present in beer are far too small to have noticeable effects on human beings. Beers made using different yeasts, grains, waters, carbohydrates, flavoring agents, and other constituents will have different tastes and bouquets, but their chemical composition is essentially as that described here.

Draft Beer   Draft beer traditionally is sold in barrels that must be kept on ice. When properly cooled, it has a shelf life of 2 or 3 weeks before it becomes **flat** and **sour**. Flat means that all the carbon dioxide is lost, and sour means that the alcohol has been converted to acetic acid. The former occurs spontaneously and the latter as a result of growth of organisms such as *Acetobacter* or *Gluconobacter*, which **oxidize** alcohol to acetic acid. *Acetobacter aceti* can oxidize the acetic acid further to carbon dioxide and water, leaving nothing behind. Today, draft beer is kept in pressurized barrels and protected from acetobacter, extending its shelf life to 2 to 6 months.

Draft beer also may be sterilized by filtration through **filter membranes** capable of retaining all microorganisms. In this condition beer may be canned or bottled and kept in perfect condition for indefinitely long periods since it is sterile.

Pasteurization   Most of the beer sold now is placed in cans or bottles and then is heated to 60°C for a few minutes. This is the temperature needed to kill acetobacter and gluconobacter and to inactivate the enzymes produced during fermentation. Because the acidity of most beers is high (pH 4 to pH 4.5), surviving organisms and their spores will not grow. The shelf life of this product is indefinite under any condition, and it may be stored for many years at room temperature.

Today, beer is produced in every part of the world. It is produced in 500,000-liter fermenters in commercial breweries and in 5-liter jars in private homes (see Figure 11.5). Beers have an infinite variety of tastes, colors, and aromas, to please even the most delicate palate of the beer connoisseur.

Other Beers

Some beers contain approximately 2% alcohol and are called **near beer**. Others contain 8% alcohol and are called **stouts**. Beers with varying characteristics are called **ales, Porters, Pilsners, Bocks, malt liquors, Weiss beers, lights**, and many others. Specialty beers are flavored with pine needles, juniper berries, strawberries, licorice, anise, ginger, clove, and all else the brewer

Figure 11.5

Traditional aging of home-made beer.

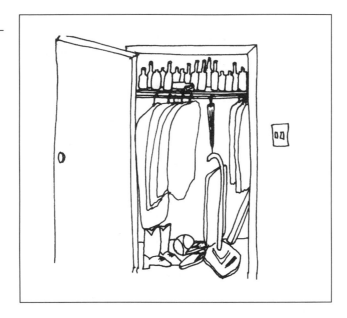

may contemplate as long as one's imagination does not spill over into noxious and poisonous substances.

The art of making beer changes constantly in small ways, and the product almost certainly improves with each change. On the other hand, significant changes in the art of making beer do not occur frequently. One has to consider that the use of hops was introduced during the 14th century, pasteurization in the 19th century, and **cold sterilization** of keg beer in the 20th century, making a total of three significant changes in 600 years.

It may be in keeping with the nature of human beings in the closing years of the 20th century that a new type of beer has been developed. Several major breweries are stocking grocery store shelves with beers that have no alcohol at all. Perhaps this beer is to be drunk while one listens to music that has no tone, meter, or melody and contemplates art that has no figure, color, or form.

WHISKEYS AND OTHER DISTILLATES

During the 11th century it was discovered that if beer were distilled by intense heating, the distillate, or **spirit**, would contain much more alcohol than the beer itself. Although the amount of alcohol in the distillate varies, it is generally in the vicinity of 50% alcohol and 50% water. This **spirit liquor** may be consumed as raw alcohol, commonly called **white lightning** in the Southern parts of the United States, or it can be aged under varying

Figure 11.6

In this 1929 photograph, the serial number is imprinted on a just-filled and sealed barrel of Glenmore bourbon prior to its going to the warehouse for aging. Photo courtesy of Glenmore Distilleries Company, Louisville, Kentucky.

conditions to produce a variety of strong, **dry drinks**. Dry alcoholic beverages are those that contain less water than beer does.

If the distillation of barley beer were performed skillfully, it would contain approximately 50% alcohol; it would be **100 proof**. The taste of such distillates is too harsh for immediate use, but if it is stored in wood casks for several years, a pleasant but strong and intoxicating **whiskey** is produced. English religious orders began producing such spirits in quantity in the 11th century.

Today, most distilled beverages are made by similar methods, using a bewildering variety of fermentable materials, yeasts, fermentation conditions, and aging procedures. Modern whiskeys generally are made by fermenting barley, distilling to an alcohol level of 80 proof (40%), and aging in oak barrels that have been charred on the inside. The aging usually takes 5 years, but some may be aged for 10 or 12 years.

Hard liquor is so much a part of life in the United States, that seven distilleries were permitted to produce limited quantities of whiskey for medicinal purposes during the era of Prohibition. The Glenmore Distilleries Company (Figure 11.6) produced medicinal bourbon throughout the prohibition years and was ready for repeal with a stock of 31,000 barrels when that noble experiment came to an end.

Bourbon is made by fermenting a mixture of corn, rye, and barley and aging the distillate for 5 years or more in charred oak barrels. The products from different fermentations and distillations are **blended** to obtain a particular character. Scotch whiskeys are made by kilning the mash to a temperature of 79°C or higher before fermentation, Canadian whiskeys get

Figure 11.7

14th century distillation apparatus on exhibit in Lerida, Spain.

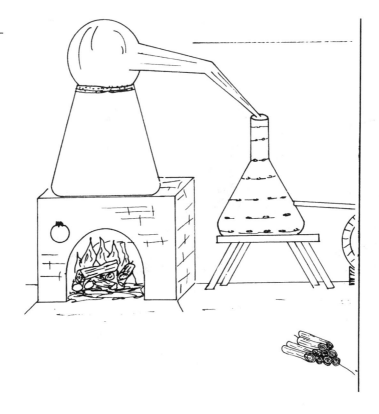

their characteristic bouquet by fermenting rye, and Irish whiskeys are made from rye, barley, and wheat.

Figure 11.7 depicts a 14th-century distillation apparatus in Spain. Other spirit drinks are produced by special processes. Some examples are:

- *Gin*, made by soaking juniper berries in distillate. It is not aged. The final taste depends on the grain fermented, the distilled fraction, and the berries used.
- *Anise*, made by soaking star anise seed in distillate until the desired flavor is obtained.
- *Absinth*, prepared by soaking wormwood, a poisonous material, in distillate. This preparation is no longer made because it affects the brain function of those who consume it. A popular drink in Europe, it was made famous worldwide by Edouard Degas in his painting, "The Absinthe Drinker," portraying a young woman with a glazed look in her eyes and a vacuous stare indicative of the absence of mental attention.
- *Aquavit*, a favorite drink of the Scandinavians, is made by steeping various seeds and berries of evergreen trees in distillate.

- *Kahlua*, is made in Mexico and other coffee producing countries. Coffee beans are steeped in distillate until the desired characteristics are obtained. Ironically, people order decaffeinated coffee to avoid sleeplessness after consuming a liberal amount of Kahlua. The latter probably contains the caffeine equivalent of many cups of regular coffee.

Homeowners often become involved in this aspect of the brewer's art by adding spices, raisins, herbs of various kinds, and fruit to homemade distillate to yield a variety of liquors. These often are used for medicinal purposes, to keep a chill away, or simply to scare the neighbors.

OTHER FERMENTATIONS

Other distilled drinks are made from the distillates of diverse starting materials. Tequila is a distillate of **pulque** (a traditional drink of Mexicans, which has been around at least 2000 years), vodka a distillate of fermented potatoes or other starchy foods, rum the distillate of fermented sugar cane, and sake of fermented rice.

In remote parts of Mexico and Central America, the outer covering of the palm or maguey plant leaf is removed and the heart tissue is placed in a crock with water. A small amount of sugar and yeast inoculum are added, and the crock is kept in a cool place for 15 to 30 days. The plant fibers imbibe the water and sugar, where the sugar is converted into alcohol along with plant carbohydrates. When finished, the plant tissues are chewed to express the alcohol and the fiber is discarded. The fermented fiber of the plant is called **sotol**.

A household fermentation called **tepache** is produced by placing mixed fruits, sugar, water, and the starter yeast in a jar, and allowing it to ferment. The finished product contains 5% to 10% alcohol (depending on the amount of sugar) and is used as a topping for ice cream, as filler for layer cakes, or as a starting material for puddings.

COMMERCIAL ALCOHOL

Next to water, ethyl alcohol is the most valuable solvent in the chemical, medicinal, cosmetic, dye, plastics, paint, and explosives industries. It also is indispensable in medical, biological, and chemical research and in many other aspects of modern technology.

In many of these uses, the alcohol must be very clean or even pure. This is accomplished by repeated distillation until a product containing 95% alcohol and 5% water is obtained. This alcohol has only pure water and pure alcohol, and all traces of the original fermentation materials are absent. This is called **pure** or **industrial alcohol**. The remaining water can be removed by chemical treatment yielding 100%, or absolute alcohol.

THE WINE FAMILY

The special art of fermentation devoted to the production of wine is probably the resolution of an ancient quandary that exists to this day. It may be stated as follows: When grapes mature, they come in such large amounts that they cannot be consumed before they spoil. Approximately 10,000 years ago a practical alternative to this quandary was found, which is still praised roundly today. If grapes that cannot be eaten are treated in a certain manner, they will ferment instead of spoiling and the result of fermentation will be a wine far more valuable than the grapes. The alcohol produced in the fermentation serves as a chemical preservative, making the wine a product with a long shelf life in comparison to the grapes whence it came.

The Cult of the Wine Drinker

In this view then, what began as an expedient method for salvaging grapes from spoilage eventually developed into a sophisticated technology that in modern times evolved into a cult of unknown dimensions. Adherents of this cult may be heard to assert the phrase "One dines with wine but feeds without it." And they are certainly in accord with the poet whose vision of paradise was expressed in the following words.

> A loaf of bread, a jug of wine, and thou,
> Oh! Wilderness were paradise enow.

In the time before it was known that water carried microorganisms, it became well established that the drinking of wine instead of water protected the individual from epidemic intestinal disorders, whereas those who drank water succumbed in large numbers. Unfortunately, it was not recognized at that time that the avoidance of water was what protected the individual and that simply boiling the water before drinking it would have had the same effect. To the contrary, it was assumed that disease was avoided because wine had curative properties and that these properties depended to a large extent on the quality and kind of wine. Wine was prescribed for many diseases, and even today many people place great faith in the medicinal properties of certain wines.

Wine is produced in almost all parts of the world, but the best wines are made in France, Italy, Spain, and Portugal. The major wine-producing countries in the world, in order of volume exported, are: Italy, France, Spain, Argentina, Algeria, Portugal, Russia, the United States, Germany, and Yugoslavia. In the United States, the largest production is in California, followed by New York, New Jersey, Illinois, Michigan, and Washington.

The quality of the wine depends on many variables, but its geographical source stands out in importance. This is in part because the soil where the vineyards grow determines to no small extent the characteristics of the wine

Figure 11.8

Vintner samples a "barrel tasting" in the famous cellars of Anjou in the Loire Valley. Note the traditional barrel syphon, called a "wine thief," and the shallow, flat "tastevin" cup. Photograph courtesy of the French Government Tourist Office

produced. The correlation between the characteristics of some wines and the soil in which they were grown is so great in some cases that, in essence, that particular kind of wine cannot be produced unless the grapes are grown in a given location. Champagne, Cognac, Bordeaux, Jerez, Rhine, Moselle, and Chianti are wines identified with their place of origin.

The production of wine is different from that of beer in that fruit sugars rather than cereal starches are the starting material for fermentation. This eliminates the malting step, as the fruit sugars are available immediately for fermentation. The grapes used in wine production are selected for their high sugar content and also for the acids, tannins, and aromatic substances they contain. Special varieties of desirable grapes have been selected and cultured for many generations according to characteristics deemed desirable and important to the **vintner**. The characteristics of the grapes often are considered the major asset of the vineyard. Many European vineyards have been in production for more than one thousand years, and their products cannot be readily duplicated. (Figure 11.8)

All American grapes were brought from Europe by Spanish and other settlers during the 16th and 17th centuries. Grape stocks not only have been maintained from that time, but many have been improved by modern techniques of selective breeding and plant genetics to yield new and more desirable plants .

Although only two kinds of wines are made from grapes, the varieties of each of these seem to be endless.

Table 11.1   Characteristics of Some Traditional Classes of Wine

Wine	Characteristics
Light wines	Contain 8 to 12% alcohol
Dry wine	All carbohydrates converted to alcohol
Full-bodied	Strong, definite bouquet
Sweet wine	Some sugar left unfermented for taste
Still wines	Little carbonation with no effervescence
Sparkling wine	Effervescent, high carbonation
Aromatic wine	Spices, herbs, etc. added for flavor
Fortified	Alcohol added to increase proof

1. *White wines* are made by removing the grape skins before fermentation.
2. *Red wines* are made by fermenting the grapes together with skins, seeds, and stems.

California vintners blurred this traditional division in the 1980s by the introduction of White Zinfandels and similar **blush wines**. Made by separating the skin from the juice of red grapes as quickly as possible, then making a white-style wine; these blush wines have become extremely popular.

The variety of wines made by these two main processes depends on the variety of grape used, the area where the grapes were grown, the sugar content of the grapes, the growing season, the yeast employed for fermentation, conditions of fermentation, substances added to the wine, length and conditions of storage, type and quality of barrel, and many other elements of the particular viniculture. This explains the nearly infinite variability among wines produced by large and modest-sized wineries. The same **varietal** wine may differ dramatically in quality when produced in different wineries, and from the same winery in different years. A few of the most common classes of wines are listed in Table 11.1.

An Infinite Variety

French vermouth and Dubonnet are aromatic wines that contain sugars and other ingredients, whereas Italian vermouth contains brandy, herbs, spices, and 15% to 20% candied sugar. Champagne is a light, white, sparkling wine named for the region of France where it is produced. According to international agreements, only wine from that region of France may be named champagne regardless of any other characteristic. Madeira, Marsala, Porto, and Tokay are wines fortified by adding alcohol or brandy along with spices, herbs, and other flavoring agents. These contain 18% to 20% alcohol, and color, flavor, and bouquet are determined in large part by the substances added.

Each manufacturer of each kind of wine may have a unique and secret formula for substances to be added and the quantities used. Jerez (sherry) is

Table 11.2 Wines Corresponding to Various Foods

Food	Appropriate Wine
Appetizers	Sherry (Jerez)
Soup	Dry sherry
Fish	Dry, light, white wine
Fowl	Burgundy or Rhine
Meat, game	Dry, full-bodied, still, red wine
Cheese	Port, Madeira, or Tokay
Dessert	Champagne

made by fermenting the juice of the palomino grape with a selected strain of *Saccharomyces cerevisiae*, and after the yeast breaks, the wine is allowed to age for 10 to 12 months. The wine then is inoculated with a second yeast and again allowed to age for many months. In the second fermentation the yeast floats on the surface, as the **flor** or flower, and the second aging takes place under the yeast.

Wines are so important a part of elegance in dining that a correct or proper wine corresponds to each kind of food served during a formal meal.

Grapes used throughout the world for wine making are generally obtained from *Vitis vinifera* or closely related stocks such as *Vitis labrusca* or *V. riparia*. The grapes to be used are watched carefully as the day of harvest approaches. Given lots of grapes to be fermented are examined daily for color, size, aroma, and sugar content (traditional wine grapes contain 10% to 25% of their weight in sugar) and they are selected for qualities deemed desirable for wine making. When at the peak of condition, the grapes are harvested, treated with sulfur dioxide to kill naturally occurring yeasts, and then washed in cold water. Unlike the days of old (Figure 11.9), they now are gently crushed in a wine press without breaking the seeds or stems. Seeds, skins, and stems contain acids, tannins, and other strong and pungent substances that impart harsh, undesirable tastes to the finished wine. As it emerges from the wine press, the juice is filtered and sodium *meta*-bisulfite is added to control bacterial growth. The juice then goes to the fermenter.

In the production of red, gold, or amber wines, filtration is omitted and fermentation is carried out in the presence of skins, seeds, and stems. Most of the substances in grapes, regardless of color, that give the wine its hue are in the skins and stems.

The Must The juice, or **must** now can be left to sit still in a cool place to give the sulfur dioxide or sodium *meta*-bisulfite time to act. Special wine yeasts rather than chance organisms from the field are used in the fermentation. If sodium sulfite at a concentration of 1 gram per 1,000 liters of wine is added,

Figure 11.9

Production of wine starts by crushing the grapes to release the fermentable juice. Photograph courtesy of Sebastiani Vineyards

all the natural yeasts, bacteria, and other microorganisms are killed or inhibited and only the selected, **pitching yeast** will grow. Yeast is carefully selected (Figure 11.10) to complement the character of the grapes used. The must contains approximately 15% to 25% sugar plus acids, tannins, pectins, alcohols, alkaloids, resins, aldehydes, phenolics, vitamins, and other plant substances.

Fermentation   The prepared must is transferred to a fermenter, where it is checked for quality and suitability for fermentation. The temperature is brought to about 10°C and the pitching yeast is added at the rate of 1 liter of starter for each 100 liters of must. Almost all commercial production of wine is fermented by strains of yeast derived from *Saccharomyces ellipsoideus*. Some European vintners have propagated their cultures for more than one hundred years and in that time have selected subcultures that are related strongly to that winery's product.

The pitching yeast grows rapidly, converting the grape sugars into ethyl alcohol, carbon dioxide, and minute amounts of other fermentation products such as acids, alkaloids, tannins, esters, and alcohols. In 10 to 12 days the fermentation will be finished and generation of carbon dioxide will cease. The yeast cells and other solids are allowed to settle, and a clear, bright, slightly yellowish wine will result.

In the production of red wine, the skins, seeds, and stems are removed at this point and the result is a deep, dark purple wine if red grapes were used. If green or yellow grapes were used, the resulting wine will be a pink or deep amber color.

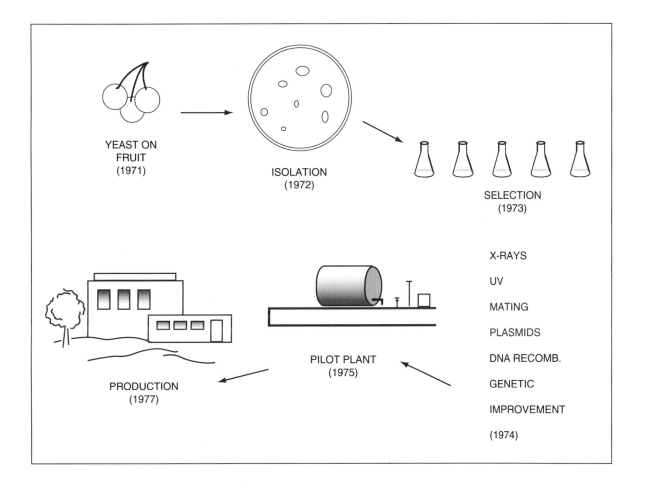

YEAST ON
FRUIT
(1971)

ISOLATION
(1972)

SELECTION
(1973)

X-RAYS

UV

MATING

PLASMIDS

DNA RECOMB.

GENETIC

IMPROVEMENT

(1974)

PILOT PLANT
(1975)

PRODUCTION
(1977)

Aging   The cleared wine is transferred to aging casks or **tuns** and sodium sul-fite is added again if necessary to bring the concentration back to 1 gram per 1,000 liters of wine. It is allowed to age for several months, during which time it changes from a harsh, bitter wine to one with a soft, mellow flavor. Wines made from the best grapes are aged for a year or more, and the finished wine may be of the finest quality.

Aging wine is checked constantly, as its flavor keeps improving with the passage of time until the peak flavor is reached. When flavor, aroma, and bouquet are at their peak, the wine may be treated with finely ground char-coal, **diatomaceous earth**, or special clays to remove acids and other undesir-able substances and also to aid in further clearing. The finished wine is allowed to settle for a final time, then filtered and blended with other wines to produce the quality desired—but this will not be the final taste and bouquet of the wine.

Figure 11.10

Selection of yeasts for com-mercial production of wine.

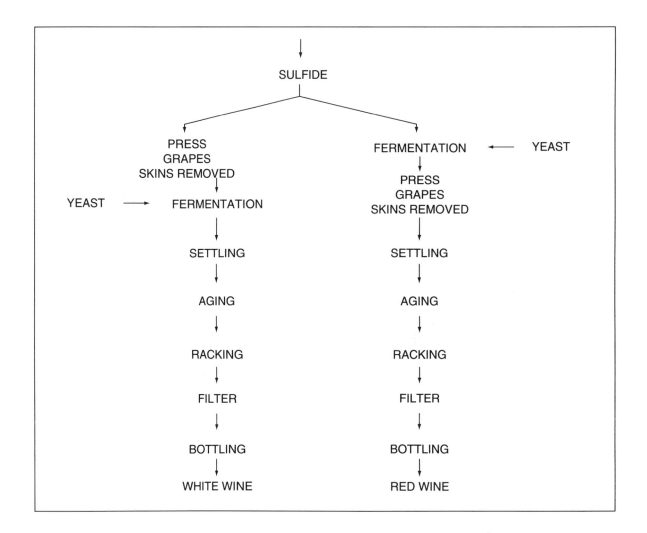

```
                                    │
                                    ▼
                                 SULFIDE
                         ┌──────────┴──────────┐
                         ▼                     ▼
                     PRESS             FERMENTATION  ◄───  YEAST
                     GRAPES                  │
                  SKINS REMOVED              ▼
                         │               PRESS
 YEAST  ──►         FERMENTATION         GRAPES
                         │            SKINS REMOVED
                         ▼                  │
                      SETTLING              ▼
                         │               SETTLING
                         ▼                  │
                       AGING                ▼
                         │                AGING
                         ▼                  │
                     RACKING                ▼
                         │               RACKING
                         ▼                  │
                      FILTER                ▼
                         │                FILTER
                         ▼                  │
                     BOTTLING               ▼
                         │               BOTTLING
                         ▼                  │
                   WHITE WINE               ▼
                                         RED WINE
```

Figure 11.11

Flow diagram of modern winery.

Bottling   The finished product is bottled, or stored in barrels. If it is to be a sparkling wine, it is carbonated either naturally or artificially after bottling. Modern wineries collect and save the carbon dioxide produced during fermentation and then add it to the bottled wine at high pressure to produce the desired amount of carbonation. Bottled wines often are pasteurized or filtered to prevent souring. The entire process is diagrammed in Figure 11.11.

Ripening   Once bottled or placed in barrels, the wine is aged for a period ranging from 6 months to 2 years (see Figure 11.12), depending on the wine and the desired traits. If it is not pasteurized, enzyme reactions continue and the character of the wine changes drastically with the passage of time. The bottled wine improves as it ages, and its characteristics mellow as time passes.

Finally the wine **matures**, at which time flavor, bouquet, and aroma are at their peak and the wine is ready to drink. The wine now will keep its character for many years—6 to 8 for lesser wines and 75 to 100 years for the better red wines. After this time the quality of the wine begins to decrease and it finally spoils.

Figure 11.12

Wine aging in wooden casks.

Other Tastes, Other Wines

Almost any fruit can be fermented to produce alcohol, and this opens the door to a large variety of wines. Many of these wines are produced commercially; others exist only in the homes or garages of home brewers. Wine is made from blackberries, elderberries, apples (cider), peaches, pears, and many other fruits and some vegetables as well.

Special wines were developed in different communities, and many were sufficiently interesting to be appreciated worldwide. The following list includes only a small sampling of the most common specialty wines.

- Cider: made by fermenting apples
- Perry: made by fermenting pears
- Chicha: made in South America by fermenting corn, manioc, plantain, or potatoes
- Mead: made from honey, originally by the Germans
- Sake: made from rice and originated in the Orient
- Pulque: made from maguey leaf and comes from Mexico
- Kifir: results from an alcoholic fermentation of milk and comes from Asia
- Pombe: an African alcoholic product made by fermenting sprouted millet

Amateur wine makers, free from concerns such as the tastes and preferences of the public, profits, or production quotas, are free to be imaginative in creating wines. Wines can be made from any fruit, vegetable, or other substance by increasing the amount of sugar, called **ameliorating**, to produce the desired quantity of alcohol and carbon dioxide. Other wines are made from fruits or vegetables that have small amounts of sugar by extracting the flavor from that material and adding sugar for the alcoholic fermentation. A typical recipe for this kind of wine follows.

> Boil 1 pound of freshly picked and washed dandelion flowers in 1 gallon water with the juice of $\frac{1}{2}$ lemon. Cover and let sit for 10 days in refrigerator.

> Filter and place clear liquid in deep crock, add 0.6 pounds sugar, and inoculate with a small package of dried wine yeast. Ferment 5 days, clear, bottle, and age for 3 months.

Other kinds of wine can be made by boiling cereal grains, extracting the starch, adding enzymes to free the sugar, and inoculating with a wine yeast. Still another kind of wine is made by fermenting granulated sugar with wine yeast, separating the clear wine, and flavoring with herbs, extracts, or anything else. This method can be used to make cinnamon, clove, rose, pine, and many other kinds of wine. Obviously, a complete list of all the wines that may be made by the imaginative wine maker cannot be compiled. A few of the possibilities, however, are:

Apple	Carrot	Dandelion
Apricot	Cherry	Elderberry
Barley	Clove	Elder flower
Beet	Coltsfoot flower	Fig, green
Blackberry	Cowslip flower	Fig, ripe
Butter squash	Currant	Fig, dried
Carnation	Damson	Ginger root

Gooseberry	Orange	Rhubarb
Grapefruit	Parsnip	Rice
Grapes, unripe	Peach	Rose hip
Lemon	Pear	Strawberry
Lemon peel	Plum	Sugar beet
Lime	Potato	Sunflower seed
Loganberry	Prune	Sugar beet
Mangold flowers	Quince	Tea
Marigold	Raisin	Tomato
Mulberry	Raspberry	Turnip

Torpedo Juice

Generations of American sailors have always known that torpedo juice is not the fuel for torpedo engines but, rather, the potent drink obtained by putting a handful of table sugar inside a loaf of uncooked, leavened bread and letting this sit in a jar with sufficient water to cover the bread. Six to 10 days later the water is squeezed from the bread and cleared by filtering through a folded towel.

Although not much for flavor, aroma, or bouquet, the alcohol produced need not take a back seat to even the most esoteric product of the brewer's art. It should go without saying that this kind of reckless brewing is not without its dangers.

Wine Spirits

Brandies   If a good grape wine is distilled to yield a spirit at an alcohol level of 140 proof or higher, some of the volatile substances of the grape wine distill over too. The spirit has a strong, subtle taste and sufficient alcohol to flare when touched by a flame. This distilled spirit may be flavored with other substances until a given character is obtained, making the final product a brandy.

Brandies are made by this process, and different kinds are produced by using different grapes, different water, distilling conditions, aging conditions, *et cetera*. Brandy is produced in many countries, but it may be called **cognac** or **armagnac** only when grapes grown in France in the wine districts known as Cognac and Armagnac, respectively, are used to produce the alcohol.

Brandies of all kinds may be made by fermenting various fruits. Among the favorites are peach, apricot, apple, pear, and cherry. Distilled fruit wines with given characteristics have earned names peculiar to the product. Some of these distilled wines are:

Raspberry	Coconut milk	Kirschwasser
Apple	Barley	Sloe gin
Cherry	Framboise	Toddy
Plum	Hard cider	Arak

Liqueurs, Cordials, and Aperitifs The distillate from grape wines, or any other wine, may be changed in character by adding various flavoring agents. Caraway seeds soaked in a wine distillate will give pernod. Eggs, sugar, nutmeg, and other spices will yield eggnog. Mixing whiskey with honey and several spices results in Drambuie. Grand Marnier is made by blending whisky with orange peel, sugar, a trace of honey, and spices. Chartreuse is made by mixing wine with various herbs, sugar, and some coloring agents.

Creams are made by adding various substances to distillates. Irish cream is a mixture of Irish whiskey and milk, creme de menthe is made by mixing distillate with sugar and mint, and creme de cacao results from mixing distillate with vanilla, sugar, and cacao beans.

VINEGAR FERMENTATION

Vinegar, or acetic acid, is produced by bacteria that oxidize alcohol and give acetic acid as the waste product of their metabolism. These organisms are named *Acetobacter xylinoides*, and they have the capacity to grow in the presence of 10% to 15% alcohol. Vinegar is a 4% to 5% aqueous solution of acetic acid. If the beginning alcohol is in the form of wine, the final product will be wine vinegar whereas cider vinegar will result if the beginning alcohol is cider.

These bacteria and this process are responsible for the souring of wine. In reality, not all spoiled wines become sour. Strains of *A. xylinoides* convert alcohol to acetic acid and the acetic acid to carbon dioxide, thereby spoiling the wine but leaving no acid behind.

Vinegar for the food industry and acetic acid for countless industrial processes are produced by many different methods, but they all depend on the metabolism of *A. xylinoides*, *A. acetigenum*, and *A. aceti*. There are slow methods and fast methods, air methods and submerged methods, and also batch processes and continuous processes for producing vinegar.

The slow method is traditional for home production of vinegar for the table. Apples, pears, or a combination of many fruits is mixed with sugar, mashed into a puree, and suspended in water. It is started with a small package of commercial, dried *Saccharomyces cerevisiae*, placed in a covered, deep crock, and left for 10 to 15 days in a cool place. When the fermentation ceases, the liquid is transferred to a large, shallow container and the incubation continued for another 10 to 15 or until the desired quality of vinegar is obtained. The liquid should be stirred at least once per day by whipping with an egg beater for a few seconds.

When the taste is satisfactory, the liquid is filtered through a folded cloth to obtain a clear, pale liquid with the strong but pleasant flavor of good wine vinegar. This must be kept in the refrigerator as there are bacteria that can convert acetic acid to carbon dioxide and water. If it becomes cloudy after

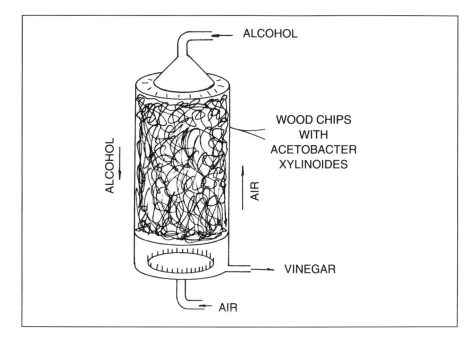

ALCOHOL

WOOD CHIPS
WITH
ACETOBACTER
XYLINOIDES

ALCOHOL

AIR

VINEGAR

AIR

Figure 11.13

*Wood chip acetic acid
generator.*

being in the refrigerator for a day or two, let it sit still until the cloudiness settles to the bottom of the container and filter again. The finished vinegar should be kept in the refrigerator.

In the days before wine was pasteurized, vinegar was not manufactured but simply was obtained from spoiled wine. With the advent of pasteurization, vinegar had to be produced to satisfy an increasing demand for both vinegar and commercial acetic acid. An efficient and reliable vinegar generator similar to the one depicted in Figure 11.13 was developed at the turn of the century.

Another method of producing vinegar depends on a double fermentation. A source of sugar is inoculated with *Saccharomyces cerevisiae*, and fermentation is allowed to proceed anaerobically until the maximum amount of alcohol is produced. When the sugar is exhausted and the fermentation is complete, the culture is made aerobic by blowing in air and inoculating with *Acetobacter aceti* for acid production.

Because France was the major wine-exporting country in the world during the middle of the 19th century, it also was the country that suffered most from the problem of wine souring. In 1852 the French government commissioned Louis Pasteur to find a solution to the problem. He discovered the bacteria now named *Acetobacter* and the role they played in wine souring. He also found the means by which the bacteria could be eliminated. If wine were bottled, sealed, and then heated just enough to kill the acetobacters but not so much as to affect taste, aroma, and color, the wine would not be affected. This made it possible to ship French wines to all parts of the world and give the French an advantage in the wine market that they have not lost to this day.

Pasteurization also prevented the growth of "vinegar eels." Vinegar eels are nematodes, visible to the unaided eye, which, though not dangerous, certainly will discourage anyone from drinking the wine.

Other acids produced by microbial fermentation that are used as food or in the preparation of food include propionic acid, butyric acid, succinic acid, and lactic acid.

OTHER FERMENTATIONS

Breads

Leavened bread is produced by the fermentation of part of the flour starch from which the bread is made. After the dough is made, a small amount of yeast, leavening, is added and the dough is kneaded to mix the yeast with the dough. The leavened dough is allowed to sit in a warm place where the yeast can grow (Figure 11.14). As the yeast cells multiply, they produce large amounts of alcohol and carbon dioxide gas. As the gas is trapped in the dough, it makes the dough rise in much the same way that a balloon filled with gas will become larger and larger.

The yeast also produces many other fermentation products that serve as flavoring agents. These mix with the ingredients of the dough, reacting chem-

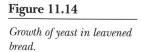

Figure 11.14

Growth of yeast in leavened bread.

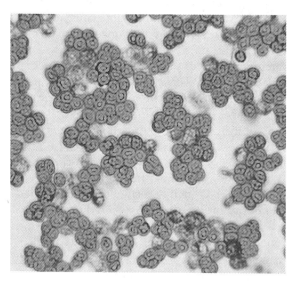

ically with them and remaining in the bread even after the bread is baked. The fermentation products enhance the flavor of the bread and give it the characteristic taste and texture of leavened bread. The alcohol and carbon dioxide also react with the dough, producing many substances that enhance the flavor of the finished product while all excess alcohol and carbon dioxide evaporate during baking. The nutritional quality of the bread also is improved, as yeast cells are an excellent food themselves, containing vitamins, lipids, proteins, and carbohydrates.

History   Historically, the starter for leavening bread came from the fermentation vat, where alcoholic fermentation was in progress or from the stored mother liquor used for producing alcohol. Eventually, bakers developed their own starters, and these came to be known as baker's yeast although it is the same yeast as brewers use. Even today, it is called brewer's yeast by brewers and baker's yeast by bakers, although everyone knows all were originally derived from *S. cerevisiae* or *S. ellipsoideus*.

Because of the nature of bread dough, it is impossible to sterilize it for the purpose of inoculating with a pure culture of a given organism. As a result, most breads are leavened by mixed cultures of bacteria and yeasts, often with a predominant organism that gives the fermentation its character. Many different kinds of breads are produced by making smaller or larger proportions of the same chemicals and using different microorganisms. Two of these are rye and sourdough.

Rye   Rye bread is produced by a sour fermentation brought about by bacteria that produce carbon dioxide and acids, including a major proportion of lactic acid. Traditionally, this fermentation was started by kneading rye flour with buttermilk and allowing the dough to rise overnight. In modern bakeries rye flour is mixed with a large inoculum of *Lactobacillus plantarum* or *Leuconostoc mesenteroides* and incubated at 25°C. A high temperature fermentation, 35°C, is carried out when *Lactobacillus bulgaricus* or milk bulgaricus are used.

Sourdough   San Francisco sourdough bread is produced by fermenting wheat flour with a combined culture of *Torulopsis holmii* and *Lactobacillus sanfrancisco*. The first organism produces gas and alcohol, while the second produces gas and acids.

Sauerkraut

Sauerkraut is produced by the fermentation of the sugars in shredded cabbage. Cabbage is cleaned, shredded into thin strips, and placed in water containing sufficient salt to make it unpalatable (approximately 3%). The cabbage and salt water are placed in a deep vat or crock and covered to protect the contents from dust and other impurities. A weight, often another crock

Figure 11.15

Progress of sauerkraut culture. The bacteria grow in numbers, pH and sugar drop, and acidity increases; then as acidity increases, the bacterial count decreases.

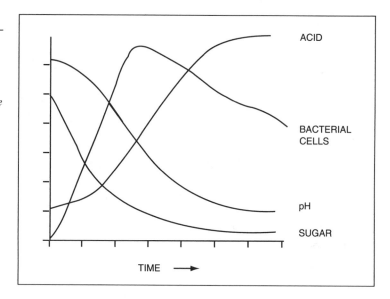

of smaller diameter filled with water, is placed on the cabbage mass lest it rise and spill out of the fermentation vat when gas begins to evolve. The preparation then is allowed to stand undisturbed at room temperature for some 10 to 30 days or until the desired flavor is attained.

Carbohydrates and other fermentable substances in the cabbage tissue make up approximately 5% of the cabbage weight and are attacked by most of the bacteria that can grow in the presence of salt. Soon *Leuconostoc mesenteroides* become predominant, and acid and gas production begins. When the amount of lactic acid reaches about 1% (vol/vol), the population of these bacteria is succeeded by *Lactobacillus plantarum*, an organism that produces several different acids but no gas.

To increase the amount of acid produced, a small amount of sugar may be added at this point. Eventually these organisms also die out and are succeeded by *Streptococcus faecalis*. These produce some acids and also gas but, most important, they remove some of the harshness of the flavor. Harshness in flavor is produced by **mannitol**, dextrans, aldehydes, and alcohols. The finished product will have a strong, pungent aroma and a strong but pleasant taste.

Under normal conditions sauerkraut contains approximately 1% lactic acid, as well as a total of 2% of other acids including acetic, propionic, and small quantities of caprylic.

In commercial production the organisms named, or closely related ones, are inoculated one after the other to produce the required series of fermentations (see Figure 11.15). The quantity of acid is monitored closely and the fermentation stopped when it reaches approximately 1.7%. This can be done by refrigeration or by pasteurization. The shelf life of sauerkraut is longer than 6 months, depending on storage conditions.

Other Fermented Vegetables

Like cabbages, almost any other kind of vegetable can be fermented for the purpose of increasing its shelf life and, often, improving its flavor and other qualities of palatability. In general, fermentation may be accomplished by washing, cutting into thin slices, placing in salt water, and inoculating with *Lactobacillus plantarum* or simply leaving the inoculation to nature. These fermentations generally are brought about by bacteria and yeasts in the air and require several weeks to mature.

In some of these cultures, different bacteria grow in successions, each leaving behind its own fermentation products that contribute to the final flavor of the product.

Preservation   Because the fermentation proceeds until the combined amount of salt and acid is so great that spoilage bacteria can no longer grow, there is little need for refrigeration or other means of preservation. These fermented foods may be kept at room temperature for prolonged periods, although changes in flavor occur because microbial action persists, albeit at a slow rate. Vegetables that can be preserved by this means include cucumbers, celery, onions, garlic, carrots, beets, cauliflower, green beans, okra, squash, peppers, mushrooms, green tomatoes, watermelon rinds, and almost any other edible plant material.

Taste and Quality   The art of fermenting foodstuffs to improve taste and quality is more highly developed in the Oriental countries than anywhere else. Many fermented foods are exported and rapidly become adopted in other countries. The best example of these special fermentations is shuyu, or soy sauce. Soy sauce is the liquid that results from fermentation of soybean meal inoculated with the fungus *Aspergillus oryzae* or other fungi. The fermentation takes place in 4 or 5 days and may be followed by growth of bacteria of the genus *Lactobacillus* and/or *Pediococcus halophilus,* yielding a variety of acids such as acetic and lactic. The fermentation is stopped by adding salt to a final concentration of 30%. The stabilized ferment is allowed to age for 8 to 10 months before use. Aflatoxins have been reported in soy sauce, although the occurrence is rare.

Other Food

In Indochina, cooked soybean meal that is fermented by *Rhizopus oligosporus* or *R. oryzae* is called tempeh; it is cooked before eating and often is fried with meat or vegetables. Soybean meal fermented by *Actinomucor elegans* is called sufu. When sufu is wrapped in rice straw mats and fermented by the bacterium *Bacillus subtilis* it is called natto. Each of these fermentations begins with soybean meal prepared in different ways and inoculated with different organisms or with koji, a mixed culture of bacteria and fungi that is common in the Orient. Each product is incubated under different conditions and for different lengths of time.

Often, other ingredients such as herbs and spices are added to give different, regional, characteristics to the product. In some formulations sugar is added to promote the production of small amounts of alcohol or fermentation acids intended to enhance taste and aroma. In many countries individual families have their own formulations for the fermentation of certain vegetables or fruits.

Olives   Olives are an inedible berry with a strong, foul smell and a noxious, bitter taste. They can be turned into an appetizing, desirable product, however, by fermentation. The major objectionable odor and taste is caused by a group of chemicals collectively called **oleuropins**. If the olives are washed and steeped in alkaline water (sodium hydroxide) for several days, much of the oleuropin will be extracted into the alkali. After the alkali is discarded, the olives are washed several times with fresh, cold water to remove all trace of the steep liquor.

The olives then are immersed in fresh water with approximately 5% to 10% salt and allowed to ferment for several weeks. After this time, salt is added to reconstitute the strength of the brine, and the fermentation is continued for 10 to 12 months. During this time there are several successions of bacterial and fungal populations including bacteria of the genera *Pseudomonas*, *Bacillus*, *Leuconostoc*, and *Lactobacillus*. Each organism contributes, in its time, various fermentation products that blend and, in unison, determine the final flavor of the product. When the fermentation is finished, the olives are washed repeatedly and finally placed in hermetically sealed glass jars or cans in brine of the same strength as that used for fermentation.

Olives prepared in this fashion keep for several years under normal conditions because most bacteria and fungi cannot grow in brine as concentrated as that used for the fermentation of olives. Olives are prepared in many different ways. The one described here is one of the oldest of the traditional methods for preparing Spanish green olives. The ripe, or black, olives also are prepared by this classical method, but other methods using spices and herbs are quite popular as well. Pitting and stuffing with almonds, pimientos, onions, nuts, or fish are common methods of preparing olives. Figures 11.16 and 11.17 depict the olive tree and its product of fermentation.

Meats   Sausages generally referred to as summer sausage (e.g., Lebanon and salami) are traditionally prepared by fermentation with *Pediococcus*

Figure 11.16

Fermented olives, a major food staple in many countries.

Figure 11.17

Olive tree and green olives, Vizcaya, Spain

acidilactici. Sugar is incorporated into the sausage meat and the production of fermentation acids imparts a favorable taste and aroma to the sausage while the acids produced inhibit microbial growth and prevent spoilage. Many other meats are enhanced by fermentation to lesser or greater degrees. While today the fermentation of meats is looked upon as a means of enhancing flavor, aroma, and texture, in the past, i.e., prerefrigeration times, fermentation was an essential method for preservation.

Fish   In some countries small and unappetizing fish are cut into small pieces and fermented in water and salt mixtures. The fish are fermented by a complex and mixed bacterial flora. The fermentation takes several months, and the product is a rich, palatable, high-protein food that can be eaten alone or in combination with vegetables and other food items. Fish prepared by this means will keep for several months to a year. If the product is placed in jars and pasteurized, it will keep for 1 to 3 years without loss of quality.

Iao-chao is a liquid containing a small amount of alcohol. It is produced in China by the fermentation of boiled rice with a starter that contains yeast and bacteria. The Iao-chao is used to ferment small fillets of fish. A similar product is produced in Japan, but there the finished fermented fish is dried in thin strips and stored in the dry condition.

Eggs   Fermented eggs produced in China since time immemorial are prepared by coating washed eggs with a paste of sodium bicarbonate, charred straw, and salt. The eggs then are placed in clay jars, and lime water is added to cover them. They are allowed to stand for 30 days or more and may be eaten any time after that. There are many, unproven claims that such eggs will keep for more than 100 years. In modern "Chinese" restaurants in the United States, they are called "thousand-year eggs" which they obviously are not.

Coffee   Coffee berries picked from the plant are not ready for the percolator. First they must be cured. This may be done by drying in the sun or under artificial heat until the outer husk falls away. The alternative is to place the berries in water and to allow the pulpy husk to ferment. **Pectinolytic bacteria**, *Pseudomonas, Bacillus,* and *Erwinia* destroy the husk while *Leuconostoc, Lactobacillus,* and *Streptococcus faecalis* ferment the remainder of the husk with production of acetic, lactic, and propionic acids. When finished, the berries are washed, cleaned of adhering husk material, and roasted to taste. Residual fermentation products yield a coffee of superior bouquet, aroma, and taste.

The word coffee may have come from the area of the Red Sea now occupied by the country of Yemen. Legend says that shepherds noticed that when their goats fed on the berries of a small bush with "flowers like jasmine and leaves like laurel," they became energetic, tireless, and playsome. When the berries were presented to their Immam, he had them boiled in water and found that the rich, dark tea produced had the same effect on him and on others who drank it. It was obvious immediately that the tea stimulated both mind and body—it was a qahveh.

By the year 1500 Yemen had a well-developed trade in berries and leaves from cultivation of the bush, which was now called kahveh, because both produced a highly desirable, and addicting, aqueous extract. The plant, also called kaffa, also was exported to the Far East and cultivated there but with only modest success. A single plant brought from the East Indies island of Java to Amsterdam in 1710 gave rise to extensive plantings in Europe. These were minimally successful, and kaffa continued to be imported from Yemen, Ethiopia, and other parts of the southern Arabian peninsula. Special places, kaffa houses, for consumption of the boiled beverage, were established all over Europe. These became favorite meeting places for businessmen, politicians, revolutionaries, and intellectuals. In some places, kaffa houses were forbidden but these continued to operate clandestinely.

Amsterdam coffee plants probably were the progenitors of most of the world's modern commercial coffee plantations. The production of *coffea* (scientific name given the plant by Linnaeus) remained an Arab monopoly until late in the 18th century. In the early 1700s it was introduced into India and Ceylon, where commercial production began to rival Arabia's. In 1727 it was introduced into Brazil, tropical South America, and Central America. Within 100 years Brazil became the major producer of coffee and by 1950 produced more than half of the world's crop. Other major coffee-producing countries are Colombia, Angola, the Ivory Coast, Mexico, Uganda, El Salvador, Guatemala, Indonesia, Costa Rica, and Jamaica.

Coffee probably was used first as a medicinal herb because of its stimulating effect on human beings. And the berries were chewed by travelers on foot journeys, because they provide the double benefit of maintaining the flow of saliva in the mouth and staving off weariness. Caffeine is used as an ingredient in many medicinal preparations because of its effect on blood pressure, mental acuity, and stimulatory effect. It also is a self-prescribed "pick-up" by every American worker in the traditional coffee break. At one time coffee was used as a fermented drink not unlike a wine of low alcohol content.

QUESTIONS TO ANSWER

1. What is the difference between fermentation and spoilage?
2. How is cereal grain prepared for fermentation?
3. What role does the cereal grain embryo play in fermentation?
4. Where does the carbon from glucose go in the production of beer?
5. Why does bacterial metabolism convert alcohol to acid?
6. How is dark beer made?
7. What is the difference between beer and wine fermentation?
8. What is the difference between white and red wine?

9. What is fortified wine?
10. What is a liqueur?
11. Why is it possible to get a high alcohol content in a distillate?
12. What is the relationship between percent alcohol content and proof?
13. How can alcohol be produced from milk?
14. How can meat be fermented?
15. Why is leavened bread not intoxicating?
16. Why does leavened bread rise?
17. What is torpedo juice?
18. Can fish be fermented?
19. Can eggs be fermented?
20. What keeps "thousand-year eggs" from spoiling?

FURTHER READING

Beuchat, L. R. 1987. FOOD AND BEVERAGE MYCOLOGY. Van Nostrand Reinhold, New York.

Brevery, H. E. 1965. HOME BREWING WITHOUT FAILURES. Gramercy Publishing Company, New York.

Frazier, W. C. and D. C. Westhoff. 1988. FOOD MICROBIOLOGY, 4th ed. McGraw -Hill Book Company, New York.

Hays, P. R. 1985. FOOD MICROBIOLOGY AND HYGIENE. Elsevier Applied Science Publishers, New York.

Hahn, P. A. 1968. CHEMICALS FROM FERMENTATION. Doubleday & Company Inc. Garden City, New York.

Jay, J. M. 1978. MODERN FOOD MICROBIOLOGY. 2nd ed. D. Van Nostrand Company, New York.

Miller, B. M. and W. Litsky. 1976. INDUSTRIAL MICROBIOLOGY. McGraw-Hill Book Company, New York.

Rose, A. H. 1959. BEER. Scientific American, 200:90–100.

Microbiology of Water

12

Water, water, everywhere,
Nor any drop to drink.

Coleridge

Human beings, as all other living things, require a constant supply of water to remain alive. The human body is made of a great variety of chemical substances that can be divided into two major classes:

1. 90% of the total weight of the body is water
2. 10% includes all the other constituents

LACK OF WATER CONSERVATION

The bodies of human beings and most other animals are not made to conserve water, as water is lost constantly in various forms and for various purposes. These are listed in Table 12.1. Even so, the healthy human body has the ability to conserve and reuse large amounts of water from metabolic reactions, but these are not sufficient to replenish the entire amount lost by the essential processes listed in the table. A daily deficit of approximately 2½ liters of water must be supplied without delay. The chemical and physiological balances of the body can take place only when the **affluent** and **effluent** water of the body are in balance. Replenishing lost body water cannot be postponed because water deficit makes normal physiologic body functions impossible.

Table 12.1   Water Loss in Human Beings

Form	Purpose
Perspiration	For temperature control
Vapor in breathing	For excretion of carbonaceous wastes
Urine	For excretion of nitrogen and mineral waste; control of water balance
Tears and moisture on mucous membrane surfaces	To maintain function of eyes and membranes; to protect against infection
Feces	To maintain semi-solid state for evacuation

WATER NEEDS OF THE BODY

The exact amount of water needed depends on the physical condition of the individual, on environmental conditions, and on the level of physical activity. For example, a person in a good state of health, with an adequate, well-balanced intake of solid food, in an environment at 22°C and 50% relative humidity who performs sedentary, nondemanding work, requires much less water than one who has diarrhea and who performs hard physical labor in an environment where the temperature is 40°C and the relative humidity is 20%.

WATER CONTENT OF THE WORLD

The world contains a large quantity of water—an estimated 1.454×10^9 cubic kilometers. This water is distributed over the surface of the globe, covering approximately two-thirds of it. The water is present in three different forms, concentrated in different areas of the world:

1. Liquid water, covering approximately two-thirds of the surface of the planet
2. Ice, located at the poles and very high mountain ranges
3. Water vapor, found in the atmosphere to an altitude of approximately 100 kilometers above the surface

The three forms of water exist in a state of equilibrium, and they change in proportion to the changing of the seasons and temperatures. Figure 12.1 depicts the cycles of water on Earth.

Water in nature is seldom pure, but the contaminants it bears generally are not harmful to man and other living things. On the other hand, as the population of human beings increases on Earth, the amount of water that is

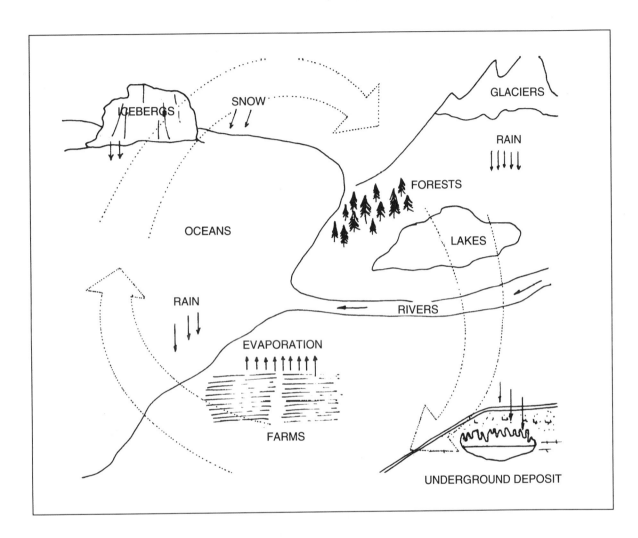

Figure 12.1

The cycles of water on Earth.

contaminated with hazardous and toxic substances is increasing constantly. The spilling of 11.5 millon gallons of oil from the Exxon Valdez in a year with 6,800 other reportable spillings of oil into the ocean waters indicates the extent of the problem.

In addition to this, adding insecticides, herbicides, and toxic industrial wastes into the fresh water systems of the world soon may overwhelm the ability of these waters to regenerate themselves. Much of the water used for human consumption now carries residual levels of carcinogenic and toxic substances. In certain places some of the toxic materials are always present in the water in quantities barely below the limits of acceptability. Because these levels increase yearly, the limits soon will be superseded. Inevitably, when safe levels of toxic substances in the water are surpassed, the water available to

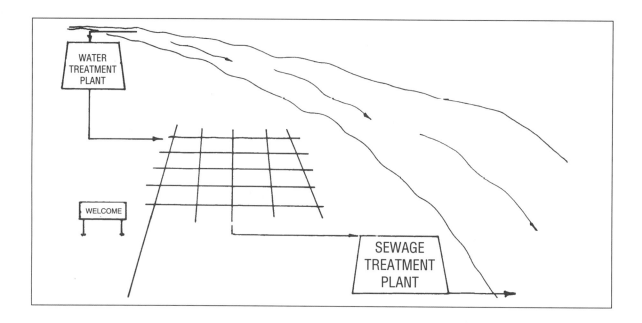

Figure 12.2

Schematic water system in modern city.

human beings for their needs will be both carcinogenic and toxic. The politics of the nation being what they are at this time, the entire issue probably will be resolved by simply changing the laws that prescribe admissible limits of contamination.

STRATEGY FOR WATER

All modern communities are built around an extensive and complex water system. Approximately half of this structure consists of the sewage collection and treatment system designed to remove waste material from the community. The heart of the sewage system is a sewage treatment plant, where waste material is rendered harmless and innoxious. The other half of the water stream on which the community lives is the pure water supply. The objective of this system is to provide safe, pure water for drinking and for preparation of food and to satisfy the many other needs that modern human beings have for clean, pure water. Figure 12.2 is a schematic drawing of a modern municipal water system.

These two streams must be kept scrupulously separated. Although microorganisms are essential for the sewage treatment process, they must be eliminated or at least reduced to small numbers in the drinking water system. Whereas the sewage system contains a representative sample of all the pathogenic organisms that inhabit the community, the potable water supply must be totally free of these agents of disease. The task is extremely difficult, but all modern, industrialized societies have the technology necessary to carry it out with little evident effort. On the other hand, many developing countries find this task impossible to perform successfully.

DRINKING WATER

Providing water that is safe for human consumption is one of the major responsibilities of all modern societies. Protecting water from contamination so it does not become a source of contagion when it is served at the table or used for the preparation of food is essentially the responsibility of the community.

THE PHYSICAL STATES OF WATER

Water is an odorless, colorless, tasteless liquid with a freezing point of 0°C and a boiling point of 100°C. It has a relative density of 1.000000 g/ml at 4°C and 0.998230 g/ml at 20°C. Water is made from the electrochemical reaction of hydrogen and oxygen:

$$2H_2 + O_2 \rightarrow 2H_2O$$

Once formed, it can be separated into the two component gases by electrolysis. Its formation by the interaction of hydrogen and oxygen and its decomposition into these two gases both adhere to strict **stoichiometry**.

The three states of water—ice, water, and steam (solid, liquid, and gas)—are discussed separately next.

Solid

When it is free of contaminating substances, water exists as a solid at all temperatures below 0°C. As **solutes** accumulate in water, the freezing point is **depressed** to an extent determined by the nature of the solute and the quantity present. When pure water is frozen into ice, the ice has some characteristics different from those of ice produced from water containing contaminants.

Because food is spoiled by microbial metabolism, ice is the perfect agent for preserving food. The most effective use of ice for preserving food is in a mixture with water at 0°C. Foods placed in such a mixture will attain a temperature of 0°C, and little or no microbial activity occurs at that temperature. In effect, the refrigerator is a substitute for this ice-water mixture, but it is much better because it operates continuously and cools the food directly without the intermediation of ice.

Because ice is present at the dining table in all climates, it is considered a part of the comestibles of human beings in the United States and a few other countries. As such, the microbiology of ice is of major importance to this treatment of the subject. Microorganisms cannot grow in water when it is in the solid state because they require the liquid medium for hydration of cell walls and cell membranes. Also, all the enzymes and other cell constituents function only when they are in aqueous solution. Nutrients must be in solution before they can be used by bacteria, algae, and fungi. Metabolic wastes, including gases, can be excreted from the cell only if they are in solution in water. As the temperature approaches the freezing point of water, microorganisms that function normally when water is in the liquid state begin to slow down. When the water becomes solid ice, all function ceases and the organism enters a state of suspended animation. Because of cell size, physical constitution of cytoplasm, pliable lipid cell membranes, and other unknown factors, bacterial and fungal cells (but not plants or animals) survive freezing, and consequent thawing, quite well. As a result, ice is an ideal vehicle for the storage and transmission of pathogenic microorganisms.

The addition of energy in the form of heat causes the ice to change to the liquid phase as the temperature becomes higher than 0°C. Ice also can go directly to the gaseous state: a process called **sublimation**.

Liquid

In the temperature interval between 0°C and 100°C, pure water exists in the liquid state. This is the brief interval of temperature in which life originated some 3½ billion years ago and in which it exists today.

All chemical substances are soluble in water to a larger or lesser extent, and this property is what, to a great degree, makes life possible. The dependence of life on the availability of water is so great that theoreticians assume that life cannot exist in its absence anywhere in the universe. All tenable, modern theories describing the origin of life on Earth specify water as the medium that supported the first chemical reactions, which eventually led to living things. Water is the major component of all living things and life fades rapidly in its absence. Not only does life depend on water, but it must be water in the liquid condition.

If energy in the form of heat is applied to the liquid, it will change into a gas. If the temperature reaches 100°C, all the water becomes vapor, or steam, and none will be in the liquid form. The changes can be reversed by removing energy from the water by refrigeration.

Gas

Liquid water evaporates and becomes water vapor which forms part of the Earth's atmosphere. Liquid water and water vapor always exist in an equilibrium that depends on several variable parameters. One of the variables that exerts a major effect on the equilibrium between water and water vapor is temperature. Another parameter that affects this equilibrium is pressure. When the temperature is low and the pressure is high, the air contains little vapor and most of the water is in the liquid form. As the temperature increases and the pressure decreases, water evaporates and the amount of vapor increases. Chemicals added to pure water affect this equilibrium in well-known and predictable ways.

Water vapor in the air is essential for the well-being of all living things because the absence of water vapor results in evaporation of water from the cells of the organism. Loss of water from the cells results in **dehydration** and subsequent impairment of physiologic activity.

SOURCES OF WATER

Water is found in many different forms and in many different places on the planet Earth. Much of the water on Earth contains salt and therefore is useless for many living things. Living things can be divided into two categories on the basis of their ability to live in salt water:

1. **Marine** forms can thrive in water containing 3 grams of salt per 100 milliliters of water.
2. Fresh water forms will be injured or killed by salt water.

Few organisms have the physiological ability to live in both salt and fresh water, and those that do have access to two different worlds. Most of the evidence available on this subject indicates that life originated in the oceans and some organisms moved from there to dry land and then, following many evolutionary stages, adapted to life in unsalted water.

Groundwater

Water that is trapped in the ground within layers of soil that prevent its movement is called groundwater. Some groundwater is very old, having collected in its present site thousands or even hundreds of thousands of years ago. Other groundwater is said to be new because it accumulated in its present site from rainwater or other surface water that recently percolated to the collection site. Groundwater travels long distances through many layers of the soil before it reaches an impermeable layer, where it collects. As it passes through layer after layer of soil, it is filtered and all the bacteria and other microorganisms are removed, rendering the water sterile.

Particulate contaminants also are filtered out of the water, and chemical contaminants are removed by chemical reaction with soil substances as the water passes through. When trapped in an underwater deposit, water is still for prolonged periods, allowing all suspended materials to settle to the bottom of the impoundment, making a perfectly clear water.

Figure 12.3

*Schematic view of deep
aquifer irrigation system.*

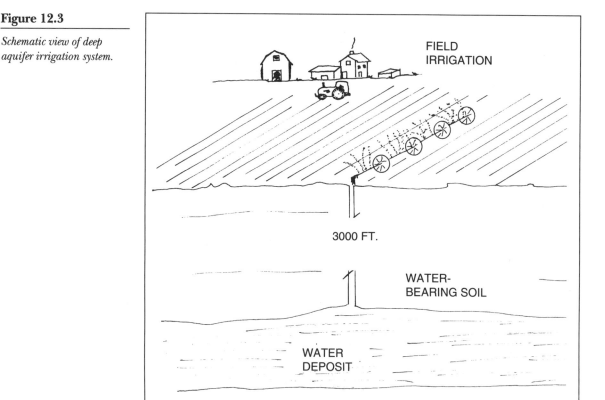

If deep deposits of water are large, they are called **aquifers** (Figure 12.3).
The more valuable aquifers are the older, and deeper ones, because they contain no modern pollutants and few, if any, microorganisms. In this regard,
"older" refers to several hundred years of age. Water is obtained from these
aquifers by deep wells, using existing water pressure or mechanical pumps.
The former are called **artesian wells**, and the latter simply **deep wells**.

Many of the aquifers in highly industrialized communities such as the
United States are becoming contaminated because of the extremely large
loads of contaminants added to the groundwater supply each day by unscrupulous industrialists working with complacent or ignorant public officials. In
addition, much of the good water from aquifers is used for agricultural irrigation, leaving nothing but contaminated surface waters for human use. As
they emerge from the aquifer, irrigation waters are free of contaminants but
they are immediately contaminated with fertilizers, herbicides, and insecti-

cides, rendering them unsafe for use by human beings. Although the soil bacteria remove large amounts of these toxic substances, a residuum remains that is not biodegradable and that accumulates in the water. Some evidence indicates that almost all the water human beings use is now contaminated with toxic residues. In spite of all the concern and effort expended to reduce environmental pollution, all natural water supplies seem to become more and more contaminated as time passes.

Many kinds of surface waters can be found, each with its own characteristics and unique properties. Each of these presents a different universe with regard to the microorganisms that exist there and those that can grow in it. The discussion necessary to describe this complex and extensive topic lies outside the scope of this text. Further discussion is limited to information pertinent to the food and beverage industry. Instead of considering each type of water independently, a generalized description of surface water is considered to show the principles involved.

Surface waters are in contact with air, soil, and all the products and waste products of modern, industrial societies. The designation *surface water* includes all water in surface collections, rivers, shallow wells, and all others in contact with the atmosphere. Contact with soil and air result in contamination of water with a great variety of chemical substances and with living organisms. Because the majority of these organisms grow in the water, vast populations of bacteria and other organisms develop there. A partial list of sources of surface waters used by human beings is given below.

Springs	Creeks	Canals
Bogs	Lagoons	Rivers
Holes	Tanks	Swamps
Seeps	Pools	Waddies
Ponds	Sinks	Paddies
Lakes	Reservoirs	

Oceans

The oceans represent a vast storage of water that is the source for all other water on Earth. Vast amounts of water vapor rise from the oceans to form clouds that move inland and drop the water on land. Evaporation also results in **desalination**. All fresh water comes from cloud vapor and generally contains only small amounts of contamination.

Part of the ocean water is frozen into the ice caps at either pole. Much of this water is frozen permanently and therefore is not available for plant or animal use. Because much of the polar ice comes from vapor frozen from the air, it contains microorganisms that became permanently entrapped there. Some microbiologists see the polar ice caps as a record of the microflora that existed in the air during the last few million years.

Atmosphere

The atmosphere is a reserve of water in the form of vapor. This vapor is in equilibrium with liquid water in the oceans and on the surface of the land. Water vapor is the means by which salt water is converted into fresh water and, because clouds formed from fresh water travel to the sea, also the means by which fresh water is converted into salt water. The atmosphere contains water vapor, liquid in the form of rain, and ice in the form of snow, hail, and sleet.

PURE WATER

Natural Water

Neither microorganisms nor other forms of life are able to grow in pure water. In a general sense, water serves as the suspending and dissolving medium in which living things grow, but the dissolved contaminants in the water are what make life possible. Nutrients and all necessary substances for maintaining cell life must be dissolved in the water to be available to cells. Since water is a general solvent for the majority of chemical substances, all natural waters are more a solution of many chemicals in the environment than a pure substance.

Clean, sparkling water actually contains all the nutrients necessary for the growth of many microorganisms. The presence of even minute quantities of **ammonium sulfate**, **magnesium nitrate**, **ferric chloride**, **sodium phosphate**, and carbon dioxide plus a source of energy such as sunlight are all that is needed to support large communities of microorganisms including many different kinds of algae, photosynthetic bacteria (see Figure 12.4), and saprophytic organisms. Because water contains no carbohydrates, **cyanobacteria** are the first to grow.

Figure 12.4

Growth of photosynthetic bacteria in water.

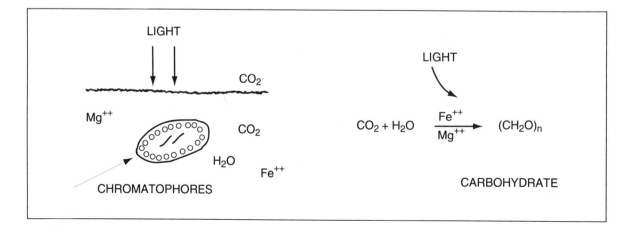

Once these are established, heterotrophic bacteria begin to grow, using the waste products of the first organisms as nutrients. Other generations now grow, using the waste products from the second generation, and so on and on as long as light and minerals are available.

Bacterial populations, though varied, do not become large because the amount of contaminants generally is small. Clear waters protected from air and light support small communities of bacteria. In general, the number of organisms present is too small to cause turbidity in the water or to impart tastes and odors. For these reasons, such waters are called pure and, in most cases, are deemed suitable for drinking.

When the populations of bacteria and algae reach large numbers, protozoa and small animals feed on them and complete a large and complex food chain. The mixed community may become stabilized, and the population of each constituent organism may bloom and ebb in a regular period as long as the source of energy, sunlight, lasts. The same principle applies to the growth of plants and animals in a closed system. In both cases some organisms give off carbon dioxide and others use the carbon dioxide to produce oxygen. In both cases the reactions are cyclic—the chemical elements are used again and again—but the energy that drives these reactions must be supplied in a constant stream.

Contaminated Water

In a different situation, with no light but organic chemicals such as carbohydrates, lipids, proteins, or other complex chemicals, growth occurs at the expense of the energy in the chemicals. In this case bacteria, fungi, protozoa, and small animals grow, utilizing the energy present in the molecules that make up the contaminating materials. As the energy is extracted from these substances, they become oxidized and their energy content is lowered by the amount that was removed by bacterial metabolism. Other microorganisms use the remaining material, and the process is repeated again and again until all the energy is extracted from the original chemicals. The original substance is **mineralized**, and many generations of microorganisms result.

Waters contaminated with human or animal wastes contain all the microorganisms present in the intestinal tracts of those that produced the waste. Municipal wastes represent an entire array of all the microorganisms present in the members of the community. Under normal conditions pathogenic microorganisms are found in contaminated waters in proportion to the incidence of each disease present in the population.

DRINKING WATER SUPPLIES

Each community obtains water for its residents from the nearest and most convenient source available. If good quality groundwater is available, this is apt to be the first choice, as it requires the least amount of treatment. Surface waters may be used after adequate treatment and purification, processes which are time consuming and expensive. The three major phases of water preparation are:

1. Collection
2. Treatment
3. Distribution

The methods different communities use to accomplish water preparation are basically the same, with allowances for differences in water source, community health standards, volume of water required, and other unique needs of the community. The description of one system suffices to illustrate the nature of the treatment required to convert contaminated water into potable water. Water treatment is a dynamic process in that dirty, nonpotable water enters the system at one end and clean potable water exits at the other end with a residence time of 24 hours or less.

An example of water supply logistics is given below.

The City of New York has a population of 7½ million people, and the municipal water system must provide 750 liters of water each day for each member of this population. This is a daily volume of 5.625×10^9 liters of purified water. To bring this quantity of water to New York City each day would require some 17,500 large tanker trucks like those that carry gasoline on the highway. Consideration of the time required for loading, transporting, and unloading this amount of water, plus the cost of unloading 17,500 tank trucks, readily shows the impossibility of such a task.

In actuality, the City of New York acquires its water from reservoirs as far as 150 miles away through a system of canals, aqueducts, and large-diameter tunnels, some of which were placed in service as long ago as 1820 and others as recently as 1965. The entire system is coordinated perfectly to accomplish the necessary task and do so with great efficiency.

The water must be handled twice within a 24-hour time period. First it must be purified and made safe for drinking and then distributed through a system of water mains and lateral pipes to every building, house, and apartment in the city. Second, after use it must be collected, treated at the sewage treatment plants, again purified, and then removed from the city to make room for the continuing torrent.

Water Treatment

Water treatment begins when the water is brought to a reservoir, where it normally is held as long as possible. If sufficient reservoirs exist to hold more water than is needed in one day, the water may be held for several days, weeks, or even months. During this time many of the pollutants are removed by bacterial metabolism, others precipitate and become part of the sediment in the bottom of the reservoir, and still others are changed from a harmful form to a harmless or even beneficial form.

The reactions that bring about these changes may be complex and involve both chemical and biological processes. Surface waters contain large populations of microorganisms including many pathogenic bacteria and viruses, but after detention for even a brief time, the populations of these organisms are depleted severely. Normally the pathogenic organisms disappear quite rapidly, as most cannot survive the harsh conditions of nature.

Water from the storage reservoir is conveyed to the water treatment plant, where a coagulant such as aluminum sulfate, ferric sulfate, or ferric chloride is added. When these chemicals react with water, they form a thin, continuous gel, or hydrate. Because the gel is slightly heavier than water, it "falls" through the water very slowly, and as it does, it traps debris and other particulate materials suspended in the water. In 4 to 8 hours, approximately 90% of the particulate material in the water is trapped in the gel and **flocculates** (Figure 12.5). The floc contains soil, organic debris, worm eggs, protozoan cysts, bacteria, fungi, and viruses, plus the majority of particulate debris in the water.

Chemicals that were dissolved in the water react with the floccing agent to form particulate products that also will be enmeshed in the **coagulum** and removed from the water. As the coagulum sinks slowly, it leaves a clear, pure water free of particles. The clear water then may be filtered through a slow sand filter to remove residual suspended material, and then passed through activated charcoal filters to remove remaining traces of chemicals that may impart tastes or odors to the water.

The final step in preparing water for municipal uses consists of adding chlorine to a level sufficiently high to prevent the growth of bacteria and other microorganisms. Chlorine reacts with many chemicals that escape the water treatment process in small amounts. The products of these reactions often are toxic or carcinogenic. As these carcinogens have been identified, the use of chlorine is being reconsidered in some communities. Many European countries have reduced the use of chlorine and have turned to **ozone** because ozonization is more effective than chlorination in killing microorganisms, and it produces no toxic materials in the water. Ozonization also is used in public hot-tubs as an alternative to chlorination.

Water emerging from the treatment plant enters the distribution system and is conveyed to the community. Because vapors and gases dissolve in water, bacteria and fungi can grow in the treated water using these substances as growth substrates. **Residual chlorine** in treated water prevents the growth of microorganisms and also kills those that escaped the treatment. This continuing antibacterial activity is important because it assures that pathogenic bacteria that survive treatment do not persist in the finished water.

The procedure just described is employed in all modern, industrialized countries for the purification of drinking water supplies. It has been the

Figure 12.5

Flocculation of particulate material in purification of drinking water.

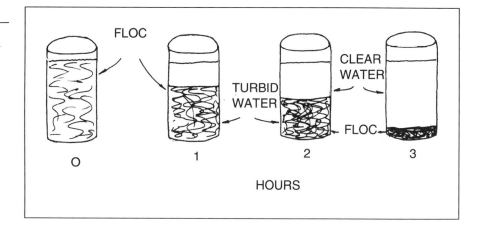

primary measure for controlling the epidemics of intestinal disease that plagued human beings throughout history. However chlorine does not totally purify water. Some viruses survive such as *poliovirus*, and recently it has been shown that cryptosporidium (a microscopic protozoan) can survive chlorination and cause epidemic disease. Children and immuno-compromised individuals are particularly vulnerable.

Constant and continual care of the water-processing system is essential for success. This assurance comes only from adequate and continual testing for the presence of pathogenic organisms. A public water system such as the one described here represents both the first line of defense against epidemics of enteric bacteria and, at the same time, the greatest threat. When the system breaks down, such as in times of war or natural calamity, it becomes the perfect vehicle for distributing pathogenic organisms to every member of the community.

Indicator Organisms

To test water for the presence of microorganisms that cause enteric diseases is difficult, but because the most likely source of these is fecal matter, water generally is tested for fecal contamination. The bacterium *Escherichia coli* is present in the fecal matter of human beings and many other animals with such great regularity that it can be reasonably stated: Water that contains *Escherichia coli* almost certainly was contaminated with the fecal matter of human beings or of animals capable of harboring microorganisms pathogenic for human beings.

Based on millions and millions of tests performed during a period of more than 70 years, the reverse also is a reasonable statement: If water is free of *Escherichia coli*, it is also free of enteric pathogens.

Since water that does not contain *E. coli* or closely related organisms is not likely to contain enteric pathogens, it is taken as an article of faith that water free of *E. coli* is safe to drink. As a consequence of the foregoing, *E. coli* is used as the organism that indicates the presence of fecal pollution. Other organisms that resemble *E. coli* and whose natural habitat also is the intestinal tract of human beings are called **coliforms**. Together with *E. coli*, coliforms are used as indicators of fecal contamination of water and food in the United States.

In many European countries *Streptococcus faecalis* instead of *E. coli* is the indicator organism of fecal contamination. In Europe *S. faecalis* is thought to be a better indicator of fecal contamination because this organism is not commonly found in the intestinal tract of animals other than human beings. The tests needed to show the presence of the *streptococcus* in water, food, or other substance are not more difficult than those used to find *E. coli*. In spite of many reasonable arguments, microbiologists in the United States remain convinced that the indicator organism of choice is *E. coli*. For example, gastroenteritis went from one of the most common causes of death in the United States to one of the less common in 80 years (Table 12.2). Because the incidence of enteric disease in the United States is as low as that in any country that uses *S. faecalis,* the choice of which indicator organism to use must, therefore, remain a moot question.

Most Probable Number (MPN)

The oldest and one of the most reliable tests for detecting low levels of fecal contamination in water is based on a multiple tube test for coliforms. It is a

Table 12.2   Leading Causes of Death, 1900 and 1980

1900 **Most Common** **Causes of Death**	1980 **Most Common** **Causes of Death**
1. Respiratory	1. Heart Disease
2. Tuberculosis	2. Cancer
3. Gastroenteritis	3. Vascular
4. Heart Disease	4. Accidents
5. Vascular	5. Respiratory
6. Nephritis	6. Diabetes
7. Accidents	7. Cirrhosis, liver
8. Cancer	8. Suicide
9. New born	9. Emphysema
10. Diphtheria	10. Homicide

quantitative test called the **most probable number test** because it employs a statistical calculation to produce an estimate of the population of coliforms in the water. It is a number based on the fact that many coliforms ferment the sugar **lactose**-producing acid and gas in the growth medium. This medium also contains **selective inhibitors** such as beef bile and the dye brilliant green (other dyes also are used), which prevent the growth of bacteria other than coliforms.

The test is based on the assumption that one organism will produce a large population of its kind when it is placed in a suitable growth medium. It is assumed that each bacterial cell in the sample to be tested exists apart and separated (not clumped together) from every other cell, and that such cells are distributed equally throughout the sample. Therefore, the presence of cells in an aliquot of sample will give a statistical value which relates to the total number of cells in the sample.

The MPN test, as it is commonly used in many water analysis laboratories, has three parts:

Presumptive test: If the Brilliant Green Lactose Bile Broth (BGLBB) tubes inoculated with water, milk, food, or other sample show bacterial growth together with acid and gas, it is assumed that the sample contains coliforms.

Positive test: If subcultures from the BGLBB culture tubes yield colonies on **Eosin Methylene Blue** agar that are typical of bacteria that ferment lactose, it is said that the sample is positive for coliforms.

Confirmed test: If Gram-negative bacteria isolated from the above cultures give positive tests for **indole** and **methyl red** and negative tests for the **Voges-Proskauer** and **citrate** tests, it is said that the presence of *E. coli* has been confirmed. Newer tests depend on **serological typing** and even on **DNA probes** for final identity.

Since three sets of five tubes each are inoculated with three proportionate quantities of sample, the number of tubes that give a **presumptive** test form a statistic which indicates the number of **coliforms** in the sample.

Results from the imaginary MPN test described in Figure 12.6, (5, 2, 0) show that statistically the most probable number of coliform bacteria is 49 organisms per 100 ml of the water tested.

Filter Membrane Test

Plastic membranes of standard pore size were developed during World War II. They are highly efficient in retaining bacterial cells and can be used to collect all the microorganisms present in a sample of water. In this test (Figure 12.7) a sample of 100 ml of water is filtered through a sterile filter membrane, and the membrane then is placed on the surface of a Petri plate containing some selective growth medium. If Eosin Methylene Blue Agar is used, the colonies of *E. coli* can be easily recognized and selected for identification by other criteria.

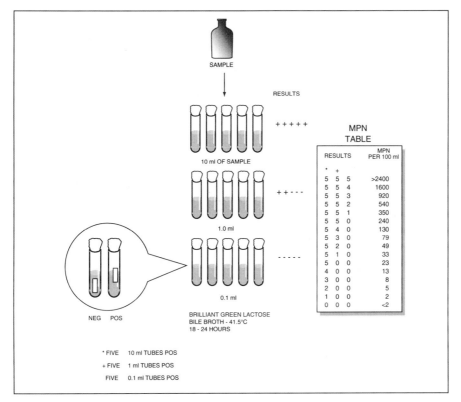

SAMPLE

RESULTS

+ + + + +

10 ml OF SAMPLE

+ + - - -

1.0 ml

- - - - -

0.1 ml

NEG   POS

BRILLIANT GREEN LACTOSE
BILE BROTH - 41.5°C
18 - 24 HOURS

MPN
TABLE

RESULTS			MPN PER 100 ml
*	+		
5	5	5	>2400
5	5	4	1600
5	5	3	920
5	5	2	540
5	5	1	350
5	5	0	240
5	4	0	130
5	3	0	79
5	2	0	49
5	1	0	33
5	0	0	23
4	0	0	13
3	0	0	8
2	0	0	5
1	0	0	2
0	0	0	<2

* FIVE   10 ml TUBES POS
+ FIVE   1 ml TUBES POS
   FIVE   0.1 ml TUBES POS

Figure 12.6

Graphic presentation of MPN test.

Figure 12.7

Detection and counting of fecal coliforms by the filter membrane test.

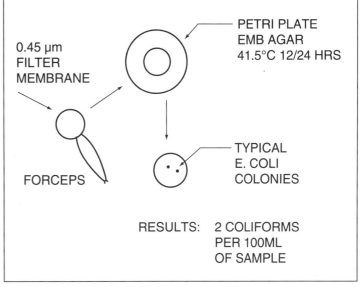

0.45 μm
FILTER
MEMBRANE

PETRI PLATE
EMB AGAR
41.5°C 12/24 HRS

FORCEPS

TYPICAL
E. COLI
COLONIES

RESULTS: 2 COLIFORMS
PER 100ML
OF SAMPLE

Figure 12.8

Effect of filtration and chlorination on typhoid fever in the United States.

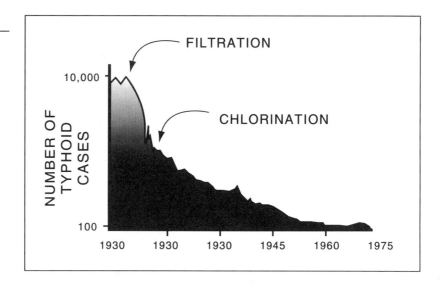

If the plates with membranes are incubated at 42.5°C, the growth of most enteric organisms except *E. coli* will be inhibited. When these cultures are incubated at 42.5°C, the test is called a **Fecal Coliform Test**. The presence of fecal coliforms is taken as definitive proof of fecal contamination of the water.

Water can be examined in many other ways for *E. coli* and coliforms, but the methods are all basically the same as the two described. These tests are used to monitor the condition of drinking water. Only water with no, or few, coliforms is considered adequate for consumption. Control of enteric epidemics in modern communities is attrributable to the purification of drinking water. The value of these procedures lies specifically in the final steps of filtration and chlorination, as this is the point at which pathogens are removed from water. Figure 12.8 shows the effect of filtration and chlorination on typhoid in the United States.

Permissible Levels of Contamination

Even after extensive treatment, water delivered to the user still may have many contaminants, both chemical and microbiological, albeit in quantities considered safe by regulatory agencies. These permissible levels of contaminants are established by public health and other agencies to safeguard the health of consumers. The standards vary from agency to agency and from place to place, but ample agreement exists on what the permissible levels should be. The Environmental Protection Agency (EPA) has established maximal allowable concentrations of chemicals and microorganisms for drinking water in U.S. communities. These are given in Table 12.3.

Table 12.3 Maximal Permissible Levels of Chemicals and Microorganisms in Drinking Water

Material	Concentration	Material	Concentration
Arsenic	0.05 mg/L	Copper	1.0 mg/L
Barium	1.0	Iron	0.3
Chromium	0.05	Manganese	0.05
Fluoride	2.4	Zinc	5
Lead	0.05	Chloride	250
Selenium	0.01	Sulfate	250
Silver	0.05	Solids, dis.	500
Lindane	0.004	Color	15 Units
Methoxychlor	0.1	Corrosivity	none
Coliforms	<1/100 ml		
Bacteria	300/ml		

The objective of all water treatment is to render a finished water that is free of tastes, odors, colors, and pathogenic microorganisms. It must be safe to drink in terms of carcinogenesis, teratogenesis, and radioactivity over and above safety from chemical and microbiological hazards. These standards must be satisfied not only at the treatment plant exit but also at the point of use. Many criteria are applied to the grading of waters. The majority of these are concerned with the use to which the water will be dedicated. In addition, industrial users develop criteria suited uniquely to their purposes, and the variations on this point are, in effect, infinite.

For example, almost all waters that harbor Gram-negative bacteria contain endotoxins. Normally, these are in such low concentrations that they are of no consequence, but if the water is to be used for intravenous injection, the presence of endotoxins creates problems. Water destined for this use must be treated by methods designed specifically to eliminate these contaminants.

A common method of rating waters for public use in the United States is shown in Table 12.4.

SEWAGE TREATMENT

Domestic Sewage

Water that has been used and contains domestic, agricultural, animal, or other waste is generally called sewage. In most communities, sewage is mixed with rainwater, street washings, river and creek overflow, and any other water that courses through the community before it reaches the sewage treatment plant.

Table 12.4   Rating System for Waters in Public Use

Rating	Criteria
Pyrogen-free	Water free of endotoxins
Potable	Water free of pathogenic organisms, tastes, odors, and sediments; adequate for all uses
Safe	Free of pathogens but may have tastes, odors, and suspended materials; must be safe to drink
Recreational	Free of pathogenic organisms, toxic chemicals, and noxious substances
Contaminated	Not safe for use; may have pathogens and/or toxic chemicals

The composite aqueous waste is a mixture of many kinds of materials all dissolved or suspended in water. Normal domestic sewage will contain organic matter, minerals, dirt, tree and plant debris, trash, dead animals, and much more.

Approximately 99% of the sewage is water, and the mixture of organic substances is so great that no individual ingredient is present in large amounts. The two basic kinds of materials in domestic sewage are those that are dissolved in the water and those that are suspended in it. The dissolved material is composed primarily of minerals, carbohydrates, lipids, proteins, and other substances of plant and animal origin. These materials collectively are called the **organic load** of the sewage, and the quantitative measure of this load is made by the **biological oxygen demand** or **BOD** value.

The BOD value is expressed in milligrams of oxygen per liter of sewage. This measure reflects the amount of oxygen that bacteria require to oxidize the organic load of the sewage in a period of 5 days and is expressed as BOD_5. Sewage with a BOD_5 value of 800 is very high in organic load, whereas that with a value of 200 is approximately normal and that with a 50 value carries a very low organic load.

The mineral constituents of the sewage contribute to the BOD load of the sewage but only in small proportion to their quantity, as many of these are in the oxidized state already. Water and insoluble materials have no BOD value, so their contribution to the total value is zero. Living things, on the other hand, consume oxygen in large amounts and therefore make an important contribution to the BOD load of the sewage. Because sewage contains large populations of bacteria and other microorganisms, the oxidative degradation of organic material takes place even during formation of the sewage. Both aerobic and anaerobic organisms are present, and both kinds function at all times in oxidizing the organic load of the sewage.

Table 12.5 Pathogenic Microorganisms in Domestic Sewage

Bacteria	Viruses	Protozoans and Metazoans
Escherichia coli	Hepatitis A virus	*Entamoeba histolytica*
Salmonella typhi	Poliovirus	*Entamoeba coli*
S. enteritidis and others	Coxsackievirus	*Giardia lamblia*
Campylobacter	Echovirus	*Balantidium coli*
Shigella sonnei	Hepatitis B virus	Eggs of flukes
S. boydii and others	Hepatitis E virus	Eggs of tapeworms
Clostridium perfringens		Eggs of roundworms
Bacillus cereus		
Yersinia enterocolitica		

Waters contaminated with human feces in the United States normally contain the pathogenic microorganisms listed in Table 12.5.

Industrial Sewage

In general, industrial sewage contains fewer kinds of materials than domestic sewage does. A given industry may simply discharge one or two substances along with an appropriate amount of wash water. On the other hand, food processors may discharge all the waste from the slaughter, rendering, and cleaning of animals along with the cleaning of vegetables and other comestibles. The BOD_5 of this kind of sewage may reach values of 25,000 when a considerable amount of blood is in the waste.

In other situations the BOD_5 values may be near zero but the discharge may be extremely toxic, as is the case in industries that discharge phenol, iodine, lead, mercury, or any of the thousands of pesticides and other such products of modern industry. In some industries the waste is radioactive, and special processes must be developed for disposing of such materials. Almost all industrial sewage requires special treatment in the plant premises before it can be added to the community sewage.

Screening As sewage enters the sewage plant, it passes through several screening devices that remove large objects of trash and other materials swept into the storm drains and other large openings in the sewage collection system. The screening becomes progressively finer (Figure 12.9), removing smaller and smaller materials until the water is free of most of the trash that came with the sewage.

Trash and other debris recovered in the screening process are incinerated on site or conveyed to the solid waste disposal site the community uses. In

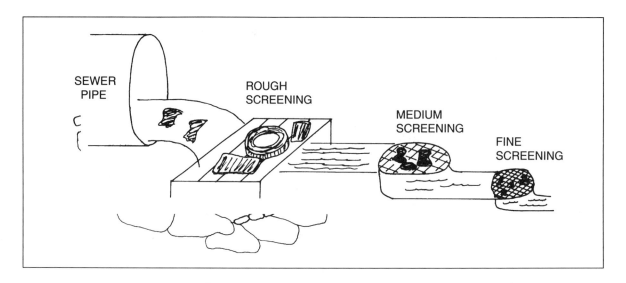

Figure 12.9
Trash removal during screening process.

many locations the quantities of trash reaped from the sewage system are so vast that a solid refuse disposal system must be constructed specifically for this material. In some cities exploratory studies are being conducted on the feasibility of burning this material for the production of electricity.

Sedimentation   The sewage is placed in a sedimentation tank of a design similar to that shown in Figure 12.10, and all solid material that settles to the bottom is removed at this time. This primary sludge contains large amounts of soil and small particles of other matter. Often, material is decontaminated by

Figure 12.10

Sewage sedimentation tank.

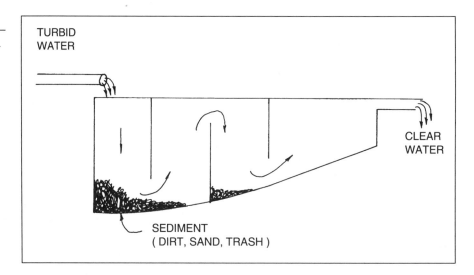

treatment with steam and used as a soil conditioner in agricultural applications. Use of this sludge as a construction material is being examined with promising results in some communities.

Oxidation The objective of sewage treatment is to remove contaminants from the water so it may be recycled and reused by the community. Because the most offensive contaminants are organic materials, water can be cleaned by converting these substances to the inorganic form, i.e., mineralization.

Before each community had its own sewage disposal plant, municipal wastes simply were run into a large body of water such as a stream, river, or lake. If the capacity of the river or lake was not overwhelmed, the natural oxidative processes, i.e., microorganisms, of the water degraded the contaminants. Lakes, rivers, and ponds dedicated to this use become polluted when the population increases beyond a given number, depriving the community of both its disposal facility and its source of drinking water.

The function of these lakes, rivers, and ponds was simple and effective. Sewage was introduced at one end of the lake, river, or pond, and pure water was removed from the opposite end. Ponds and lakes used for this purpose, however, are easily poisoned or overloaded, and often they cease to function in cold weather when they freeze.

Modern communities utilize a rapid and efficient treatment method called the **activated sludge process**. The essential aspects of this process are (a) inoculating the incoming, cleared sewage with effluent from the previous load of sewage, which contains vast populations of microorganisms already, and (b) aerating the inoculated sewage by spraying into the air or by bubbling either air or oxygen into the water (Figure 12.11). In any case, oxygen is provided to the microorganisms and organic materials are degraded rapidly.

Settling After oxidation, the sewage water is allowed to settle, and a floc containing suspended solids and microbial cells forms rapidly. Part of this sludge, which bears a vast load of microorganisms, is used to inoculate the activated sludge tank, and the remainder is transferred to an anaerobic digester, or the entire sludge may be decontaminated by steam treatment and used as a soil fertilizer or soil conditioner. The water that emerges from the activated sludge tank is clear and clean, and it generally has a BOD_5 value lower than 10.

Anaerobic Digester Organic materials that cannot be degraded by aerobic oxidation may be degraded by anaerobic bacteria. The anaerobic digester (Figure 12.12) serves to encourage the growth of anaerobic bacteria, many of which have metabolic processes quite different from those of aerobic organisms. Anaerobic bacteria give off methane, hydrogen, ammonia, hydrogen sulfide, and other odoriferous substances as waste products.

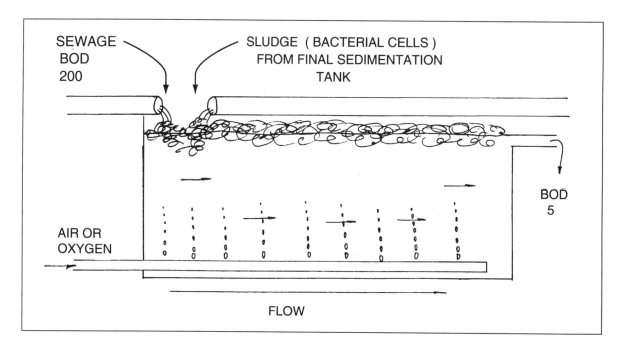

Figure 12.11

Activated sludge aerator in typical sewage treatment plant application.

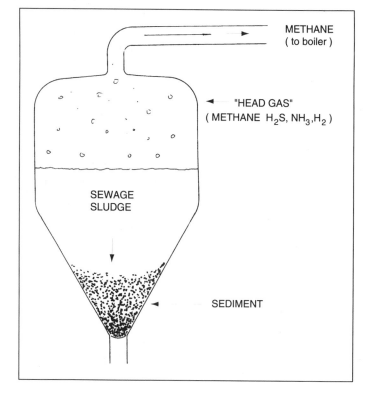

Figure 12.12

Schematic view of Imhoff anaerobic digester.

In many sewage treatment plants the methane is collected and burned for the production of steam, which can be used to generate electricity. Some sewage treatment plants obtain their entire energy requirements by such processes. Sludge from the anaerobic digester is removed and treated the same as aerobic sludge, and the water is added to the activated sludge tank for a second pass through the aerobic digester.

Chlorination As the water leaves the sewage treatment plant, it is chlorinated, ozonized, or treated with ultraviolet radiation to kill pathogenic microorganisms that may have survived the process. Under some conditions the water may be discharged without final treatment, but this represents a serious hazard to adjoining communities. Finally, the water is discharged into a stream, river, or lake, and thus returned to nature. The entire transit through the sewage treatment plant may have been of the order of 15 to 20 hours. The discharged water should be no worse than the water it is discharged into; that is, its BOD_5 value should not be greater than that of the receiving water. Figure 12.13 presents a schematic drawing of a modern sewage treatment plant.

Growth of microorganisms does not occur if the organic load of the water is low or near zero. No pathogens should be present in treated water, as it should not serve as a source of environmental contamination.

QUESTIONS TO ANSWER

1. What is steam?
2. What is water vapor?
3. What are the characteristics of a liquid?
4. What are the characteristics of solid water?
5. What living things survive frozen in ice?
6. What is sublimation?
7. What is an artesian well?
8. What is organic material?
9. What is inorganic material?
10. What is mineralization?
11. What is ozonization?
12. What is finished water?
13. What does the word coliform mean?
14. What does MPN mean?
15. What is lactose?
16. What is endotoxin?
17. What is activated sludge?
18. What is an Imhoff Digester?
19. What is BOD_5?
20. What is potable water?

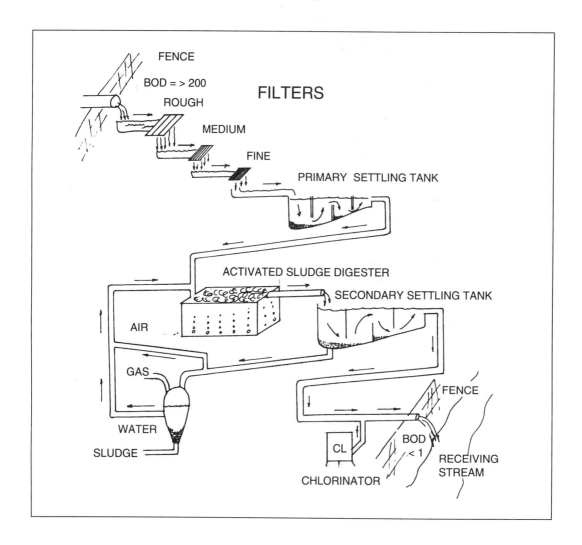

Figure 12.13

Scheme of modern sewage treatment plant.

FURTHER READING

Brock, T. D. and M. T. Madigan. 1988. BIOLOGY OF MICROORGANISMS, 5th ed. Prentice Hall, Englewood Cliffs, New Jersey.

Cliver, D. O., R. A. Newman, and J. A. Cotruvo, eds. 1987. DRINKING WATER MICROBIOLOGY, Journal Environmental Pathology, Toxicology, Oncology 7:1–365.

Iwan, G. R. 1987. DRINKING WATER QUALITY CONCERNS OF NEW YORK CITY, PAST AND PRESENT. Annual New York Academy of Science 502:183–204.

Riemann, H. and F. L. Bryan. 1979. FOOD BORNE INFECTIONS AND INTOXICATIONS. 2nd ed. Academic Press, New York.

Microbiology of Milk

13

Look at them milkers, ain't they grand,
Make you want to hold 'em, one in each hand,
They's four of them suckers. Don't be dumb,
You on'y got two hands, and on'y one thumb.

Shiaashooh

Milk is a complete food for many animals and an excellent medium for the growth of almost all microorganisms. It is produced by all females of the class *Mammalia* for the nutrition of their progeny and may serve as the sole source of food and water for the developing young animal for many months or even years. The milk of all female mammals is rich in all necessary carbohydrates, fats, proteins, vitamins, and mineral substances. It also provides certain hormones essential for the normal development of the young and all the antibodies necessary for protection of the offspring during the time that its own immunological system is deficient.

Humans have intervened dramatically in the milk-producing abilities of certain domestic mammals such as cattle and goats. These animals are made to produce much greater quantities of milk than necessary for their young. The surplus milk is intended for use by human beings. Milk is considered an essential food and, in some cases, bottle-fed babies may reach childhood never having tasted their own mother's milk.

ASPECTS OF MILK

Contamination

Milk in the mammary glands is sterile and remains so as long as it is not contaminated from the outside. It becomes contaminated with bacteria, viruses, and fungi as it passes through the milk ducts in the teat, as all these organisms generally are present there. The process of spoilage begins at this time, and proceeds until the waste products of microbial metabolism change the taste, aroma, and consistency of milk. Microorganisms grow only in the distal extensions of the milk ducts, because cells lining the canal further to the interior prevent bacterial growth and stop invasion of the inner, milk-secreting tissues.

When bacteria overcome the body defenses and proliferate in the milk-secreting cells, a disease condition results and all the milk produced under those conditions is contaminated with microbial cells. When bacteria grow in the mammary gland tissues, the disease they produce is called **mastitis**. One of the most common causes of mastitis in dairy cattle is infection with the Gram-positive bacterium *Staphylococcus aureus*. This and many other diseases can be transmitted by drinking contaminated milk or even by just coming in contact with it, as in the preparation of cheese and other dairy products.

Fresh Milk

Milk is said to be fresh immediately after it is collected from the animal. In the production of milk from cows, goats, yaks, camels, or other animals for use by human beings, local custom determines the amount of care devoted to controlling microbial contamination.

In the United States, where cow's milk is used almost exclusively, milk quality is synonymous with bacterial content. By maintaining scrupulously clean milking machines, receiving containers, and all auxiliary equipment, as well as clean, disease-free animals, the extent of contamination is kept to a minimum. In addition to this, all personnel involved adhere to a strict code of personal hygiene lest post-production contamination defeat the entire process. To prevent growth of the contaminants that enter into collected milk, the milk usually is chilled and then stored at a temperature of 10°C or less.

Fresh milk may be drunk immediately after it is procured, and in the past this was the case. In prerefrigeration times the shelf life of fresh milk was limited to a few hours during the summer and a few days in cold weather. After this time it was consumed as sour milk, was used for making butter, or was separated into curds and whey.

Figure 13.1

Modern milk collection, storage, and processing.

Bottled Milk

The selling of milk in sealed containers in the United States started at the end of the 19th century and became popular during the first decade of the 20th century. Heavy glass bottles were universally used at first, but these were replaced by cartons during World War II and by plastic in the 1970s. Formless plastic bags are used in many countries, but these present problems of storage once the container is opened. The major reason for bottling milk came as a marketing strategy, but many other benefits accrued from this practice, including:

- more efficient refrigeration,
- protection from contamination,
- longer shelf life,
- identification of producer,
- responsibility, and
- safer transportation.

At the turn of the century, fresh milk was placed in a clean bottle and maintained at a temperature of 10°C or lower. This was a significant improvement over the older methods of dispensing milk from the seller's large container to the buyer's small, open container. That practice offered ample opportunity for contamination previous to and during this transfer, as well as during subsequent storage and use. Sealed bottles offered protection in this sense and their use revolutionized the marketing of fresh milk and cream.

The most significant aspect of placing milk in a container and sealing it against microbial invasion came after pasteurization was introduced. In the early days of this practice, the milk was placed in the container and sealed before pasteurizing it, effectively guaranteeing that it did not bear disease-producing organisms. Today, milk is pasteurized as it emerges from the blending area and then is placed in clean containers. When the preparation area of dairies is kept scrupulously clean, the possibility of introducing pathogenic microorganisms at this point is very small.

Sterile/Evaporated/Condensed Milk

The shelf life of bottled milk kept at 4°C to 10°C is counted in terms of 2 to 3 weeks at most. After this time the milk sours and spoils. A strategy for keeping milk suitable for drinking for prolonged periods was introduced in the 1920s. This was accomplished by placing the milk in metal cans, sealing them, and heating to 100°C or higher. Milk prepared by this method will remain useful for an indeterminate time and may be transported over large distances. Because a large part of the milk is water, transportation was deemed more efficient if the milk was condensed. This is accomplished by evaporating the water but leaving all nutrients behind before canning. This product, **evapo-**

rated or condensed milk, became the mainstay in urban areas from the 1920s to the 1940s.

Approximately 60% of the water is removed, and in some cases sugar is added (**sweetened condensed milk**), making the osmotic pressure very high. The only organisms that survive the heating process are in the spore form, and these will not germinate in such milk. Condensed milk now is a more common staple of food in Third World countries, and in the United States it is used primarily for cooking.

Powdered Milk

If all the water is removed from milk or skim milk, by heating at 50°C to 90°C, a fine, white powder, containing all the ingredients and nutrients of whole milk, can be produced. Adding the appropriate amount of water gives a product of the same nutrient quality as milk, but its taste and appearance bear no resemblance to the original. Many men in the armed forces of the United States, including this author, probably suffered more from having to drink powdered milk during World War II than from enemy action.

Although powdered milk is not sterile, it will not spoil, since there is no water for microbial growth. On the other hand, since only milk of low microbiological quality is used to make powdered milk, it may contain microbial toxins. Staphylococcal toxin will remain active for long periods of time and will result in intoxication when consumed. Many outbreaks of powdered milk intoxication occurred in the period from 1930 to 1950, but stringent regulations have made such problems a thing of the past.

Microbiological Quality of Milk

The microbiological quality of milk is determined by counting the number of viable bacteria and fungi present in a representative sample. The assumption is that milk with low microbial populations is less likely to harbor pathogens than milk with high populations. In addition, the freshness of the milk is very much a function of the number of microorganisms present. In general, the most important aspect of milk microbiology is the total count or the number of all organisms present in the sample. Although counting the total number of organisms in a sample of milk is impossible, the tests recommended are designed to yield the highest number of organisms possible.

Tests for pathogenic bacteria or coliforms normally are not deemed necessary since all dairy products are pasteurized before they are sold. When such tests are prescribed, they are the same as the ones used for water and other food items. None of the pathogens of importance to human beings survive pasteurization when this procedure is performed properly, but in practice epidemics of gastroenteritis caused by pasteurized milk do occur. Some of these epidemics have been caused by bacteria that survived pasteurization, by pre-formed bacterial toxins, and others by products contaminated after the pasteurization process.

Since spoilage is brought about by microbial action, the larger the number of bacteria, the closer the milk is to spoiling. This consideration is so important that the bacterial content of the milk being sold often determines the price paid for milk. If the microbial assessment of milk is to be useful, methods to rapidly estimate bacterial populations in milk must be available. Some of the procedures for estimating bacterial populations in milk can be carried out in just a few minutes. These often are used at the time milk is being purchased. Four of these are discussed briefly here.

Dye Reduction Test   Bacterial cells, as all other living things, obtain energy by **oxidizing** food molecules and, as a consequence, reducing other molecules inside of the cell. Chemical substances can enter bacterial cells and intervene in the natural, metabolic process. Some of these substances are dyes that may be one color when oxidized and another when reduced. Because of this characteristic, substances such as **methylene blue**, **resazurin**, and **triphenyltetrazolium chloride** are called *oxidation-reduction indicators*. The first two named here are blue in the oxidized state and colorless in the reduced condition, and triphenyltetrazolium chloride is colorless when oxidized and changes to red when reduced.

The oldest and most common **dye reduction test** for assessment of bacterial load employs methylene blue. The test is performed by placing a predetermined quantity of methylene blue in a sample of milk, incubating the mixture, and recording the time required for the color to change from the oxidized to the reduced form. The number of organisms present in the milk at the beginning of the test can be estimated by knowing the time it takes for the color to change. For example, if 1×10^5 viable bacterial cells per milliliter of milk, are present, the time for the color of the **oxidation-reduction** indicator to change will be 3 hours. A population of 1×10^8 cells will produce the color change in approximately 10 minutes. By comparing the time required to decolorize a given amount of methylene blue and the number of bacterial cells (determined by the plate count method), the dye reduction test yields a fairly accurate estimate of the bacterial population.

In addition to this, the sensitivity of the test can be changed by increasing or decreasing the amount of methylene blue. The dye reduction test is a field test that requires equipment no more complex then a watch and two test tubes. It is used to grade milk as shown below:

Time for Color Change	Grade of Milk
60 min	high grade
30 min	acceptable
10 min	poor
(no change)	(control; milk heated)

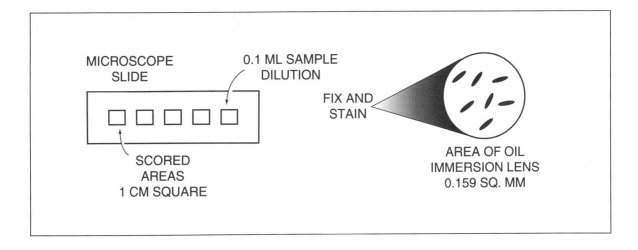

Table 13.1   Typical Results from Direct Cell Count Slide

Dilution	Total in 25 Fields	Average per Field
no dil	TNTC*	—
1:10	TNTC	—
1:100	1997	80
1:1000	406	16
1:10,000	87	3
1:100,000	12	<1
1:1,000,000	0	0

*TNTC = Too numerous to count

Figure 13.2

Breed slide for direct count of total number of cells.

Direct Microscopic Count   A useful estimate of the total number of bacterial cells in milk also can be made by the direct microscopic count of cells. In this procedure a **ten fold dilution series** of the milk sample is prepared and aliquants are placed on a special microscope slide. A sample of 0.1 ml of each dilution is placed in a measured (one square centimeter) area of the slide. The slide is allowed to air-dry, fixed in 95% alcohol for 30 seconds, washed in distilled water, stained with methylene blue, and allowed to air-dry (Figure 13.2). The bacteria in 25 **oil immersion fields** are counted, and the average number of cells for a field determined from the results. The results shown in Table 13.1 are from an actual determination.

The rule to follow in this kind of determination is to select the dilution that yields an average of 50 to 100 cells per microscopic field. Because each field from the 1:10 dilution has an average of 80 bacterial cells, this datum will be used. Each field on the slide seen with the oil immersion lens covers an area of the slide that can be measured as shown in the following.

$$
\begin{aligned}
\text{Diameter of oil immersion lens} &= 0.45 \text{ mm} \\
\text{Area of field} &= \pi r^2 \\
&= 3.1416 \times (0.225)^2 \text{ mm} \\
&= 3.1416 \times 0.050625 \\
&= 0.159 \text{ sq mm}
\end{aligned}
$$

The area inside each square of the special slide is 100 mm². The amount of milk placed in this area was 0.1 ml of a 1:100 dilution or 0.001 ml of the milk sample. If 0.159 mm² has an average of 80 cells:

then:

$$
\begin{aligned}
80 \text{ cells} : 0.159 \text{ sq mm} &= x \text{ cells} : 100 \text{ sq mm} \\
80 \times 100 &= x \times 0.159 \\
x &= (80 \times 100) \div 0.159 \\
&= 50{,}314 \text{ cells in } 0.001 \text{ ml of sample,}
\end{aligned}
$$

therefore:

$$
\begin{aligned}
50{,}314 \text{ cells} : 0.001 \text{ ml} &= x \text{ cells} : 1 \text{ ml} \\
50{,}314 \times 1 &= 0.001 \times x \\
x &= (50{,}314 \times 1) \div 0.001 \\
&= 50{,}314{,}000 \text{ cells per milliliter}
\end{aligned}
$$

or:

$$5.03 \times 10^7 \text{ cells per ml of milk}$$

Like all other methods of estimating the number of bacterial cells, the direct microscopic count has many shortcomings, including errors in dilution, measuring the area of the square, and washing cells off the slide. Also, both living and dead cells are counted. The results are useful, however, because it has been shown that they are consistent in expressing the quality of milk. The direct microscopic count is a rapid field test that can be performed in a few minutes in any location where a microscope can be used.

Total Count   Estimates of the number of **viable** (live) bacteria in milk are made by the pour plate method. The milk sample is shaken vigorously to mix milk and cream, and 10.0 ml are transferred to 90.0 ml of sterile water in a bottle with stopper. This mixture is shaken vigorously, and 10.0 ml is transferred to another 90.0 ml **water blank**. Subsequent dilutions are made in the same manner to yield a series of tenfold dilutions labeled 1:10, 1:100, 1:1,000, 1:10,000, 1:100,000, and 1:1,000,000. In practice, these are labeled 10, H, T, TT, HT, and M or, using the logarithms to indicate the amount of milk in 1 ml of dilution mixture, i.e., –1, –2, –3, –4, –5, and –6. In an everyday situation, the labels used are 1, 2, 3, 4, 5, and 6.

After the dilutions are prepared, 1.0 ml of each dilution is placed in each of three sterile Petri plates, and melted agar at a temperature of approxi-

Table 13.2 Typical Data from Total Plate Count of Milk

Dilution	Counts	Average
1:10	TNTC, TNTC, TNTC*	—
1:100	TNTC, TNTC, TNTC	—
1:1,000	TNTC, TNTC, TNTC	—
1:10,000	287, 344, 258	314
1:100,000	66, 78, 75	73
1:1,000,000	10, 32, 24	22

*TNTC = Too numerous to count

mately 45°C is poured into each plate. The agar and diluted milk are mixed carefully to disperse the cells evenly throughout the melted agar medium. The plates are allowed to cool and when the agar gels, the plates are inverted and placed in the incubator at 37°C.

Generally, Plate Count Agar or Nutrient Agar is used for this purpose, but many others will give suitable results. The colonies that develop are counted after 24 and 48 hours of incubation. Only plates that have more than 30 but less than 300 colonies are used to determine the plate count. Colonies must be separated from each other and be easily countable if the estimate is to have significance. Typical colony counts may appear as shown in Table 13.2.

The 1:100,000 dilution gave an average of 73 colonies per plate and is the only dilution that falls within the 30 to 300 range. The dilution of 1.0 ml of milk to 100,000 parts by mixing with water means that each ml of the dilution mixture contains $(1 \div 100,000)$ 0.000,01 ml of milk. Therefore, if each 0.000,01 ml of milk has an average of 73 bacterial cells, there must be:

$$73 : 0.000,01 = x : 1.0$$
$$x = (73 \times 1.0) \div 0.000,01$$
$$x = 7,300,000 \text{ viable bacteria per ml of milk}$$

This number is better expressed as 7.3×10^6 and is considered an estimate of the size of the population rather than a count of the total number of cells. Estimates of bacterial populations determined by this procedure are subject to many errors, but they are quite consistent within a framework of the procedures used, and the information obtained is valuable in determining the quality of milk. The greatest value lies in allowing comparisons among different samples of milk. A graphic description of the plate count method is shown in Figure 13.3.

Limulus Lysate Test   *Limulus* is the genus to which the horseshoe crab, *Limulus polyphemous*, belongs. If the blood cells from this crab are lysed in pyrogen-free water and the lysate is preserved in specially cleaned (**pyrogen-free**) glassware, it can be used to identify lipopolysaccharide from

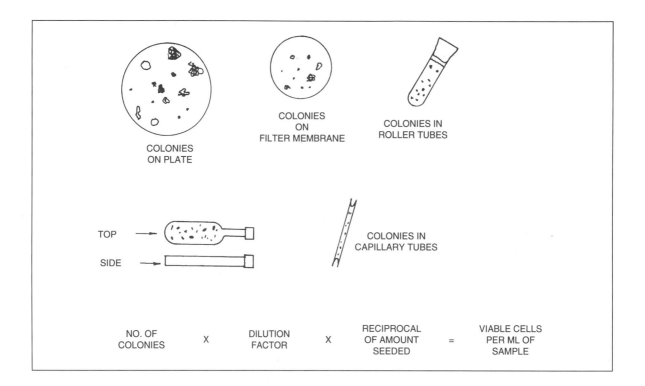

Figure 13.3

Different devices for obtaining total colony counts.

Gram-negative cells. To accomplish this, the lysate is mixed with the lipopolysaccharide fraction of the bacterial cell wall and is allowed to react. The lysate will gel in a few minutes. If the limulus blood cell lysate is standardized and if a dilution series of the milk sample is used, the amount of lipopolysaccharide in the milk can be estimated.

Lipopolysaccharide is found primarily in Gram-negative bacteria, but experience has shown that the ratio of Gram negatives to Gram positives is quite constant, and from this ratio the total bacterial load of the milk can be calculated. Limulus lysate is the most sensitive test for lipopolysaccharide and coincides well with other estimates of milk quality. The test can be performed in the field with a precalibrated **kit** of materials and reagents in a period of one hour, and the results are consistent with those obtained in the laboratory.

Grading of Milk

Milk for sale to the public is graded according to the bacterial population estimates obtained by the pour plate method. The counts shown in Table 13.3 are those obtained at the time the milk leaves the production plant, and it is assumed that if the milk is refrigerated, they will not change significantly until after the milk is sold.

Table 13.3   Bacterial Count Standards for Milk Leaving Production Plant

Grade	Criteria
Certified raw milk	Not to exceed 10,000/ml total; not to exceed 10 coliforms per ml
Certified pasteurized milk	Not to exceed 500/ml total; not to exceed one coliform per ml
Grade A raw milk	Not to exceed 100,000/ml total; not to exceed 10 coliforms per ml
Grade A pasteurized milk	Not to exceed 20,000/ml total; not to exceed 10 coliforms per ml
Industrial	Not acceptable for drinking

Milk can be graded in many other ways but those in Table 13.3 are the ones most commonly used for fresh bottled milk. Milk that contains larger populations of bacteria and coliforms generally is designated as **Industrial Grade** and is used for the production of powdered milk, condensed milk, or evaporated milk. Milk with high total counts and large populations of coliforms usually is employed in cooked food preparations, candy, or pastries. If the total counts are high but the coliform counts are within limits, milk usually is fermented to produce one of hundreds of cheeses or yogurts, or be used for the production of ice cream.

Butter

Fat can be separated from milk by churning or other method that allows the fat droplets to collect. Once separated, butter can be preserved for long periods because most bacteria cannot grow on it. On the other hand, lipolytic bacteria and fungi can split the fatty acid molecules and cause **rancidity** or spoilage. Like other microbial processes, lipolysis is retarded by refrigeration.

NORMAL FLORA

Microorganisms enter milk through contact with the animal that produces it, milkers and grooms, utensils used for collection, and contact with water, soil, and air. On the other hand, bacteria of the genera *Lactobacillus, Leuconostoc, Streptococcus,* and *Pediococcus* plus various fungi and viruses normally are present in the milk ducts of the udder and enter the milk as it leaves the cow's teats. The combination of organisms in the milk represents a universal inoculum of all the microorganisms present in the environment where the milk is obtained and where it is processed. With few exceptions, most microorganisms grow or survive in the milk. If the temperature is suitable, growth is

Figure 13.4

Increase in total count of bacteria in milk as a function of time.

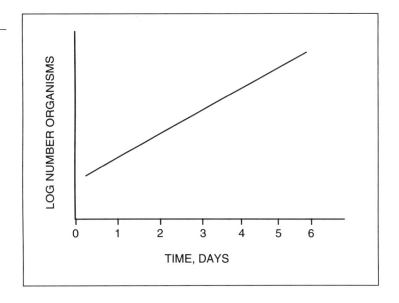

rapid, but even at refrigerator temperatures, growth proceeds, albeit at a slower pace.

The organisms that become predominant and create the largest populations are of the genera *Lactobacillus, Streptococcus, Pseudomonas, Arthrobacter, Flavobacterium, Alcaligenes, Leuconostoc,* and *Pediococcus.* With the exception of the viruses, the populations of these organisms increase with the passage of time if temperature and other conditions are suitable for growth. Growth of the total population can be shown graphically, as in Figure 13.4.

DISEASES OF MILK

When the growth of a given organism becomes predominant, the contamination bears the characteristics of the contaminant. For example, if pseudomonas serves as an inoculum, the growth of *Pseudomaonas* gives the characteristics of this organism to the spoilage process. In the milk industry these phenomena are regarded as **milk diseases**. Table 13.4 shows some of the most common of these diseases and the organisms that cause them. Normally one or more of these organisms will grow together to produce mixed "diseases." Combined characteristics resulting in, say, ropy-blue milk or red-rancid milk.

SPOILAGE

As the number of microorganisms increases, the flavor, texture, color, and aroma of the milk change, and the milk is perceived as in the process of spoil-

Table 13.4   Diseases of Milk and Organisms Most Often Associated
with Them

Disease	Associated Organisms
Bitter milk	*Flavobacterium, Acinetobacter, Proteus*
Fruity taste	*Pseudomonas, Flavobacterium, Alcaligenes*
Discoloration	*Chromobacterium, Pseudomonas*
Bloody milk	*Serratia marcescens* or animal blood
Red milk	*S. marcescens* or *Brevibacterium erythrogenes*
Blue milk	*Pseudomonas aeruginosa*
Ropy milk	*Bacillus subtilis*, yeasts
Slimy milk	*Bacillus, Alcaligenes*, micrococcus
Rancid milk	*Pseudomonas, Staphylococcus, Alcaligenes*, yeasts and other fungi

ing. Normally, milk spoilage proceeds in a predictable, orderly sequence that
begins with souring and ends up with liquefaction. Under certain conditions
mixed populations of bacteria and fungi become predominant and spoilage
begins. Spoilage results from the growth of many extraneous organisms, making the milk useless. Proteins in spoiled milk will not coagulate, thereby eliminating the possibility of cheese production. Figure 13.5 shows a culture from
spoiled milk.

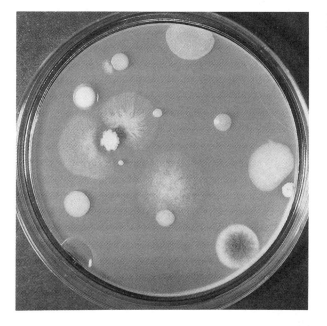

Figure 13.5

*Culture from spoiled milk;
note large number of fungi
and yeasts.*

SOUR MILK

A normal bacterial flora that grows rapidly develops spontaneously if the temperature is higher than 10°C but lower than 40°C. The major part of the growth takes place at the expense of milk sugars such as lactose. Milk contains little oxygen and the sugars are **fermented** with liberation of various acids including butyric, lactic, and acetic. These acids make the milk sour, and the acid has a selective effect on the subsequent development of the bacterial flora.

The first step in fermentation is brought about by *Streptococcus lactis*, *S. cremoris*, and organisms of the genera *Leuconostoc* and *Lactobacillus*. These organisms perform the primary metabolism of milk sugars and release acetic and lactic acids, which bring the milk pH to a value of 6 or less. Taste, consistency, and bouquet of the sour milk are determined by the amounts and kinds of acids produced. The process is called a **mixed acid fermentation**, in contrast to single-acid fermentations, and the major impact on the taste and appearance of the milk comes from the ratio of the acids produced by organisms from the three genera named above.

As the sour milk ages and the fermentation proceeds, the taste and quality of sour milk changes accordingly. Sour milk fanciers may gauge the quality of sour milk quite accurately by measuring the time allowed for fermentation and by calculating the mixture of acids present.

Buttermilk

As the number of fermentation acids increases, pH falls to values between 4 and 5. At this point, **casein**, the major milk protein, precipitates and the milk becomes buttermilk. This product, clabbered milk, has characteristics quite different from those of sour milk. Also, **renin**, an enzyme produced by *streptococcus* and other bacteria, causes all other milk proteins to precipitate. The milk becomes thick and viscid and shows small flecks of butter throughout.

Buttermilk has a sufficiently high concentration of acids to inhibit the growth of most bacteria, giving buttermilk some stability and a fairly long shelf life. Some strains of *Streptococcus lactis* and *S. cremoris* produce esters, lactones, ketones, high molecular weight acids, and diacetyl in small amounts. These substances have strong, pungent odors and give buttermilk its characteristic flavor. The final bouquet of finished buttermilk depends on the kinds and quantities of fermentation products given off by the bacteria present.

Liquefaction

Acid-tolerant bacteria and fungi continue to grow in the buttermilk, using the fermentation acids as growth substrate. As the production of acids diminishes, the pH rises slightly and **proteolytic organisms** such as *Pseudomonas*, *Bacillus*, and *Clostridium* begin to grow. This leads to degradation of proteins and liberation of large amounts of amino acids, which allows a great variety of bacteria and fungi to grow.

The remainder of the fermentation acids are consumed by the expanding populations of bacteria and fungi, and **lipolysis** commences. As this proceeds, the milk takes on a yellowish, watery appearance, the pH becomes neutral, and the remainder of the organic substances are consumed. The milk is now **mineralized**, and few organisms can grow in the remaining solution. Most of the microorganisms have perished by this time for lack of nutrients, although large numbers of bacteria and fungi survive in the resting stage. Because many spores are present in this residue, it would take many years for this fluid to become sterile.

Cultured Milk

Fresh, pasteurized milk inoculated with large populations of *Streptococcus lactis*, *S. cremoris*, and *Leuconostoc mesenteroides* and incubated at 20°C to 22°C for 12 hours becomes **buttermilk**. Much of the conversion of the milk to buttermilk results from the production of lactic and acetic acids, but enzymes also take part in the precipitation of milk proteins in the same manner as in normal fermentation. Varying the ratio of the organisms added as inoculum will yield variations in taste, aroma, and consistency of the buttermilk. The variations are attributable not only to the differences in the fermentation acids but also to the production of small amounts of esters, aldehydes, lactones, and ketones as well as other products of fermentation. The amounts and kinds of substances produced are predetermined to attain the desired taste and bouquet of the finished product.

Buttermilks with varying characteristics are produced in different countries. The Scandinavian countries produce buttermilks called langfil, lattfil, ymer, skyr, villi, and filmjolk. These products are made with different organisms and under different conditions but are really only variations on the same theme.

Langfil is made using *Streptococcus cremoris* and *S. lactis*. A special strain of *S. lactis* produces **ropiness** or a stringy consistency that makes it impossible for the buttermilk to flow through pipes and valves. It must either be produced in vats, as is the traditional method in the home, or fermented in the final container, as is the case in modern commercial production.

Ymer is a traditional drink of the Russians, Finns, and Lapps. It includes strains of *Streptococcus lactis (diacetylactis)* that produce large amounts of acetaldehyde and diacetyl and relatively little acid.

Acidophilus Milk

Fresh milk that is pasteurized, inoculated with *Lactobacillus acidophilus*, and incubated for 24 hours yields a buttermilk popular in the United States and many other countries. Many devotees believe that acidophilus milk has curative properties. Others think that drinking large amounts of acidophilus milk will protect against enteric infections. This belief was held more widely in the past than it is now, especially if the buttermilk was consumed instead of water, which often was contaminated with enteric pathogens. Elie Metchnikoff (1845–1916), the great Russian scientist, not only supported the idea that cultured milk prevents enteric disease but also promulgated the theory that it promots a longer life span. Metchnikoff drank buttermilk constantly from a shoulder water bag he always carried, but, nevertheless, he lived only 71 years.

Milk Bulgaricus

This cultured milk is produced by inoculating fresh, pasteurized milk with a combination of *Lactobacillus bulgaricus* and *Streptococcus thermophilus*. Commercially, the culture first is incubated at 42°C for 24 hours and then at 22°C until the desired taste and bouquet are attained. In the home, taste and quality depend entirely on the "starter" used, as incubation temperature cannot be controlled easily. Much of the taste of Bulgaricus milk comes from acetaldehyde and other minor products of fermentation.

Yogurt

Yogurt is made by adding *Lactobacillus bulgaricus* and *Streptococcus thermophilus* to milk that has been heated to 90°C for 30 minutes. Heating destroys most of the enzymes and the bacteria present in the milk and permits the flavors produced by the added organisms to dominate. Acids and enzymes produced by *L. bulgaricus* and *S. thermophilus* cause precipitation of the milk proteins, giving yogurt its characteristic consistency. The consistency is variable and is determined in large part, by the fermentation acids, lactones, ketones, and enzymes excreted into the milk by bacteria.

Yogurt has been used in one form or another since approximately the 12th century. It is made by many different formulas and under as many different conditions. In earlier times milk was allowed to ferment spontaneously by action of the bacteria normally found therein. Incubation probably was accomplished by placing the milk container in a cool corner of the kitchen. Although most yogurt produced commercially today is from cow's milk, traditionally the milk of many animals has been fermented. In India, water buffalo milk was used in the past and still is in use in rural areas. In North Africa, Spain, Lapland, and Scotland, yogurt is made from the milk of the camel, goat, reindeer, and sheep, respectively.

Kifir, Koumiss, and Laban

Fermented milk in which bacteria or fungi produce carbon dioxide and alcohol is common in many parts of the world. Organisms used for the alcoholic

Figure 13.6

Commercial production of curds using bacterial enzymes and dilute acid.

fermentation of milk include *Lactobacillus acidophilus, L. kefir, L. brevis, Leuconostoc mesenteroides, L. cremoris, Streptococcus lactis, S. thermophilus, Saccharomyces cerevisiae, Candida kefir,* and many others. In many of these fermented milks, the precipitated proteins, or curds, hold carbon dioxide and float to the surface. The yogurt so produced contains large amounts of carbon dioxide and approximately 3% ethyl alcohol, giving the product a light, frothy taste and an alcoholic punch not found in other milk products.

CHEESE

Curds and Whey

Microbial action beyond the point where buttermilk is produced results in accumulation of greater quantities and varieties of acids and other fermentation products. As a result, the milk continues to change in appearance, flavor, and aroma. *Streptococcus lactis* and other bacteria that produce renin and other enzymes that cause milk proteins to coagulate, promote the formation of stable, well-formed curds, and a clear liquid commonly called whey.

The curds can be collected by passing the coagulated milk through a cloth strainer. Curds are made of the milk proteins, lipids, and some of the high molecular weight carbohydrates. Whey contains vitamins, amino acids, some carbohydrates, and most of the minerals from the milk. In the home, whey is used as a liquid stock in preparing other foods. Commercially whey is added as a mineral and vitamin enrichment to foods for human or animal use. Figure 13.6 shows the commercial production of curds, using bacterial enzymes and dilute acid.

Figure 13.7

Variety of cheeses made from the same milk.

Variety

Curds are the point of beginning for the production of cheeses of a great and fantastic variety. There are two major kinds of cheese. One class is called **unripened** cheeses because these are not aged and the other **ripened** cheeses since they are aged for periods of time varying from a few days to more than two years.

Aging means that bacteria and fungi grow in the curd and impart different flavors, appearances, and aromas to the cheese as fermentation products accumulate. Thus, the method by which fermentation is carried out determines the kind of cheese that results. The variety is infinite, and many households in many countries make cheeses unique to them. On the other hand, commercial production is limited to approximately 400 different kinds of cheeses worldwide. Only a few of these will be described here as examples of the potential of this industry (see Figure 13.7).

Cottage Cheese

Cottage cheese is undoubtedly the first cheese made by human beings—some 5,000 years ago. Today it is made by coagulating the proteins from skim milk with a commercial preparation of renin and acid. It also can be made by adding commercial lactic acid to skim milk until precipitation of proteins is complete.

A third method of forming the curd from skim milk is by introducing a heavy inoculum of *Streptococcus lactis* so that the curdling is completed in 4 to

6 hours. The process includes stirring the coagulum and adding texturizing agents to obtain a soft curd with a smooth, creamy consistency. This process involves no aging or fermentation, and the finished product has no significant taste or bouquet, but it does have a desirable texture and consistency.

Cottage cheese is a great favorite of Americans. In other countries it is known by many different names including **queso casero** in Mexico and **quesillo** in Colombia. In home production various substances are blended in with the curd to give specific flavors. These range from chopped cherries, to nuts, to Jalapeño peppers, according to personal taste.

Cheddar/American Cheese

Cheddar cheese is a classic English cheese but is now so much of a favorite in the United States that it commonly is called American cheese. It is a fully ripened cheese with a hard, heavy consistency and a deep yellow color.

Cheddar cheese is produced by fermenting whole milk with a combination of *Lactobacillus lactis*, *L. cremoris*, and *Streptococcus faecalis*. The curd is formed by the action of renin from *Streptococcus*, or by adding the isolated enzyme from either bacterial or animal origin. After the curd is formed, it is separated from the whey and placed in wooden forms called **hoops**. After drying for several days, coloring, e.g., **annatto**, a coloring agent derived from the plant *Bixa orellana* is added to produce the characteristic yellow-orange color of authentic cheddar cheese. When thoroughly dry, the forming cheese is moved to the **curing room** and kept there at a temperature of approximately 15°C for 2 or more months. Mild, soft cheddar cheese is obtained after 2 to 3 months of curing, while sharp, hard, dry cheese is produced after a curing time of about 2 years. It requires a longer time and much more care and attention to produce the sharp, dry cheese than it does to produce the mild variety. The extra care and effort are reflected in the price of the finished product. The consumer is rewarded, however, with a cheese of unique, full flavor.

Swiss Cheese

Originally from Switzerland, Swiss cheese is now a universal favorite. It is made from pasteurized, whole milk fermented with *Lactobacillus bulgaricus* and *Streptococcus thermophilus*. When the curd is formed, it is placed in wooden hoops and pressed several times to remove residual whey. Late in the fermentation process, *Propionibacterium shermanii* is added to supply the finished cheese with propionic acid, other desirable acids, flavoring agents, and carbon dioxide. The carbon dioxide trapped inside the heavy, tenacious curd has no means of escape and produces the eyes, or holes, that characterize this cheese (see Figure 13.8).

The fermenting curd then is moved to the curing room and is turned daily for several weeks. Each time it is turned, the surface is rubbed with salt

Figure 13.8

Swiss cheese with large (left) and small (right) eyes.

water or salt until a desirable surface consistency is attained. At this point it is transferred to a fermentation room and kept there at 20°C for 2 to 3 months. While in the fermentation room the surface is again washed with salt water at least once per week and the cheese turned each time it is salted. The fermented cheese then is transferred to the curing room and kept at a temperature of 10°C for 10 to 12 months. Some Swiss cheeses are aged even longer, but the care required raises the price so much that little of this kind of cheese is marketed in the United States.

Camembert Cheese

Camembert cheese is made from pasteurized whole milk fermented with *Lactobacillus bulgaricus* and other lactic acid producers until a curd is formed. The curd is separated from the whey and beaten to a soft, homogeneous consistency. It then is placed in hoops and allowed to drain for several days until all the whey has been removed. When the cheese is formed, the hoops are removed, the surface inoculated with the mold *Penicillium camemberti,* and placed in the curing room at 10°C to 12°C for 2 to 3 months or until the desired characteristics are attained. This process is called **surface curing** because the mold grows from the outside toward the center.

The taste, aroma, and soft, creamy consistency characteristic of Camembert cheese are introduced into the cheese by the mold and by bacteria and other fungi found normally in the milk and in the curing room. These organisms grow on the surface and produce a **mat** or **pellicle** that seals in moisture and keeps the cheese from hardening. Bacterial and fungal enzymes break down the original curd to produce the creamlike texture of the interior portion of the Camembert cheese. This last process takes approximately 6 months, but the time can vary to satisfy the maker's expectations. The pellicle formed naturally will protect the cheese from drying out for 6 to 18 months if the cheese is kept in a cool, dry place. Commercially produced Camembert cheese may be wrapped in paper or cloth to extend its shelf life.

PASTEURIZATION

Pasteurization is designed specifically to prevent the transmission of disease by milk and milk products because it is effective in killing all microorganisms that cause disease in human beings. When first introduced, pasteurization of milk was not altogether popular because heating causes undesirable changes in the taste and appearance of milk and also adds to the cost of production. Today, however, with the universal pasteurization of all milk and milk products, the taste of pasteurized milk is regarded as the natural taste and that of raw milk as the "off taste."

Pasteurization was first accomplished by heating milk to 61.7°C for 30 minutes and then cooling and holding the milk at approximately 10°C. This treatment did not coagulate the milk proteins, and changes in taste, smell, and consistency were minimal. This is called the **holding method** and was first proposed for use in the United States in 1899 by Theobald Smith.

Today, **flash pasteurization** is used because the changes produced in the milk are less pronounced and, consequently, less objectionable. High temperature-short time pasteurization requires heating to 72°C for 15 to 17 seconds, followed by rapid cooling to 10°C and holding at this temperature until used.

The method called **Super Pasteurization** is in use in many places today. It requires a temperature of 82°C applied for 3 seconds, with subsequent cooling to 10°C. Pasteurization destroys all pathogens in the milk and reduces the microbial load, but it does not kill all the organisms therein.

Pasteurization of milk originally was designed as a method of heating milk to kill the organism that causes tuberculosis in human beings, *Mycobacterium tuberculosis*. Since the organism that causes Q fever, *Coxiella burnetii*, is more resistant to heat than is *M. tuberculosis*, it now has replaced the latter as the target organism. Since the objective of pasteurization is to destroy a specific organism, the characteristics of the medium determine, to a large extent, the temperature required to accomplish the objective.

The temperatures required to pasteurize buttermilk (pH 4.5) are much lower than those required for milk, as *C. burnetii* is killed easier in an acid medium than in neutral media. On the other hand, cream, yogurt, and some cheeses require exposure to different temperatures for different periods of

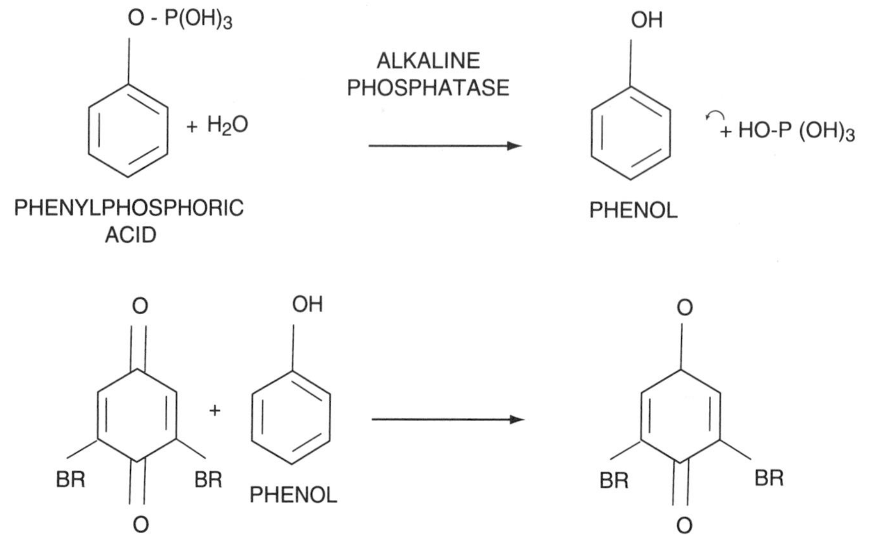

Figure 13.9

Alkaline phosphatase test.

time because of the different degrees of acidity and other physical characteristics of these products.

Pasteurization is such an important aspect of the dairy industry in making
milk safe for human consumption that tests had to be developed to confirm
the fact that it had been, indeed, properly carried out. Culture methods used
to test for the presence of *Escherichia coli* and coliforms are of little value, as the
entire test procedure may take as long as 48 hours and cultures to detect the
presence of *M. tuberculosis* take as long as 6 to 8 weeks.

Alkaline phosphatase is an enzyme present in all milk, regardless of its
source. This enzyme is destroyed when milk is heated to pasteurization temperatures. Since the presence of the enzyme can be detected by a simple col-
orimetric test, it is possible to state whether a given milk sample was or was not
pasteurized by performing the color test. Other tests based on the heat inac-
tivation of lactoperoxidase and *gamma*-glutamyl transpeptidase have been proposed. Also, the presence of alkaline phosphatase may be detected by the use
of Fluorophos, a proprietary substance that yields a fluorescent product when
it is acted upon by the enzymes.

If the enzyme is present in the milk, the reaction shown in Figure 13.9 will
proceed, terminating in the formation of **indophenol**. This will be evident by
the appearance of a blue color in the reaction tube. If, on the other hand,
the milk was pasteurized adequately, the reaction will not proceed as shown in
the figure and the tube will remain colorless.

DISEASES TRANSMITTED BY MILK

Many diseases are transmitted by milk. Some of these are diseases of the milk animal and pass to human beings when the microorganisms that cause them are conveyed in the milk. In others the microorganisms are transmitted from one human being to another using the milk as the vehicle of transmission. Milk is also commonly contaminated by insects and vermin such as mice and rats so that the organisms these animals harbor may enter the milk and thereby be transmitted to human beings.

Zoonosis

Animal diseases transmissible to human beings are called **zoonotic diseases**. In some cases, e.g., tuberculosis, the pathogenic microorganisms cause essentially the same disease while in others, e.g., undulant fever, the infection produces completely different results.

Since milk is an excellent growth medium for many pathogenic microorganisms, even a few organisms introduced into unprotected milk may result in large populations when conditions are appropriate. Many pathogenic organisms are eliminated from milk products by the production of acids and other fermentation products, especially when stored at room temperature. To the contrary, survival of pathogens is extended at lower temperatures and is at maximum in ice cream as long as it remains frozen.

Tuberculosis

At the turn of the century, stomach tuberculosis was a common illness of human beings, with a predominance among children who lived on farms. The bacterium *Mycobacterium tuberculosis variant bovis* causes tuberculosis in cattle and various other animals, including human beings. In the active phase of the disease in dairy cattle, the bacteria travel via the bloodstream to the animal's udder and there infect the glands that produce milk. Shed into the milk in large numbers, these bacteria remain viable for long periods in unpasteurized milk.

When consumed by human beings, the bacteria become enshrouded in the mucus produced by the cells of the stomach and escape the action of the digestive processes. They invade the cells of the small intestine, then the cells of the mesenteric lymph nodes, and finally produce lesions on the exterior tissues of the stomach wall. When these lesions break through to the interior of the stomach, a tubercular ulcer develops.

This form of tuberculosis occurred early in childhood as a result of drinking the unpasteurized milk of tuberculous family cows. It was not uncommon for two or more generations of children to be infected by the same animal during its milk-producing life. Butter, cheese, and all other foods made from the contaminated milk also served as vehicles for transmitting tuberculosis both within the family and to other members of the community.

Three accomplishments in the science of bacteriology led to the eradication of tuberculosis from dairy cattle. Although tuberculosis had been recognized as a disease transmissible among human beings since biblical times, not until the late 19th century was it recognized as a zoonotic infection. The relationship between milk and stomach tuberculosis was not established until the early part of the 20th century.

First Accomplishment   The fact that tuberculosis was caused by a living microorganism was discovered in the laboratory of Robert Koch in 1882. From his work it also was established that tuberculosis could not be caused by means other than invasion of the body by the etiologic agent. In the past it had been assumed that tuberculosis was caused by demons, by **miasmas**, by malnutrition, and by many other agencies that, in reality, had no relation to the disease. The significance of Koch's discovery of the causative agent of tuberculosis was that the disease could be prevented by avoiding the causative organism.

Second Accomplishment   The finding that heating milk, pasteurization, destroyed *M. tuberculosis* made it possible to break the chain of infection that had always existed. To render milk safe from tuberculosis, it was heated to 63°C for 30 minutes and then protected from recontamination. A law requiring that all milk be pasteurized was introduced in New York state in 1910. This resulted in an immediate decrease in the rate of stomach tuberculosis in urban areas, but the same was not attained in rural settings until much later. The continued use of the family cow served as a source of human tuberculosis until the 1940s and it was only the enactment of stricter laws that finally led to eradication of stomach tuberculosis in the United States.

Third Accomplishment   In 1891 Robert Koch found that the presence of tuberculosis in human beings could be detected by injecting a small amount of a specially prepared extract from the culture medium where *M. tuberculosis* had grown. The injection was made just under the outer layer of the skin, **intradermally**, and the presence of active infection was noted by a reddened, swollen area. The active material was called **tuberculin** (now purified protein derivative) and it could be used in cattle as well as in human beings. This made it possible to test not only human beings but also dairy herds and family cows for the disease.

A law was enacted making it unlawful to use milk for human consumption from cows that tested tuberculin-positive or those that had not been tested. The combination of education, tuberculin testing, and milk pasteurization made zoonotic tuberculosis a terror of the past.

Death Rate   There is a striking contrast in the incidence of stomach tuber-culosis between populations where pasteurization is practiced and those where raw milk is consumed. Six cases of milk-transmitted tuberculosis were reported in the United States in the period from 1950 to 1960. In an equal time period previous to pasteurization, the number of deaths would have been in the thousands. The death rate from all types of tuberculosis prior to pas-teurization and tuberculin testing in the United States was approximately 400/100,000; now it is approximately 2/100,000 while the rate from stomach tuberculosis is near zero.

Undulant Fever

Brucella abortus and *Brucella melitensis* cause an infectious disease transmissi-ble to human beings by cattle, sheep, and goats. *Brucella abortus* causes **infec-tious abortion** in cattle, and *B. melitensis* causes a systemic, and often chronic, infection in sheep and goats. The bacteria invade the bloodstream of these animals and emerge in their milk.

When human beings ingest the raw milk or milk products, the bacteria grow in the intestinal tract and invade the cells of the intestinal wall. From there, they enter the lymphatic system and then the blood circulation, which carries them to all parts of the body. The result is a generalized infection.

The first symptoms appear after an incubation period of 15 to 30 days and are characterized by malaise and low fever with characteristic "spiking" tem-peratures. The bacteria are Gram negative and release endotoxins when killed by the phagocytic cells of the body. Destruction of large numbers of **phago-cytic cells** results in release of large amounts of endotoxin from bacteria which had been ingested by the phagocytic cells. The consequence of this killing off is a temporary relief from the infection as the number of bacterial cells is decreased dramatically. The endotoxin, however, causes a severe rise in tem-perature (spiking). When the temperature falls again in 2 to 3 days, the bac-teria grow again and the cycle is repeated. These episodes are repeated at regular intervals and are characteristic of **brucellosis**.

As a result of the undulating characteristic of the fever, the disease is called **undulant fever**. Other names used to describe the infection are **Malta fever**, **Mediterranean fever**, and **milk fever**. The etiologic agent of undulant fever was discovered in 1887 by David Bruce (1855–1931), a British army bac-teriologist stationed in Malta.

Although brucellosis was a serious problem in the United States in the past, it was brought under control by pasteurization of milk and milk prod-ucts. Brucellosis was reduced further as an infection in the U.S. population by introduction of the Bang's test in 1931. This is a blood test that reveals the presence of *B. abortus* antibodies in the serum of cattle or human beings. It now is used universally in dairy herds as the result of a law requiring testing of all animals used for the production of milk for human consumption. The

Figure 13.10

Incidence of undulant fever in the United States, 1945–1985.

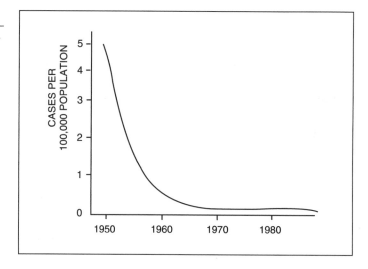

infection also can be acquired by ingesting meat that is not cooked thoroughly and even from butchering or handling infected animals. There were 6,200 cases of undulant fever in the United States in 1946, fewer than 1,000 in 1957, and 200 in 1983. This trend is shown in Figure 13.10. Of the 200 cases of brucellosis diagnosed in the United States in 1983, 84 occurred in Texas while 22 states had none. More than 95% of the 200 brucella infections were found in **abattoir** workers and animal handlers while the remainder were caused by drinking nonpasteurized milk or eating unpasteurized dairy products.

Q Fever

Q fever was first recognized as a human disease by E. H. Derrick in 1935 in abattoir workers in Australia. Q signifies "query" because the symptoms of the infection are diffuse, ambiguous, and do not represent a recognizable, well-defined syndrome. Symptoms first appear 3 to 4 weeks after exposure to the infectious agent and vary from mild to severe. Many cases are asymptomatic, but others include headache, myalgia, chest pains, chills, and malaise. When symptoms are intense, death may occur but fatal infections are rare.

Q fever is caused by *Coxiella burnetii*, a member of the group of bacteria called **Rickettsia**. These organisms cause mild disease in many animals including cattle and goats. The Rickettsia are present in all the tissues, feces, urine, saliva, and milk of infected animals. Human beings become infected by contact with infected animals, and milk seems to be an excellent vehicle for transmitting the Rickettsia.

Coxiella are highly resistant to extreme temperatures, desiccation, and chemical agents. As a result of this resistance, they may be found in dairy products that are free of all other pathogens. They survived pasteurization as it was specified in the laws of 1910 and are difficult to kill with disinfectants. For pasteurization to be effective in preventing milk-borne Q fever, the temperature used now must be 79°C for 20 seconds with subsequent cooling to 9°C. These temperatures now are employed in the United States, and the incidence of Q fever has decreased dramatically. Only 18 cases of the infection were reported in the United States in 1983, and no fatalities.

Staphylococcal Infections

Staphylococcus aureus causes many diseases in animals, including mastitis in cattle, goats, and human beings. The most common problem associated with mastitic cows resides in the fact that the contaminated milk can cause several different diseases in human beings. These include mastitis, septic sore throat, pneumonia, septicemia, impetigo, meningitis, scalded skin syndrome, carbuncle, and several other skin infections.

When cows have frank mastitis, the infection is obvious, as the udder is inflamed, painful, and may even shed blood and pus. Under these conditions the milk from infected animals should be avoided and not put to human consumption. On the other hand, when the symptoms are absent but the infection present, milk carries vast populations of undetected staphylococcus that continue to grow after the milk is harvested. Drinking milk with large population of *S. aureus* causes staphylococcal food intoxication when there is sufficient time for toxin formation in the milk. Milk also serves as an ideal source of infection for workers who handle it or who handle products that contain the unpasteurized milk.

ECONOMICS

An economic hazard also is involved, because adding the milk of one asymptomatic infected cow to the milk produced by an entire herd of healthy animals will taint the entire milk harvest. Pasteurization will destroy the bacteria as it does those that cause tuberculosis, brucellosis, and Q fever, but it does not destroy the staphylococcal enterotoxin. In the past, milk containing staphylococci often was the source of diseases so different one from the other that connecting them to a common cause was difficult or impossible. In most cases finding the source of the contamination also was impossible.

Today, modern techniques for culturing staphylococcus, bacteriophage typing, and daily record keeping allows epidemiologists not only to find the milk causing the infections but also to find the specific animal producing the tainted milk. Staphylococcus infections go freely from human being to animals and back again, making eradication of the organism from dairy herds difficult.

NONZOONOTIC ILLNESSES

Milk has the ability to harbor infectious agents of all kinds. If it is not protected from contamination, it becomes a vehicle for the transmission of many diseases. Contamination of milk by the fecal matter of human beings or animals has resulted in epidemics of typhoid fever, dysentery, cholera, poliomyelitis, hepatitis, echo viruses, and amebic dysentery. Contamination by saliva and nasopharyngeal mucus may result in the transmission of diphtheria, pneumonia, meningitis, staphylococcal and streptococcal infections, and many viral diseases. Contamination of milk with soil may result in fungal diseases, anthrax, **perfringens food poisoning**, and *Bacillus cereus* food poisoning.

RELATED FOODS

All puddings, creams, mousses, custards, cream pies, cream sauces, cake fillings, malted milks, milk shakes, and ice cream are potential sources of infectious microorganisms because these foods contain large amounts of milk. The same cautions and alarums that apply to fresh or cultured milk apply to these. Each of these food items has found its place in the epidemiological literature as the vehicle by which some epidemic was started in the human population.

The history of **Typhoid Mary** (Mary Mallon) would suffice to prove the ease with which milk products in the kitchen lend themselves to the transmission of diseases. Mary made a specialty of cream desserts and ice creams, which she served to admiring employers who then died of typhoid fever. Typhoid Mary (profiled earlier in the book) left a trail of more than 5,300 dead in the 15 years she worked in the New York City area during the first quarter of the twentieth century.

QUESTIONS TO ANSWER

1. How is milk maintained sterile in the mammary glands of animals?
2. When does milk become contaminated?
3. Why do bacteria grow in milk?
4. Why does milk spoil when microorganisms in it become numerous?
5. How does the dye reduction test work?
6. What is a tenfold dilution series?
7. What is the calculation for direct microscopic count of bacteria by the Breed slide method?
8. What is the calculation for total plate count?
9. What is the basis for grading milk?
10. How is cottage cheese made?

11. How is cheddar cheese made?
12. What makes the holes in Swiss cheese?
13. Why is Camembert cheese creamy and Parmesan hard?
14. How is milk involved with tuberculosis?
15. What is undulant fever?
16. How can cows get staphylococcal mastitis from human beings?
17. What was the effect of pasteurization on the incidence of tuberculosis in human beings?
18. What was the effect of pasteurization on the incidence of undulant fever in human beings?
19. What other diseases may be transmitted by milk?
20. What diseases may be transmitted by cheese?

FURTHER READING

Banwart, G. J. 1979. BASIC FOOD MICROBIOLOGY. AVI Publishing Company, Inc., Westport, Connecticut.

Brock, T. D. and M. T. Madigan. 1988. BIOLOGY OF MICROORGANISMS. 5th ed. Prentice Hall, Englewood Cliffs, New Jersey.

Hays, P. R. 1985. FOOD MICROBIOLOGY AND HYGIENE. Elsevier Applied Science Publishers, New York.

Jay, J. M. 1978. MODERN FOOD MICROBIOLOGY. 2nd ed. D. Van Nostrand Com-pany, New York.

Riemann, H. and F. L. Bryan. 1979. FOOD-BORNE INFECTIONS AND INTOXICATIONS. 2nd. ed. Academic Press, New York.

Microbiologic Quality of Food

14

I wonder if the cabbage knows
He is less lovely than the rose;
Or does he squat in smug content,
A source of noble nourishment;
Or if he pities for her sins;
The iris who has no vitamins;
Thinker, philosopher, wise, eh wot!
But, does he know how soon he'll rot?

Anonymous

The process of spoilage of all foods begins either at the time they are harvested or slightly before harvest. Microorganisms present on or in all food substances have little effect on these as long as the processes of life are active, but once harvested, microbial action accelerates and diversifies. Many fungi and bacteria produce enzymes that degrade the structure and texture of all vegetal and animal tissues. When this process is taken to the point that the color, odor, taste, and appearance of the food changes noticeably, human beings judge it to be spoiled and inedible.

Fruits and vegetables, as well as all other plant materials, harbor large populations of microorganisms on surfaces exposed to soil and water, but these are kept at comparatively low levels by the plant's defense mechanisms while it is alive. When the plant dies, or when part of it is harvested, chemicals that inhibit microbial growth—such as organic acids, phenolic acids, alkaloids, resins, and waxes—are no longer produced and the microorganisms begin to proliferate rapidly. Bacteria and fungi also invade those parts of the plant that had been sterile during its life. Microbial populations grow by using the substances of the plant for nutrition, and eventually the entire plant tissue is converted into microbial cells and waste products of microbial metabolism.

The largest part of the flesh of animals is generally free of bacteria, fungi, and other microorganisms during life, but when the animal is butchered, the defense mechanisms are destroyed and large microbial populations develop rapidly. To the contrary, if viruses are present in the animal's tissues, reproduction stops when the animal dies, because the cells cease to function. In the absence of the body's defense mechanisms, microbial growth on animal tissues occurs at maximal rates, and spoilage begins in a matter of hours. The microbiological processes involved in food spoilage make an interesting chapter in microbiology and have been studied extensively.

SPOILAGE OF BEEF

Bacteria begin the first metabolic actions that eventually lead to spoilage. These pioneer organisms are the ones normally present in the body of the animal itself. They generally are found in the **lymph nodes** and surrounding tissues during life, when their growth and metabolic activities are kept under control by the actions of the animal's white blood cells. At death, when the controlling activities of the white blood cells fail, the microorganisms break out of the tissues they inhabit and begin to reproduce rapidly. Other microorganisms are introduced into the flesh (meat) by the hands of abattoir workers and the utensils they employ. By the time it reaches the market, fresh meat carries large populations of microorganisms (see Figures 14.1 and 14.2).

Figure 14.1

Fresh meats carrying large populations of microorganisms.

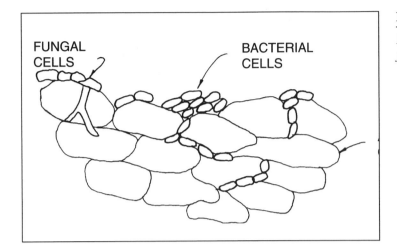

Figure 14.2

Presence of bacterial and fungal cells in fresh meat.

The Mechanism of Spoilage

Bacteria and fungi begin to grow by using substances from the flesh that are dissolved in the **serous fluids** of the tissue. As these pioneer microorganisms grow, they excrete waste products that may be used by organisms that were incapable of attacking the unchanged tissue. This second wave of organisms utilizes the waste products secreted by the first group and in turn secrete waste products of their own. All the microorganisms growing at this time also metabolize parts of the animal flesh that were changed or modified by the action of the first set of waste products, making these available to other organisms present but not yet growing on the flesh. The activities of these organisms allow a third wave to begin growing and the cycle to expand.

In addition to this, the pioneer organisms, generally bacteria, secrete enzymes that cause secondary changes in the structure and chemical composition of the flesh. This may allow other organisms to utilize the end products of enzyme degradation, and these in turn will secrete other enzymes and thus begin a different kind of cyclic growth and degradation. The effect is an ever expanding cascade of growth of new organisms until most of those present become involved.

This process continues as long as substance remains to serve as growth substrate for bacteria and fungi. As the bacteria and fungi die, they too become part of the growth substrate for other generations of microorganisms until finally all the substance of the meat, and the microorganisms that caused its degradation, have been consumed. The final result is the **mineralization** (conversion of organic substances into inorganic substances) of meat and microorganisms as well. In the end, only the minerals of the meat are left and, theoretically, the one organism that consumed the last bit of growth substrate. Theoretically also, this organism should die of starvation.

Rate of Spoilage

As succeeding waves of new organisms begin to grow, the process accelerates as the earlier populations become larger and larger. Most bacteria and fungi grow faster at **mesophilic temperatures**, causing spoilage to proceed more rapidly during the summer months, with maximal activity in the hotter parts of July and August. Spoilage is slower during the cooler parts of the year and stops altogether when the meat is frozen.

Every kind of living matter eventually decays, and all its constituents will be decomposed by microbial metabolism. If the largest ox or the smallest microorganism dies in the desert, all the different chemicals that make up the organism eventually degenerate to the mineral state. Within a few months, or at most a year, all the materials will be degraded by the metabolic activities of the microorganisms in the environment.

If the environment has sufficient moisture, even the bones of the ox will disappear since these are made of minerals held in an organic matrix; as the organic matrix is degraded by microbial action, the minerals are lost to the environment, and in a given period of time no trace whatever will be left of the ox or of the microorganisms that consumed it. All the matter that once made up the ox and the bacterium will be returned to their simplest form, minerals. These minerals mix with those of the soil only to be again converted to living matter by the metabolic activities of other living things at some future time. This is the nature of the process by which all matter on the Earth is cycled many times to support countless generations of living things. This recycling is what makes life on Earth **infinite** while the quantities of materials that living things are made from are **finite**.

Criteria for Spoilage

The guidelines most often used to describe meat quality in terms of microbial growth are subjective and may be interpreted according to local law or even local custom. In prerefrigeration times most meat was eaten past the point that we, today, would call spoiled since fresh meat was available only on the day the animal was slaughtered. Using spices, herbs, and vegetables to hide off-tastes was common, and many recipes still in use today were developed for this purpose alone.

The designation of "spoiled" or "fresh" also may vary according to the market criteria of supply and demand. The demands made of the supplier of fresh meat are more rigorous when prices are high and supplies are plentiful than when prices are low and supplies are scarce. The standards in use today in many countries are more exigent than they have been at any other time in history.

Most of the meat consumed in modern, industrial countries today is examined and graded according to established criteria of quality. The six points generally employed in grading beef are:

Table 14.1 Correlation Between Total Plate Count and
Opinion of Organoleptic Panel in Grading Meat

Days	Total Count per Gram	Opinion of Panel
0	1.2×10^4	Fresh, good
1	1.9×10^4	Fresh, good
2	2.8×10^5	Mellow, good
4	9.0×10^5	Tired, OK
8	1.0×10^8	Slimy, ugh
16	3.3×10^9	Smell, can't eat
32	9.7×10^9	Bad, won't eat

1. Odor,
2. Color,
3. Palpable growth of slime on surface,
4. Color of cut bone,
5. Presence of more than 1×10^8 viable bacteria per square centimeter of surface, and
6. Overall, subjective appearance

Judgment

Since there are no objective criteria for such traits as odor, color, and palpability, those employed by necessity are subjective measures. In the best situations a panel of **organoleptic** experts may be assembled, and by majority vote these say whether the odor or the color is "off" or whether they are "right." In the same way they indicate whether the slime on the surface of the meat is palpable or not, i.e., normal or abnormal. The panel may consist of any arbitrary number, say 10 or 12, or it simply may be one person selected as a judge. In practice, the decision commonly is made by a judge or an **inspector**, depending on the situation and circumstances. Table 14.1 compares total plate count and opinions of an organoleptic panel in grading meat. Table 14.2

Table 14.2 Sum of Opinions By Organoleptic Panel of 10 Participants

	Day 0		Day 2		Day 4		Day 6	
	Fresh	*Spoiled*	*Fresh*	*Spoiled*	*Fresh*	*Spoiled*	*Fresh*	*Spoiled*
Odor	10	0	10	0	4	6	0	10
Color	10	0	8	2	3	7	1	9
Slime	9	1	8	2	5	5	2	8
Bone	8	2	7	3	4	6	0	10
Overall	10	0	8	2	3	7	2	8
SUM	47	3	41	9	19	31	5	45

Figure 14.3

Graphic presentation of bacterial counts of meat in Table 14.2 and organoleptic opinion.

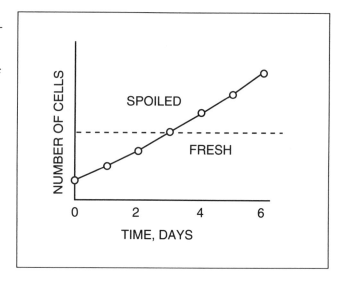

gives a summary of the opinions, and Figure 14.3 graphically presents these data.

OTHER MEATS

The term beef refers only to the muscle tissues of cattle while the animal contains many kinds of meat other than muscle. These include the organ meats such as kidney, pancreas, liver, heart, stomach, intestines, brain, and testicles. Any of these organs are popular in certain communities and may even be regarded as delicacies while they are considered inedible in others.

The standards for grading these meats are essentially the same as those used for grading muscle tissue, with the added precaution that most of these are much more vulnerable to microbial action than are the muscle tissues. The reason for this increased vulnerability lies in the higher water and carbohydrate content of organ meats in comparison to muscle meats. In addition, the action of microbial cells is more evident on some meats than it is in muscle meats.

Non-Beef Meats

The standards used for grading other kinds of meat may be different, but the criteria are basically the same. Odor, color, palpable slime, and bacterial content serve better than any other criteria for establishing the degree of freshness of any kind of meat. Pork, poultry, fish, and even specialty items such as squid, shrimp, and shellfish can be graded by the same guidelines.

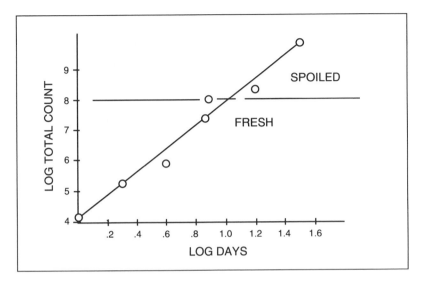

Figure 14.4

Bacterial counts of smoked snake steaks and organoleptic opinions.

Often the criteria must be made up when the question is presented. If no background of information exists, criteria may have to be established for the specific purpose of a particular test. In this case, a comparison generally is the best method by which to proceed. What is the microbiological condition of smoked snake steaks? To determine this, sufficient smoked snake steaks have to be acquired to run the following tests.

The total bacterial count of steaks purchased from Ace's Exotic Foods is determined using several growth media such as Nutrient Agar, Trypticase Soy Agar, and smoked snake steak agar. Since the steaks normally are kept in the refrigerator, it suffices to incubate the culture plates at room temperature under aerobic conditions. Several pieces from individual steaks are mascerated aseptically in 500 ml of sterile water, and a tenfold dilution series is made for plating on the media prepared. The amount of steak material per milliliter of mixture is determined by evaporating the water and weighing the residue in several 10 ml samples of the lower dilutions.

These tests were performed at 0, 1, 2, 4, 8, 16, and 32 days. The colony counts from the medium that gave the highest plate counts were used for constructing Figure 14.4. The quality of the smoked snake steaks sold at Ace's was determined by an organoleptic panel and compared to the plate counts as shown in Figure 14.4. On this basis the quality of these steaks now may be determined by the bacterial count.

Aging of Certain Meats

The quality of beef is improved by aging; the consumer considers its taste and texture more desirable. Almost all beef sold in the United States is aged.

Table 14.3   Storage Life of Various Meats

Meat	Storage Life (Weeks)
Beef	2–6
Baby beef	2–4
Pork	1–2
Organ meats	1–2
Poultry	1–2
Fish	1–3
Shellfish	1

This is accomplished by hanging the cut and prepared carcass at a temperature of 15°C for 18 to 24 hours. During this time the enzymes in the meat escape from the cells where they are found during the life of the animal and work on the tissues to bring about a certain amount of structural degradation.

This action often is enhanced by injecting the animal with meat-tenderizing enzymes just before it is killed so the enzymes are distributed throughout the body by the circulatory system before the animal dies. The enzyme most commonly employed for tenderizing meat is called **papain** and is obtained from the papaya. It has the virtue of causing sufficient changes in the chemical structure of the meat to make it soft while retaining sufficient body and strength to prevent mushiness.

In addition, a certain amount of growth of psychrophilic bacteria and fungi occur to bring about a small but significant amount of tissue degradation. The result of these combined effects improves the overall quality of the meat.

In other meat-aging processes, the meat is kept at temperatures between 10°C and 25°C for extended periods, but microbial populations are kept as low as possible by using ultraviolet radiation, antibiotics, or chemical preservatives. The objective of the hanging and aging process is to allow the blood to drain out, as this gives the meat a better color and to obtain the maximal amount of tenderizing while avoiding the build-up of slime. In addition to this, a certain amount of mellowing of harsh tastes and flavors results from aging.

When the meat has aged to the desired condition, it is moved to the storage cooler and kept there as long as 6 weeks at a temperature of 2 to 4°C. Other animal meats do not last as long in the fresh condition as beef does. The storage life of some meats is given in Table 14.3.

Special Aging

Another type of aging is accomplished by allowing growth of selected psychrophilic bacteria and fungi. In this process an ordinary piece of meat is converted into a choice cut costing several times as much as the original meat. The meat is cut, trimmed, and aged in special aging lockers where the selected organisms are found. The bacteria and fungi in these lockers reside in sawdust on the floor, adhere to the walls of the locker, and in the air.

The process involves extensive degradation of the meat and accumulation of metabolic waste products that enhance the flavor of the meat. Fungi grow into the tissue itself, and the combination of fungi and beef produce a flavor far superior to that of beef alone. Beef steaks so treated are called "**furry steaks**" because of the fungal mycelia that cover them during the aging process. When the meat is cleaned and prepared for cooking, no visible trace of the microorganisms remains, but fungal mycelia and bacteria permeate the structure of the meat. The growth of bacteria and fungi improve the texture, taste, and appearance of the cooked steak. Aging lockers are in effect, special environments where special bacteria and fungi are found but where spoilage bacteria are either absent or in very small populations.

Game Meats

Game is almost always allowed to age to attain some mellowing of the harsh or strong flavors characteristic of these meats. Aging is more common in Europe than in America and more a practice of the past than of the present. The hanging of game birds in full plumage for several days, or weeks during cold weather, was considered at one time the only way in which these choice animals could be eaten. Boar, elk, deer, mountain goat, and other game also were hung until changes in color and texture became evident. In some societies hanging to the point where insect larvae, maggots, appeared was considered the optimal point of aging and the maggots became part of the subsequent feast under the euphemism "field rice."

Consumption of Spoiled Meat

In earlier times the science, or perhaps art, of preserving meat was considerably limited. Prior to the easy availability of the home refrigerator, meat was eaten in fresh condition only on the day when an animal was killed or, in cold climates, for a few days afterward. The remainder of the animal was preserved by drying, smoking, salting, or freezing where this was possible. It was not unusual for part of the meat to be eaten even after it was in the early, and perhaps even late, stages of spoilage.

Such meat required special treatment to remove offensive tastes and odors. Often this meant soaking or boiling the meat with aromatic herbs, ashes from the fireplace, or baking soda. The use of salt, sugar, spices, beer, wine, and fruits often was intended to mask the offensive odors and tastes of spoiling meat. Some of the recipes developed to make unpalatable meat acceptable were so successful that they are used today even though the problem is now for the most part a thing of the past. Salt, sugar, and some common spices are in everyday use by almost everyone. Spices, herbs, and wine now are used to enhance the flavor of fresh meats rather than to hide the flavor of spoiled ones.

PRESERVATION

Because the quality of foods is very much a function of microbial load, anything that can limit or diminish microbial growth will be considered a preservative. Meats can be preserved by many methods that were worked out empirically long before the presence of microorganisms was known. Meat also may be preserved by modern methods designed specifically to capitalize on some special vulnerability of specific organisms. Although meat can be preserved through many means, all of them can be placed into one of three categories: (a) cold, (b) chemical, and (c) exclusion.

In modern, industrialized countries, refrigeration and freezing are the methods of choice. Meats are generally sold fresh or in the frozen state and stored in the refrigerator or freezer in the home. Cold cuts, sausages, hams, and other meats are normally preserved by addition of chemicals such as the acids that result from natural fermentation. Meats preserved in cans or jars designed to exclude microorganisms are more expensive than those that are frozen or refrigerated and, consequently, tend to be more common in industrialized societies. The consumption of freshly killed meats is more common in emerging nations.

Effect of Temperature

The effect of refrigeration is to slow the metabolic activities of microorganisms to a minimum, thereby retarding the spoilage process. As noted previously, microbial action continues at refrigeration temperatures although it is limited to that of psychrophilic organisms and minimal activity on the part of mesophilic bacteria and fungi. The data in Table 14.4 show that refrigeration is far inferior to freezing as a means of preserving meat but that it is a useful method of slowing down microbial growth.

Many microorganisms are destroyed by exposure to the temperatures generally employed in refrigeration, i.e., approximately 2°C to 6°C. The vast majority of organisms survive such exposure but may be totally incapable of carrying out metabolic activities that are harmful to meat. Others do so but at greatly reduced rates, while still others, the psychrophilic bacteria and fungi, proliferate at high speed. In general, psychrophiles have optimal growth temperatures in the area of refrigeration temperatures although, as a general rule, their rates of growth are somewhat lower than those of the

Table 14.4 Comparative Storage Life of Different Foods

	Storage Life	
Meat	*Days* *4°C*	*Months* *−20°C*
Beef steak	3–5	10–12
Ground beef	1–5	3–6
Pork	3–10	2–3
Poultry	2–3	6–12
Sausage	10–30	1–4

mesophilic organisms. Spoilage of refrigerated meat results from the growth and metabolic activities of psychrophiles and mesophiles combined.

Many variables affect the keeping qualities of meat. The manner in which the animal is killed has much to do with keeping quality. The flesh of animals killed with knife and spear with much tearing of organs and tissues undergoes heavy contamination and, consequently, rapid spoilage. This is accentuated when killing takes place in the field and refrigeration is not readily available. Animals killed in an abattoir represent the other extreme.

Meat allowed to bleed after the animal is sacrificed lasts longer because blood sugar supports the growth of microorganisms. When blood is removed from the carcass immediately after death, growth is retarded significantly. Meat dressed under clean, hygienic conditions lasts longer because the number of microorganisms introduced into the meat before refrigeration is small. The total number of organisms present at any given time is a function of the original number. It follows that the smaller the number introduced originally, the smaller the number at any subsequent time.

The time required to cool meat from that of the animal's body temperature at the time of slaughter to the time when the low temperature is attained determines the amount of microbial growth that will take place. This time is kept at a minimum with rapid chilling equipment under commercial conditions. Also, it lasts longer if it is wrapped during chilling and subsequent storage.

Freezing

Although meats have been preserved by freezing from time immemorial using climatic conditions, this can be done only when the temperature of the ice is considerably lower than 0°C. Freezing cannot be accomplished with ice in an ice box, as the temperature of ice made from distilled water is only 0°C whereas most meats have freezing points lower than 0°C. Common household and commercial freezers operate at approximately 20°C below the freezing point of water (−20°C), a temperature more than adequate for freezing meat.

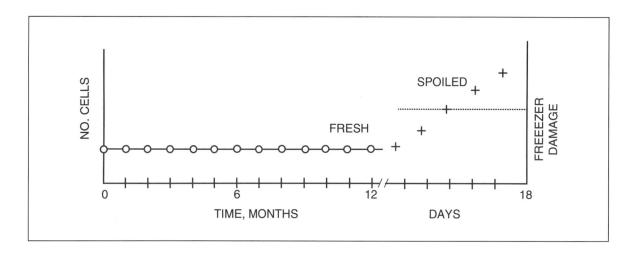

Figure 14.5

Preservation of meat in the freezer and subsequent removal to room temperature.

Meats that are frozen may be stored for prolonged periods, as indicated by the data in Figure 14.5 and Table 14.4. Since the water in frozen meat is in the solid state, all microbial activity ceases and the processes of spoilage are arrested. Although microbial degradation does not occur in frozen meats, the freezer life of all meats is finite, as Table 14.4 shows. Foods that are not packaged properly will dry out or will suffer "freezer burns." Freezer burns are areas of the meat that become dry and lose their texture and appearance. Meats that are not prepared properly for freezing will undergo tissue and cell breakdown. This results in the release of water from ruptured cells when the meat is thawed, leaving a mushy, discolored, unappetizing product. On the other hand, properly prepared meats can be recovered in excellent condition even after extended periods of time.

Freezing makes it possible to transport meat over long distances and to stockpile it in vast quantities. Institutions that use large amounts of food, such as the armed forces, schools, prisons, and restaurant chains, generally purchase frozen meat and meat products. An old, but good, joke to tell armed forces recruits is that the meat they are consuming for their supper may be older than they are. In some instances during World War II, this may have been quite true.

Exclusion of Microorganisms

A great variety of meat products are preserved by canning. In this method the meat is placed in a can or jar and then sterilized. Preservation then depends on exclusion of the microorganisms that might cause spoilage. It is well established that the meat remains edible and safe as long as the integrity of the container is maintained. In this method of preservation, there is no residual enzyme activity since all the enzymes in the meat are destroyed by the heat employed in sterilizing it. Although this is the ultimate means of preserving food, it has several major drawbacks, including the following:

1. Loss of flavor. Heating causes drastic changes in taste and odor.
2. Loss of texture. Heating changes structure and texture of meat turning it into a mushy mass.
3. Loss of color. Heating destroys color; meat becomes gray and nondescript in color.
4. Loss of nutrients. Some amino acids, vitamins, and other substances are destroyed by heat.

Many stories and popular writings attest to the preservation of canned foods for 50 years or longer. Canned meats can be shipped long distances and stored for many years without need of refrigeration as is the case with Spam. Even though the food may be unattractive and unappetizing, it will be safe and, with certain limitations, it also will be nutritious.

Spices As Preservatives

It is easy to understand the necessity of eating meat past the point of freshness in prerefrigeration times and also to understand that by developing methods that would improve the quality of the meat, a virtue was made of necessity. In another approach to the problem of rapid spoilage, spices were used not so much as flavoring agents but rather as preservatives. Herbs and spices were applied to meat while it was still fresh rather than during preparation for the table and were valued for their preservative properties rather than taste enhancement.

Spices with strong, pungent qualities were thought to prevent spoilage better than those with softer and more subtle tastes and aromas. This notion has been proven correct by microbiological studies on the antibiotic properties of salt, black pepper, clove, cinnamon, and nutmeg. The present-day flavoring of foods in southern regions as contrasted with the bland, unseasoned foods of northern lands attests to the previous dependence on spices as preservatives in warm climates. Although the home refrigerator has made spices unnecessary for food preservation, clever chefs and brilliant housewives have turned that previous need into culinary virtue.

Dried Meats

Many foods were developed for their keeping qualities, and many of these are still in use although the nature of the spoilage problem is completely different now. Early in their history all communities, regardless of their origin or location, developed ways of drying meat to keep it from spoiling. Generally, this involves cutting the meat into thin strips and laying it in the sun to dry.

Figure 14.6

List of ingredients in beef jerky.

Drying also may be accomplished by exposure to fire such as smoking or by the use of chemicals such as salt or sugar.

Once dried, meat may be ground into a fine powder for easy storage and transport. Native Americans mixed dry powdered meat with other comestibles such as nuts, berries, and cereals to make a high-calorie food that was easily transportable and had a long storage life. In some Indian communities it was called **pemmican** and was a valuable bartering item in frontier United States during the 1700s, 1800s, and even early 1900s. In the Southwest, where the temperature is high and the humidity is very low, the thin strips of meat can be stored in that form. A modern-day dried meat product, beef jerky, has become the cult food of country-western fanciers and long-distance truck drivers (Figure 14.6).

Sausage

The manufacture of sausages has two separate virtues: The first is that all parts of the animal that the community may not wish to consume in their original form may be ground into unrecognizable form and mixed together to obtain an unrecognizable composite. The second virtue resides in the fact that the sausage material may be fermented with the production of acids, which serve as preservatives, giving the sausage a fairly long shelf life. Often the sausage is spiced heavily to disguise both the flavor of the meats used and that of the fermentation acids.

Chemical and Physical Methods of Preservation

Many chemical and physical methods by which meat may be preserved for varying periods are available in modern communities. Meat may be "cured," or treated with chemicals that retard microbial growth. These chemicals may be salt, sugar, the acids and aldehydes of wood smoke, and the fermentation acids produced by microorganisms. If cured meat is protected by a container or wrapper, it lasts longer than meat which has not been so treated.

Non-Beef Meats

Poultry   Poultry remains edible for only a short time unless it is rapidly chilled or frozen. Cleaned, cut, and trimmed poultry must be cooled as rapidly as possible to avoid microbial growth. This is accomplished by

---

BEWARE! DANGER! POISON!

Each year between 2 million and 80 million individuals are infected with *Salmonella*, *Campylobacter*, and other pathogenic bacteria found in the feces of diseased chickens and 1,000 to 2,000 die from these infections. The source of this contagion is not some third world black market run by some chicken cartel. It is your friendly neighborhood supermarket. Most of the chickens (up to 65%) sold to the public in the 'hood market contain the etiologic agents of classic enteric disorders. But how can this be? The United States bristles with Nobel Prize microbiologists, super technology, miracle medicine, and all the knowledge available in the world. The answer to this chicken problem is neither easy nor simple but seems to be deeply ingrained into the fiber of the industry.

Chickens are eviscerated by machines which spill intestinal contents onto the edible parts of the carcass. The carcasses are next placed in a vat of cold water to reduce the temperature of the carcass and to wash off microbial contamination. Unfortunately, this "washing" actually results in bathing the animal in water heavily contaminated with fecal matter including that from diseased animals.

In 1993, some seven billion chickens were slaughtered and less than 10% of these were examined by even the most superficial method. As a result, 35% to 65% of fowl flesh sold to the consumer in the United States was contaminated with *Salmonella*, *Campylobacter*, *Listeria*, *Staphylococcus* or other pathogenic organisms.

Several proposals are being considered to eliminate this source of infection. They all depend on microbiological tests designed to detect fecal contamination. The use of chemical disinfectants to reduce the number of bacteria and eliminate pathogens is a major part of all the plans being discussed.

For the time being, it is prudent to treat all poultry as if it were contaminated. Any fowl meat brought into the house is best handled with disposable gloves, washed in the sink carefully so that water does not splash onto other foods or work areas, and, as quickly as possible, placed in the microwave oven or cook pot until lightly but thoroughly cooked. After this, the chicken may be safely cut, trimmed, and prepared in the desired fashion.

Bon Appetit!

immersing the prepared animal in clean water at approximately 4°C (water kept in the refrigerator overnight) or, better, in ice water. After the carcass, or parts, have been chilled, they must be dried thoroughly, then placed in plastic bags to prevent desiccation.

Prepared in this manner and kept in the refrigerator, poultry lasts 2 to 5 days in a fresh condition. Storage in the freezer at –20°C preserves the meat for 6 to 12 months without signs of spoilage. Cooked poultry products such as TV dinners and fried chicken stored at –20°C last approximately one year before taste and odor deteriorate markedly.

Fish Fish, shellfish, and marine animal meats have shorter freezer lives than beef or chicken. On the other hand, because these foods have strong, penetrating smells not unlike those of spoiling matter, they often can be comfortably consumed even after they are well into the spoilage phase. Though spoiling seafoods possibly, and quite likely, carry large populations of microorganisms, the bacteria and fungi that cause fish spoilage tend to be marine forms that lack disease-producing capacities. The finding of epidemics of disease induced by *Vibrio parahemolyticus*, *V. vulnificus*, and *V. fluvialis* in marine fish indicates to some experts that even the oceans are being polluted by the activities of human beings. Shellfish (oysters, clams, scallops, mussels) being filter feeders collect and accumulate microorganisms from water when harvested and refrigerated, the pathogens are preserved and transmitted to humans when the shellfish are consumed raw.

Some of the pathogenic organisms associated with shellfish-borne infections are marine while the rest are brought to the marine environment by raw sewage. *Vibrio vulnificus*, *V. parahaemolyticus*, *V. cholerae*, *V. fluvialis*, and *V. mimicus* are marine organisms capable of causing human disease while *Salmonella typhi*, *Salmonella spp.*, *Campylobacter spp.*, *Plesiomonas spp.*, and hepatitis and Norwalk viruses are carried in domestic sewage. New York and Florida have the highest rate of shellfish-induced infection while other coastal states, Louisiana, Texas, Alabama, Massachusetts, Connecticut, and California, contribute significantly.

SPOILAGE OF PLANT FOODS

The variety of plant foods on Earth is so vast that human beings have not yet had time to explore them all. Modern societies seem to select a few plants for use as food and to integrate them into the very fiber of their society. Examples of this phenomenon are seen in the Irish and potatoes, Americans and corn, and many Asian peoples and rice. Selection probably is based on important criteria such as ease of cultivation, nutritional adequacy, and economics of crop yield. Although many plants have been moved from their original habi-

Figure 14.7

Stages in aging of these fruits, evidenced by differences in color.

tat, others have not. Potatoes, tomatoes, and pineapples are examples of plants transported from one continent to another.

All vegetables, cereals, fruits, berries, and nuts spoil with time. Plants are protected during their life by chemical substances that inhibit microbial growth. Protection extends after harvest as long as those chemical substances remain part of the plant material and in concentrations high enough to effectively inhibit microbial growth.

In addition, plants are protected by integuments, coats, hulls, shells, and outer coverings that are impervious to microbial action. All plant foods from tomatoes to pecans will spoil completely in one or two days if the outer, protective covering is damaged to the extent that microorganisms can reach the internal tissues.

Because plants have a sessile nature and are closely associated with the soil, they generally are covered with bacteria and fungi, and these organisms immediately begin the processes of decay if they enter unprotected plant tissues. If the outer covering is not damaged, plant foods can last in a healthy, edible condition for periods ranging from a few days to many months. For example, strawberries remain edible for only two to five days under the best of conditions, whereas some varieties of pears last three to five months if they are wrapped in tissue paper and kept in a cool dry place. Nuts last 10 to 12 months if kept cool and dry, while most cereal grains remain edible for approximately one year if stored in cool, dry, well-ventilated rooms. The coconut is said to stay fresh and edible for periods in excess of 30 years. Aging in fruit is pictured in Figure 14.7.

Figure 14.8

Enzyme activity causing bananas to age.

Unlike animals, which die when their flesh is rendered, plants often give one harvest of food after another without risk to their being. In the same way, animal flesh is dead when it is converted to food, but fruits, vegetables, cereals, and nuts remain alive and their cells functioning normally until cooked or ingested.

In many plant products spoilage need not be induced by microorganisms. Bananas spoil from the actions of enzymes in the fruit itself while still sealed off from microorganisms by the peel (Figure 14.8). In like manner, nuts become rancid as a result of chemical reactions while the nut meat is still sterile and protected from microorganisms.

Foods such as nuts, seeds, and some fruits that are protected by coverings impervious to microorganisms do not spoil until the covering is damaged or becomes so old that it loses its protective capacities. Many of these coverings also are provided with protection by substances capable of resisting microbial attack. An example is the thin, dry layer of tissuelike "skin" covering onions or the covering of potatoes, which protects them from microorganisms. When spoilage begins, however, the protective covering is degraded by microbial action. Impervious coverings generally are provided for the protection of reproductive structures such as plant and animal embryos. In many cases protection of the vital part seems to be necessary for only a brief period, and then the entire structure becomes vulnerable to microbial degradation.

All plant matter, with the exception of some pollen hulls, is degraded by microbial metabolism, and its chemical components are returned to the soil for recycling in the same manner as animal materials.

PRESERVATION OF FRUITS AND VEGETABLES

Because fruits and vegetables are so varied in their biological properties and so different in their physiologic characteristics, each has different reactions to the different forms of storage. The cells and tissues of each piece of fruit or vegetable that has been harvested remain alive as long as the product remains fresh. Enzymes and other cell constituents continue to function to maintain the plant product alive and fresh. The physiologic activities of the plant material are maintained by stored energy sources. In many cases this stored energy material is what human beings consume as fruit or vegetable while discarding the seeds or embryo of the plant. Apples, figs, carrots, and potatoes are examples of this phenomenon.

When fruits and vegetables are heated, frozen, or in any other manner killed, the fruit or vegetable is no longer considered fresh. Examples of extensive biological activity after harvest are the ripening of bananas, avocados, and tomatoes, and the outgrowth of potatoes. In addition, many fruits and vegetables can heal from bruises and other injuries suffered during harvesting if placed in a cool, well-ventilated place and allowed sufficient time for cell growth. The colder the storage temperature, the longer damaged tissues take to heal since this process depends on enzyme activity and numerous, complex chemical reactions which are affected by temperature.

Refrigeration

Only fruits and vegetables that are not damaged can be preserved successfully by refrigeration. If the product is damaged, psychrophilic microorganisms will grow on the tissues and cause spoilage. Unlike meats, which are all stored at the same temperature, plant materials must be stored at temperatures that are individually suitable for each product. The storage temperature must be selected on two critical points:

1. The temperature that best retards spoilage, and
2. A temperature that does not injure or kill the plant cells.

The **optimal storage temperature** is essentially the lowest temperature at which the plant material may be stored without causing injury or death.

The home refrigerator does not provide the ideal temperature for storing most fruits and vegetables and, consequently, products do not last as long in a fresh condition, as Table 14.5 indicates. To prevent changes caused by enzymes, plant products often are **blanched** before being stored in the refrigerator or freezer.

Almost all fruits and vegetables can be frozen if they are prepared adequately before freezing. In some cases blanching is sufficient, but with other products, especially fruits, peeling, cooking, or immersing in syrups or other fluids is necessary to ameliorate the effects of freezing on plant tissues.

Table 14.5   Cold Storage of Fruits and Vegetables

Material	Optimal Refrigeration Temperature	Storage Life
Apples	1°C	2 months
Asparagus	0°C	4 weeks
Bananas	15°C	3 weeks
Broccoli	0°C	8 days
Celery	0°C	4 months
Cherries	0°C	2 weeks
Cucumbers	8°C	2 weeks
Grapes	0°C	8 days
Onions	0°C	8 months
Oranges	1°C	10 days
Potatoes	4°C	9 months
Peaches	1°C	4 days
Sweet potatoes	14°C	6 months
Strawberries	0°C	5 days
Tomatoes	12°C	10 days

Drying

Fruits and vegetables can be successfully dried and preserved for prolonged periods at ambient temperatures if kept in cool, well-ventilated storage areas. Foods preserved by drying must be protected from insects and animal vermin because these consume or destroy large quantities of product rapidly. It is estimated that 40% of all foods preserved by drying are ruined by insect and animal pests. A study conducted in 1970 showed that approximately 50% of stored wheat crops in some areas of the world were rendered useless by the actions of fungi, various insects, rats, and mice.

The method of choice for the storage of cereals, nuts, seeds, and other kinds of vegetal products that are protected by hard, water-impervious outer coats is drying. These crops are field-dried, then cleaned to remove all parts of the plant that bear moisture before storage in cool, ventilated bins. Under the best conditions these storage bins are protected from insects and other pests. Although spraying or treating with pesticides may be necessary, this should be avoided if possible.

Canning

Canning was introduced as a method of food preservation in 1810 by Nicolas Appert, a purveyor of foods in Paris during the end of the 18th and beginning

Table 14.6 Appearance and Effect of Contaminated Canned Food

Name	Appearance	Organisms and Problem
Flat sour	Can flat, no change	Contents sour, *Bacillus*
Flat sulphide	Can flat, blackening	Hydrogen sulfide
Flat aerobic	Can flat	Condensed milk coagulated
Swell	Can swollen, leaks	Putrid odor, meat liquified
Sour swell	Can swollen, leaks	Fermented, cheesy odor
Soft swell	End of can soft	Fermented
Hard swell	End swollen hard	Fermented, large amount of gas
Springer	Swollen at both ends	Fermented, can may leak

of the 19th century. Appert started by placing food in bottles intended for champagne, sealing them with the traditional stopper, and placing them in boiling water. Bottling or even canning fruits and vegetables was not an immediate success, as the product had to be sterilized in the container and consequent changes in taste, texture, and appearance yielded an unappetizing final product. Many changes have been made to the original method of canning food, but the basic problem, overcooking, is still a part of modern-day canning technology.

In the past, failure of processing equipment to sterilize the contents, or failure to seal cans properly, resulted in bacterial growth in canned foods. Deaths from consuming this food were not rare previous to the 1950s. Table 14.6 explains the terminology developed by food industry workers to describe the effects of bacterial contamination on cans and on the food content. The possibility of food intoxication tends to be small when the effect of contamination is plainly evident, because only a hardy spirit would consume food so obviously spoiled. On the other hand, when the effects of contamination are not evident, a greater likelihood exists that someone will eat the food and become ill.

Ionizing Radiation

The use of **ionizing radiation** to sterilize canned and packaged foods has been successful in rendering a more appetizing final product since this **cold sterilization** does less damage to the food than heat does. On the negative side, the cost of radiation is much greater, the equipment is more costly, and the possibility exists for serious injury and harm to workers in the area of the radiation fields used for sterilization. In addition, undesirable tastes sometimes develop in foods protected by high doses of radiation.

MICROBIOLOGICAL QUALITY

Between harvest and final spoilage lies the time during which a particular food is thought of as fresh and suitable for consumption. During this time microbial populations are increasing in number constantly and consuming the food material as they do. The processes of microbial growth and subsequent spoilage are inexorable, and all methods of preservation simply forestall but do not affect the final outcome.

Each vegetal product bears its own, unique microbial flora, and this, to a large extent, reflects its past history. Fruits and vegetables with smaller microbial populations have been better protected from contamination, and therefore, are fresher than those of the same kind that bear larger microbial populations.

With vegetal products, but not meat products, it has to be borne in mind that the cells that make up the fruit or vegetable age and die. Eventually the fruit or vegetable becomes inedible even though it does not spoil from microbial action.

Microbial Load

The relationship between microbial populations and the quality of foods can only be estimated and must be viewed with caution. For example, a fish just removed from the water, which has very few microorganisms in its flesh but is infested with the larvae of *Diphyllobothrium latum*, certainly is not safe to eat. Populations of bacteria and fungi are indicators of edibility on empirical rather than theoretical grounds. That is, bacterial and fungal populations of foods judged to be spoiled or inedible by other criteria are measured and these are compared with foods judged to be fresh. The relationship between bacterial counts and food quality, as well as the limits used to determine spoilage, in terms of microbial load and edibility, then are set on these bases in retrospect. Table 14.7 illustrates the effect of salmonella populations on the quality of meatloaf.

Determination of Microbial Load　Microbial populations can be measured by many different methods, but the most common in the food and beverage industry are based on colony counts using Petri plates or some other device that will yield comparable results. The assumption behind all such methods is that each colony that appears on the growth medium in a Petri plate grew there from a single bacterial cell that came from the sample. This assumption is valid only in very few instances. In most cases colonies result from aggregations of cells that may occur in the sample in pairs or in microcolonies of several or many hundreds of individual cells adhering to each other. In these instances the term **colony-forming units** is used rather than number of organisms to acknowledge this phenomenon. This acknowledges that in all

Table 14.7   Correlation Between the Number of *Salmonella* and the Microbiological Quality of Meatloaf

Salmonella in Meatloaf	Results	
No. Cells	Sickness	Death
5×10^9	Yes	Yes
6×10^9	Yes	Yes
6×10^4	No	No
6×10^6	Yes	No
4×10^8	Yes	Yes
3×10^8	Yes	Yes
5×10^7	Yes	No

but a few cases, determining the actual number of cells is impossible and the numbers obtained from "plate counts" are really minimum numbers of cells with no information regarding the maximal number of cells that may be present in the sample.

Another source of ambiguity resides in the fact that the microbial flora is made up of many diverse organisms and there is not one growth medium that will satisfy the growth requirements of all the organisms present in the sample. The result is that only a cross-section of the organisms present in the sample is counted on any growth medium used while the majority probably go undetected. Experts who have considered this problem contend that, as a general rule, only 10% of the bacteria in any sample are counted by plate count determinations. This contention is based on the following criteria:

1. The counting procedure is not accurate.
2. Only some organisms grow on any given medium.
3. Only some organisms grow at any given temperature.
4. Only some organisms grow in a given gaseous atmosphere.
5. Some organisms produce antibiotics that inhibit the growth of neighbors.
6. Some organisms are killed by manipulations of the counting procedure.

Measurement of the microbial populations of foods yields data that are of value in comparing one sample to another sample of the same kind of food. These data allow estimation of the degree of spoilage in a given food when the final degree of spoilage is known already from previous study of other samples. On this basis, almost all commonly used foods have been characterized according to their microbial populations. Table 14.8 shows plate counts from various food items that are considered suitable for any use at table. These are total bacterial counts obtained on Nutrient Agar, incubated aerobically at 30°C to 35°C, and the colonies counted after 48 to 96 hours of incubation.

Table 14.8   Microbiological Evaluation of Certain Foods

Product	Total Count per Gram	Coliforms	Salmonella
Beef, fresh	1×10^5	100	none
Beef, frozen	5×10^5	50	
Beef, ground, raw	1×10^7	100	none in 25 g
Crab	1×10^4	100	
Poultry	1×10^4	100	none in 25 g
Shellfish	1×10^4	100	none in 25 g
Ice cream	5×10^4	20	
Cheese, cottage	—	10	
Eggs, powdered	1×10^4	10	none
Milk, powdered	5×10^4	<3	none in 100 g
Fish, fresh	1×10^6	4	
Vegetables, frozen	1×10^4	10	
Custards	1×10^5	100	
Cream pies	5×10^4	50	

Microbiological Standards   Measuring the microbial content of foods is a very difficult task, but interpreting the results obtained in terms of the wholesomeness or safety of the food is even more difficult. The problems in this area begin with knowing what part of the food to test and how to test it. In the past the assumption was that the best approach was to determine the number of bacteria found in portions of the sample equal to those that would be consumed by a normal person eating a normal meal. This led to much confusion because no data are available on what constitutes a normal person or a normal meal. Other such methods were tried, some for long periods, but all were found to have serious shortcomings. The standard that has acquired universal acceptance now is the expression of bacterial count per gram of sample.

A second problem area concerns the nature of the sample itself. In measuring the bacterial content of fresh, healthy strawberries, all the organisms are found on the outside of the fruit and few or none in the inner tissues. In this case, 1 gram of strawberries may have high counts if that weight contains much of the integument of the berry while another, equal, portion of the same berry would give low counts if that sample contained only internal tissues.

The same dilemma exists for all foods, and only an empirical approach yields a profitable resolution. Most laws, ordinances, or specifications that describe limits and requirements in terms of microbial loads specify in some detail the nature of the sample to be used. Generally, food substances that are homogeneous present little problem in that the approach to sampling is obvious. Flour, milk, salad dressing, ice cream, crab meat, ground beef, and other such products simply are mixed thoroughly and subsamples are obtained from several different parts of the sample.

Strawberries, lemons, peanuts, radishes, apples, carrots, fish, quail, and other such foods are taken at random from a large sample and ground or "liquified" with sterile, distilled water to yield a homogeneous slurry that represents the entire sample. Beef, turkeys, pork, and other large articles may be assessed by counting the number of organisms on the surface, in the outer one inch of the sample, or in a "core" sample taken with a canula of a given size and inserted to a given depth.

Multiple samples are measured since the results obtained from identical, multiple samples indicate the degree of accuracy of the procedure. Sample dilutions, growth medium, incubation temperature, and other conditions must be standardized as much as possible so the results obtained bear relevance to other such measurements. Although the procedures used to enumerate bacterial populations are fraught with many opportunities for error, they serve well to give data concerning the microbiological quality of any food tested when they are employed within the limitations obvious to any microbiologist.

An Example   Several complaints of intestinal distress have been filed during the last week by individuals who ate from the salad bar at Dirty Dan's Awful Fast Foods Emporium. On the other hand, no unambiguous evidence has been presented which would alarm the public health authorities or even indicate a relaxation of standards at this eating establishment. Dirty Dan wishes to be clean and seeks help from a consulting laboratory. The purpose is to determine if the food in his Awful Fast Foods joint is safe or not.

A typical procedure used by many microbiologists, as well as by many governmental agencies, for the evaluation of foods in public eating establishments may be carried out as follows:

1. Pay the price of the salad bar meal and fix a plate to "take out," containing as wide a variety of items as possible without selecting on the basis of recently filled servers as opposed to those that are nearly empty.
2. If possible, obtain two other filled plates by having two associates accompany the person obtaining the sample.
3. Place the three samples in sterile protective containers brought from the laboratory for that purpose. Label each sample and place in a container with ice to transport to the laboratory.
4. Each sample is treated separately but identically in every respect.
5. In the laboratory remove each plate from its protective container, label, weigh, and record weight of plate and food.
6. Place the entire contents of the plate in a sterile kitchen blender jar, and reweigh the plate with leftover traces of food. Subtract this weight from

Figure 14.9

Graphic presentation of sample weighing.

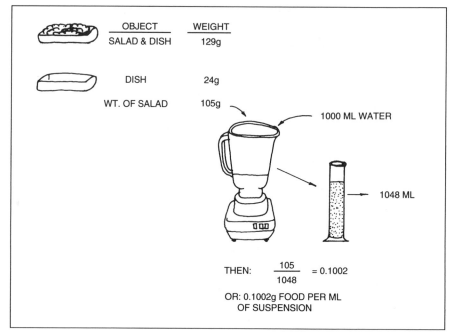

OBJECT	WEIGHT
SALAD & DISH	129g
DISH	24g
WT. OF SALAD	105g

1000 ML WATER

1048 ML

THEN: $\dfrac{105}{1048} = 0.1002$

OR: 0.1002g FOOD PER ML OF SUSPENSION

the first weight to determine the amount of food placed in the blender jar.

7. Add sterile, distilled water at 0°C to the blender jar and blend the food. Add more water if necessary until a homogeneous, thin slurry of food is obtained. Check temperature of the slurry continually, and cool by placing blender jar in ice bath if necessary to keep temperature between 0°C and 4°C.

8. Measure volume of slurry in a sterile, precooled graduated cylinder, and use sterile, cold water to wash all traces of food from blender into the graduated cylinder. Record total volume. Dividing the weight of food by the volume of slurry will give the amount of food per milliliter (ml) of slurry (Figure 14.9).

9. Transfer 10 ml of slurry ($10 \times 0.1002 = 1.002$ g of food sample) to a bottle fitted with a water-tight stopper containing 90 ml of cold, sterile, distilled water. Label this bottle 1:10. Shake contents thoroughly, and transfer 10 ml (0.1002 g of sample) of mixture to another water blank.

10. Label second bottle 1:100, shake well, and again transfer 10 ml as before. Continue as shown below, maintaining the temperature of diluted slurry at 0°C to 4°C.

11. Have ready a flask with 500 ml of melted Plate Count Agar at 45°C and 18 sterile Petri plates.

12. Shake contents of bottle labeled 1:100 thoroughly, and place 1.0 ml of dilution in each of three Petri plates.

13. Rapidly repeat Step 12 with the rest of the dilutions.

14. Add sufficient melted agar to each plate to cover the bottom, doing so without removing the plate covers completely.
15. Stack the plates and rock gently to mix water dilution and agar thoroughly. Continue mixing until agar begins to gel.
16. Allow plates to cool until agar is gelled firmly and invert plates so that agar is on top and place in the 30°C incubator.
17. At the end of 48 hours of incubation, examine the plates and select the three plates from each of the three samples that contain more than 30 but less than 300 colonies each. Count the colonies, obtain the average of the three plates, and factor the dilution to determine the number of colonies per ml of slurry. Example:

Dilution	Number of Colonies			
	Sample 1	*Sample 2*	*Sample 3*	*Average*
1:100	TMTC*	TMTC	TMTC	—
1:1,000	TMTC	TMTC	TMTC	—
1:10,000	287	379	212	—
1:100,000	128	145	139	137
1:1,000,000	31	12	16	—

*Too many to count

One ml of the 1:100,000 dilution contained viable bacterial cells that produced, on the average, 137 colonies on each plate. Therefore, the original slurry contained:

$$137 \times 100,000 = 1.37 \times 10^7 \text{ cells per ml of slurry.}$$

Since each ml of slurry contained 0.1002 grams of salad:

$$1.37 \times 10^7 : 0.1002 = x : 1$$
$$x = 1.37 \times 10^7 \div 0.1002 = 13.7 \times 10^7$$
$$= 1.37 \times 10^8 \text{ bacterial cells per gram of salad}$$

Now:

Assume that the results obtained from the other two samples (plates of salad) were as follows:

Sample 1 1.37×10^8
Sample 2 8.31×10^8
Sample 3 9.80×10^7

The average of these three counts:

$$3.5 \times 10^8 \text{ bacteria per gram of salad}$$

would be an excellent indication of the microbiological quality of the food in the Salad Bar at Dirty Dan's Awful Fast Foods Emporium.

In an analysis of this type, the same dilutions prepared for total plate counts would have been inoculated into Brilliant Green Lactose Bile Broth in quantities of 10, 1, and 0.1 ml, and the Most Probable Number assay (p. 000) conducted to determine the number of coliform organisms and also the number of *Escherichia coli* present in the three samples.

Assume that the results were:

100 ml of Slurry or 10.02 g of Salad	Coliforms	*E. coli*
Sample 1	11	0
Sample 2	6	2
Sample 3	26	10
Average:	14	4

Conclusion: Dirty Dan's is not bad, and he probably will not be closed down by the Health Department this week.

All procedures to determine the microbiological quality of food are essentially as that described above. Different foods may require special treatments and often special media. Special incubation conditions are necessary for the cultivation of special organisms. Anaerobic, thermophilic, acidophilic, and nutritionally fastidious bacteria require special treatment. The data given in Table 14.9 show the microbial load of foods known to be safe for human consumption. Any item with a microbial content less than that shown in the table may be considered safe and suitable for all food purposes unless it is shown to contain pathogenic organisms.

Indices of Microbiological Quality

Few federal or national laws state requirements in terms of microbiological load for foods sold to the public. Responsibility for the microbiological qual-

Table 14.9   Permissible Limits of Bacterial Content in Various Foods

Food	Total Count per Gram	Coliforms per Gram	Authority
Beef, frozen	5×10^6	<50	Oregon
Beef, frozen	5×10^6	50	N. Dakota
Beef, frozen	1×10^6	250	Rh. Island
All raw meats	1×10^5	100	Mass.
Smoked meats	5×10^4	10	Mass.
Ice cream	5×10^4	20	Military
Dry milk	5×10^4	90	Military
Custards	1×10^4	100	New York
Malted milk	4×10^4	10	Military

Table 14.10 Bacteriological Load of Foods Available on the Open Market

Food	Total Count per Gram, Average	Coliforms per Gram	Number of Samples Tested
Beef steak	2.6×10^6	48	56
Beef, ground	3.3×10^7	5.2×10^4	96
Beef liver	4.1×10^8	—	56
Pork	9.3×10^8	—	56
Chicken, broilers	1.6×10^5	407	72
Sole fillets	4.7×10^7	—	9
Shrimp	9.3×10^6	158	19
Fish sticks	1.0×10^4	30	78
Lima beans, frozen	1.8×10^5	28	8
Mashed potatoes	7.6×10^5	708	5
Peas, fresh	1.0×10^5	650	3
Fruits, various, dried	7.1×10^4	—	3
Beef pie, frozen	7.1×10^5	300	48
Tuna pie, frozen	3.8×10^4	115	36
Flour, bag	4.9×10^6	—	1

(data from personal notebooks)

(— = determination not performed)

ity of foods is very much left to the discretion of those who produce them and those who sell them in their final form. Because of this, certain legal entities such as city, county, and state governments, as well as many federal agencies, establish independent criteria called **indices** of microbial content or **permissible limits**. Enforcement of these requirements is in the hands of those who promulgate them and generally is on the basis of payment for goods received.

For example, the city of Chikopee, Ohio, may establish an index of bacterial counts for food items purchased for its city jail. A purveyor must provide certified total bacterial counts for items offered for sale. If the bacterial counts are higher than specified, the items delivered may be refused on this basis. The data shown in Table 14.10 describe some allowable limits of bacterial populations specified by various agencies selected at random.

When the microbiological quality of foods is specified by purchasers, the allowable limits generally are more stringent than those normally found in foods offered for sale in the open market. The findings described in Table 14.10 were obtained from routine bacteriological analyses performed on various food items available on the open market, and their suitability for human consumption are deemed acceptable. The bacteria encountered were considered normal for the kind of food, and the presence of coliform bacteria is assumed to be associated more with soil and water contamination than with fecal contamination.

Figure 14.10

Scanning electron photo-micrograph of bacteria on meat cutting board surface. Bar = 1 μm. Courtesy of Dr. Ben Tall, Food and Drug Administration, Washington, D.C.

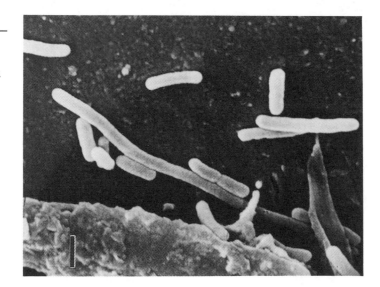

The data in Table 14.10 are an indication that almost all meats and meat products contain coliform organisms when they are sold to the public. To slaughter an animal, clean the carcass, and dress the meat without contaminating the final product with soil, water, air, implements, worker's hands, clothing, and even the contents of the animal's intestinal tract would be almost impossible. When examining such meats, it is not unusual to find bacteria of the genus *Salmonella*. Recent studies have shown that approximately 65% of all chicken meats sold to the public harbor salmonella of various serological types.

Many meat products carry other pathogenic bacteria such as *Staphylococcus aureus*, *Bacillus cereus*, and *Clostridium perfringens*. Meats containing pathogenic bacteria present little danger if they are refrigerated adequately during storage and if they are cooked thoroughly. Still, some danger of infectious disease persists because these foods serve as a transport medium for the introduction of pathogenic organisms into the food preparation area.

Meats containing pathogenic bacteria present a definite risk since they come in contact with salad vegetables and other foods that are consumed raw. Contaminated meats also are a source of pathogenic organisms that may contaminate the hands of the cook, kitchen utensils, serving ware, and the entire premises as well.

Enteric pathogens are not hardy organisms and their numbers diminish rapidly once food is removed from the refrigerator. Some food producers treat meats and even some vegetables with mild bactericides and fungicides such as nitrate and nitrite in order to maintain freshness and to avoid discoloration of the product. In some operations, poultry is treated with antibiotics to ward off spoilage. A gratuitous result of these treatments is the removal of the greater number of pathogenic organisms.

Many fruits and vegetables contain coliforms and enteric pathogens when sold to the public. These organisms typically are added to plant products when they are washed, trimmed, and wrapped in areas where meat products have been handled previously. In some cases contamination takes place in the field. Fruits and vegetables that grow on the surface of the ground and come in contact with water that may have been contaminated by animal or human excreta are particularly prone to harbor enteric organisms. The coliforms present in plant materials are of both the "nonfecal" and the fecal kind.

CONCLUSION

If one measure were to be used in grading foods and beverages, it should be "freshness," along with some assurance that pathogenic organisms are either absent or are present in such small numbers that they constitute no hazard. If more detailed knowledge were required, total microbial count together with coliform count would provide the best index of food safety. In the last analysis, however, it must be recognized that microorganisms are not only symbiotic with human beings but are commensal as well.

There should be no doubt that eating is more than an adventure, often it is also a major risk. A risk that is best minimized by adhering to the many rules, conventions, customs, and the knowledge recently acquired by scientific studies.

QUESTIONS TO ANSWER

1. Why does meat smell bad when it spoils?
2. Why do vegetables smell different from meats when they spoil?
3. What good comes from food spoilage?
4. The life of human beings depends on: (a) the availability of fresh food and (b) food spoilage. Which is more important?
5. What is the one, certain, absolute criterion of food spoilage?
6. How does bone spoil?
7. What is an organoleptic panel?
8. Why is it important for meat to bleed after slaughter?
9. How is the taste of game enhanced by aging?
10. How is the taste and texture of beef enhanced by aging?
11. How were spices first used in food?
12. Why will microorganisms not attack dried meat readily?
13. Why will microorganisms not attack frozen meat readily?

14. What is wrong with salting meat for preservation?
15. Where do halophilic organisms grow?
16. Where do osmophilic organisms grow?
17. Why are bananas not stored in the refrigerator as apples are?
18. What protects raisins from microbial invasion?
19. Why is ionizing radiation not used commonly in food preservation?
20. Can foods that contain salmonella be sold legally in grocery stores? Explain.

FURTHER READING

Jay, J. M. 1978. MODERN FOOD MICROBIOLOGY. D. Van Nostrand Company, New York.

Banwart, G. J. 1979. BASIC FOOD MICROBIOLOGY. AVI Publishing Company, West-port, Connecticut.

Hayes, P. R. 1985. FOOD MICROBIOLOGY AND HYGIENE. Elsevier Applied Science Publishers, New York.

Mountney, G. J. and W. A. Gould. 1988. PRACTICAL MICROBIOLOGY. Van Nostrand Reinhold Company, New York.

Microorganisms As Food

15

Appetizers

Sea urchins and mushrooms

Main course

Ewe's udder stuffed with mushrooms
Cow's brains cooked with milk and eggs

Side dishes

Boiled tree fungi with honey
Goose liver ground with mushrooms

Dessert

Truffles in oil

Roman menu, ca. 100 A.D.

The use of microorganisms as food staples goes back to the dawn of human existence on Earth. It is not difficult to imagine our earliest ancestors eating whatever they could find and preferring those items that were easier to find. The mushrooms fit this description and, consequently, were probably among the earliest favorites. In the earliest experiences, modern human beings ate selected mushrooms with knowledge obtained by trial and error regarding the distinction between edible and poisonous mushrooms. By trial and error, humans discovered which of the many mushrooms available were unappetizing, noxious, or toxic and which were palatable and safe to eat.

With little doubt, women were the primary providers of mushrooms for the family cooking pot and consequently the ones who determined the suitability of mushrooms and which kinds were best left alone. Because women and children performed the food-gathering tasks while men hunted, women were responsible for remembering the characteristics of edible mushrooms that were considered family favorites. The knowledge that made it possible to distinguish valuable items of food from those that caused illness or death was not only preserved by women but also was transmitted by women from one generation to the next.

Human beings developed foods that included the large microorganisms—i.e., mushrooms and algae—as food items, as well as those in which preparation was carried out by the use of microorganisms. The major example of the latter is the entire panoply of fermented foods including the fermented alcoholic beverages.

Although discovery of the microorganisms was not to come about for many thousands of years yet, early human beings were able to use the microorganisms as food and to do so successfully. The only other microorganisms besides the mushrooms to be used for food were the algae. By necessity, the use of algae was limited to coastal areas. In one form or another, then, microorganisms always have been a part of the daily diet of all human beings.

MUSHROOMS AS FOOD

Mushrooms are eaten both fresh and in the dried form. When atmospheric conditions are favorable, mushrooms spring from the ground in great profusion and may be gathered at that time and dried in the sun for later use. They may be stored safely for many months without loss of quality or nutritional value. Other microorganisms were not discovered until the 19th century and consequently were not used as food *per se*, although fungi and bacteria were used extensively in fermentations and in cheese making.

Until the early part of the 20th century, mushrooms for the table were obtained primarily by the ancient art of taking basket in hand, walking through the woods, and gathering the mushrooms that grew there. This art required an expert's knowledge to distinguish mushrooms that were safe to eat from those that were not. This method is still very much in use in many parts of the world, but mushroom hunting is viewed more as a gourmand's pastime than as a means of filling the family larder as it was in times past. As pointed out in the discussion on intoxications, the practice of mushroom hunting, whether for pleasure or out of necessity, is fraught with danger for the neophyte. Still, those among us who consider eating mushrooms as one of life's pleasures would rather do without food than without mushrooms.

Field mushrooms (Figure 15.1) offer such a rich variety of species and strains that each meal is considered a gourmet's feast. The variety of tastes, textures, and flavors that can be sampled from one spring morning's walk through the woods is such that the truly indoctrinated are not likely to be satisfied with mashed potatoes.

The favorite field mushrooms in the United States include the king bolete, *Boletus edulis*, which is also one of the favorites in Europe. The king bolete is cultivated in small amounts in many countries, but epicures pre-

Figure 15.1

Common field mushrooms of Western Kentucky.

fer the field variety. Dried, field-picked king boletes sell for approximately $100 per kilogram, whereas the same mushroom cultivated commercially brings approximately $10 per kilogram. The king bolete mushroom, has a cap approximately 4″ to 6″ in diameter but, under ideal conditions of temperature and humidity, field king boletes may develop caps more than a foot in diameter. In the latter case, the edible parts of individual mushrooms may weigh two kilograms or more.

Butterball (*Suillus granulatus*), the admirable bolete (*Boletus mirabilis*), the French mushroom (*Agaricus bisporus*), and the orange bolete (*Leccinum aurantiacum*) are among the best known mushrooms found growing in forests of pine, birch, beech, oak, and aspen. All mushrooms of the genus *Leccinum* are said to be edible. Although none of the *Leccinum* are poisonous, poisonous mushrooms of other genera may be confused with leccinum. The orange bolete is priced for its delectable taste and bouquet, but others of the same species, though edible, are hardly worth the trouble because they offer little in terms of flavor and eating pleasure. Hundreds, or perhaps thousands, of fungi are known to be safe for human consumption and are eaten regularly in most parts of the world. Some of the most common ones are listed in Table 15.1, and three are illustrated in Figure 15.2.

Mushrooms have been cultivated by amateur growers and gourmands for thousands of years and by commercial growers since the 17th century. Mushroom culture on a commercial scale was practiced in France by the 1650s. Several kinds of mushrooms were grown in caves using moistened horse manure as the cultivation medium. In more recent times amateurs in

Table 15.1   Edible Mushrooms Popular in Many Parts of the World

Common Name	Scientific Name	Remarks
Hen of the woods	*Polyporus frondosus*	May weigh 30 kg
Cauliflower fungus	*Sparassis radicata*	Parasite on tree roots
Lilac mushroom	*Pleurotus sapidus*	Easy to identify
Oyster mushroom	*Pleurotus ostreatus*	Grows only on dead or sick aspen
Shaggy mane	*Coprinus comatus*	Black ink used for writing
Chanterelle	*Cantharellus spp.*	Sold pickled
Black diamond	*Tuber melanosporum*	Truffle, may cost $500 per kg
Morel	*Gyromitra esculenta*	Contains toxin, but edible
Honey mushroom	*Armillariella mellea*	Cultivated in the field; pickled
Nameko	*Pholiota nameko*	Cultured on blocks of wood; Japan
Snow fungus	*Tremella fuciformis*	Thought to be of medicinal value

Europe and in North and South America cultivate fungi by several different methods. Of these, the simplest requires only that the mature spores of the desired mushroom be scattered in a place suitable for their growth. Mushrooms that grow naturally on grassy ground, tree roots, or fallen logs should be seeded on similar sites at the time of the year when wild mushrooms are disseminating their spores.

Spore Culture

Figure 15.2

Common edible fungi.

The spores of mushrooms may be obtained easily by picking mature mushroom caps or puffballs, placing them in a plastic bag, and shaking the bag to

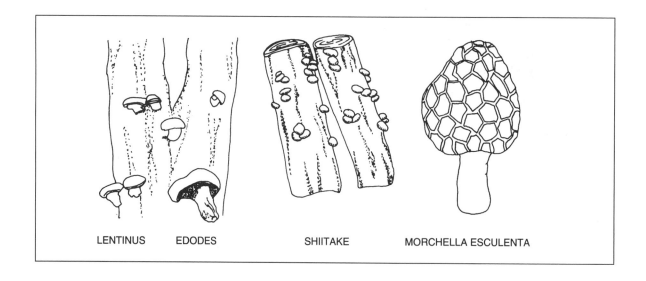

LENTINUS   EDODES   SHIITAKE   MORCHELLA ESCULENTA

break the cap or ball, thereby releasing the spores. If water is added to the plastic bag, the spores will be suspended in water. These spores now may be sown by sprinkling the spore suspension on the ground. When performed properly, this method assures a good crop of mushrooms at the same time that the wild mushrooms are maturing.

Once inoculated, the ground produces one crop after another since natural seeding occurs in subsequent years. Some mushroom gardeners encourage growth by breaking up the ground and fertilizing with horse manure in the early spring. All other mushrooms should be removed from the ground as they appear. This "weeding" process is particularly important if poisonous varieties are found to inhabit the area being farmed.

Spawn Culture

Mushrooms also may be grown from spawn. In this method the mushrooms are started from mycelia that have developed in the ground already but that have not yet broken through. For the amateur, the simplest method is to find a field that contains several buttons of the desired mushroom and gently spade up the soil to a depth of 2″ under the mushrooms. This soil contains large amounts of mycelia of the desired mushroom. It may be gently broken up and minced before scattering over ground similar to that in which the mushrooms normally grow. The ground should be spaded lightly to a depth of 1″, the soils well mixed, and the ground watered well. If weather, temperature, soil, and other conditions are optimal, a crop of the selected mushrooms may appear within a few days. If conditions are not optimal, mushrooms will appear the next time the wild mushrooms of that kind bloom in the field.

As in the cultivation with spores, great care must be exercised to prevent the growth of poisonous organisms together with the selected crop. In all cases, farming of mushrooms for the table should be attempted only by experts who are well-versed in mushroom culture. Amateurs should culture mushrooms only for aesthetic purposes and never for culinary reasons.

Tree Culture

Mushrooms that normally grow at the bases of trees often parasitize the tree's roots. These mushrooms may be propagated by transferring young saplings from the forest where the tree and mushrooms grow to the garden where the sapling and its parasites, the mushrooms, are desired. If the saplings selected were parasitized and if the transfer was accomplished properly, trees and mushrooms will grow together for many years. This technique has been tried many times in efforts to bring truffle culture to the United States. The truffle, *Tuber melanosporum*, grows in the root system of oak trees in France and neighboring countries. Success in transferring parasite and host to a new environment has been limited.

Figure 15.3

Dog finding truffles in ground in France.

Log Culture

Mushrooms that grow on fallen logs may be transferred by bringing the proper kind of log into the garden and cutting notches or drilling holes into the wood and filling these with decaying wood that contains the mycelia of the desired mushroom. Moist, shady areas where the logs do not dry out are the places of choice. An entire generation of a French family, it has been stated, can eat mushrooms from such a log and pass it on to the next generation without its production having diminished noticeably.

In general, many fungi will grow on logs and even on the dead wood of living trees, but for predictable production the logs must be selected carefully to obtain the optimal harvest of fruiting bodies. The wood, irrigation water, environmental conditions, soil nutrients, and many other variables influence the quantity and the quality of the mushrooms produced.

Living wood, and even recently felled wood, contains antimicrobial substances that make it resistant to invasion by fungal mycelia and inhibit the formation of fruiting structures. Old wood that has undergone considerable weathering and bacterial growth gives the best results in terms of fruiting body production. In many cases mushrooms flourish only on a specific kind of log and do not grow on any other.

Once the logs are selected and the holes, cuts, or indentations made on each log, they are placed in the log yard or in a log house to acclimatize. Pieces of wood from a previous log that bore mushrooms are cut to fit the crevices in the new logs and inserted tightly. The newly prepared logs are irrigated and placed in the laying yard for fungal growth to proceed.

When mature, mushrooms produce a crop of fruiting bodies that can be harvested. The logs continue to yield mushrooms for prolonged periods, giving one crop after another for many years. *Lentinus edodes* (shiitake, or forest mushroom), *Pleurotus ostreatus* (oyster mushroom), *Flammulina velutipes* (enokitake), and *Stropharia rugosoannulata* (strophaire) are all cultivated on logs by amateurs and commercial producers alike.

With few exceptions, almost any edible mushroom can be grown in the garden or basement if the gardener is knowledgeable, observant, and patient. On the other hand, many mushroom fanciers have secret formulas and methods for growing special mushrooms and often claim to be the only person who can grow some rare species that no one else can grow. This is so much a part of the art and science of mushroom culture that no one can know what is fact and what is fancy in cultivated mushrooms. If one assumes that half of the stories told are true, several hundred edible fungi are grown by amateurs in the United States alone.

COMMERCIAL PRODUCTION OF MUSHROOMS

Mushrooms are cultivated commercially in many countries, representing a valuable resource in some of these. The major species produced are described in Table 15.2. Evidence suggests that mushrooms were produced for commercial purposes in the first century A.D. but the material sold then was of a primitive nature when compared to mushrooms of the same species sold today.

Commercial production of mushrooms has increased dramatically in the United States during the last 20 years. The vast majority of it is confined to one single species, *Agaricus bisporus*. This species also is cultivated in many other countries (Table 15.3) and is the most common of the commercial mushrooms.

As production has increased, the price of the white, common mushroom has dropped. In 1985, approximately 800,000 metric tons were produced in

Table 15.2 World Production of Mushrooms in 1975

Common Name	Scientific Name	Cultivated In	Tons per Year
French; common	*Agaricus bisporus*	France, USA, Italy	670,000
Shiitake	*Lentinus edodes*	Japan, Asia	130,000
Straw mushroom	*Volvariella volvacea*	China, Japan, Asia	42,000
Enokitake; Winter	*Flammulina velutipes*	Europe, Japan	38,000
Nameko	*Pholiota nameko*	Japan, Taiwan, Asia	15,000
Oyster	*Pleurotus ostreatus*	Europe, South America	12,000
Black diamond	*Tuber melanosporum*	France, England	200

Table 15.3   Increase in World Production of *Agaricus spp.*

Country	1939	1980
France	20,000	115,000
USA	17,000	220,000
England	6,800	82,000
Denmark	400	10,000
Germany	300	155,000

the United States, and these were sold at approximately $7 per kilogram; the cost in 1976 was approximately $8 per kilogram. In contrast, only 230 tons of black truffles were produced in France in 1985, and the cost of these was approximately $600 per kilogram in the United States. The cultivation of truffles in the United States holds a great deal of interest at present. California and Florida are the major centers of activity in this effort but, to date, little progress has been made. On the other hand, the cultivation of truffles on a commercial scale has been in effect in France for the last 10 years. Previous to this time, all truffles sold commercially were obtained by finding the fruits with the aid of animals.

Mushroom House

Mass production of the common mushroom usually is accomplished in windowless, concrete buildings with both heating and cooling systems to maintain constant temperature, humidity, and illumination. The interior of the building is divided into rooms called **bays**, each fitted with a system of shelves or racks where large boxes can be placed. The large boxes, or beds or flats (Figure 15.4), are made of thick, hard wood and are approximately 30 cm deep, 1M wide, and 2M long. Before introducing the mushrooms, the bay, racks, and beds are swept clean and sprayed with a steam hose to kill all foreign fungal spores, bacteria, and animal pests in the mushroom growing area. Nematodes are particularly harmful, as they swarm in the beds and destroy the entire crop of mushrooms even before it forms. Periodically, and on a predetermined schedule, the mushroom house is emptied, swept clean, steamed, and ventilated until the walls and floors are thoroughly dry in order to prevent the buildup of "wild" microorganisms in the mushroom culture area.

Substrate

A substrate composed of various organic materials is produced in the **compost yard** and sterilized before use. An infinite variety of composts can be utilized, but each producer searches for the one that gives the best results under the conditions prevailing in that particular site. The optimal substrate can be

found only by trial and error and, consequently, may take many years to develop.

Figure 15.4

Flats of Agaricus bisporous.

Compost Yard

A starting point may be the blending of equal parts of horse manure, straw, and light sandy loam soil. Chemical fertilizers, or more commonly, chicken house sweepings often are added in measured amounts to fortify the compost. This mixture may be allowed to compost for approximately one month, during which it must be watered lightly and turned frequently so a homogeneous, light substrate results. Turning should be done carefully to allow self-aeration in the heap lest it overheat in the center.

When the heat generated in the center of the compost heap decreases, the compost is placed in beds, covered with plastic film, and steamed for several hours to reduce the number of microorganisms in the compost. The beds then are placed in their racks in the bays and allowed to cool to the bay temperature. The plastic cover is left in place until the bed is seeded with mushroom spores.

While the beds are being prepared, spawn is obtained by extracting mature spores from mushroom caps selected from a previous crop. Spawn also may be grown in the laboratory, but the traditional methods employ spores from choice caps selected from good crops. Spores or spawn are added to the surface of the prepared bed and mixed into the top layer of substrate. The bed then is moistened and left in the dark at a temperature of 24°C to 25°C. Fungal mycelia develop vigorously during an incubation period of 10 to 20 days.

When the mycelia are well developed, a shallow layer of **casing** or **fruiting soil** is added to the top of the bed to cover the developing mycelia. Casing soil can be any kind of rich garden soil or one specially formulated with specific chemicals known to accelerate fruiting. Some of the chemicals often added to casing soil are calcium, phosphorus, and nitrogen.

After adding the casing soil, the beds are wetted and incubation continues for 10 to 15 days or until the mycelia cover the casing soil. When mycelia appear on the surface of the casing soil, the temperature is lowered to 15°C and the bay is ventilated with air precooled to 15°C. Air circulation has two purposes: (a) it helps to maintain a constant temperature throughout the bay, and (b) it removes accumulated carbon dioxide that retards growth of the fruiting structures.

The mycelia begin to form **primordia** in approximately 5 days, and buttons start forming in another 2 to 3 days. Finally, almost as on a signal, there is a break, and a large number of mature buttons appear on the surface of the beds. Thereafter, as mature buttons are harvested, new ones arise and a new crop can be harvested every 5 to 10 days during a period of 30 to 60 days.

Preparation for Sale

After the mushrooms are harvested, they are washed in cold water and stored at 4°C to 5°C until sold. The storage life is limited to 7 to 10 days for fresh mushrooms, but this may be extended for many days if the mushrooms are used for other purposes. Mushrooms may be dried in dehydrating ovens at 70°C and stored at room temperature. Dried like this or dried in the sun by the amateur home grower, the keeping life of mushrooms is extended by many months. Dried mushrooms tend to be more common in the Orient than they are in Western nations. Most of the mushrooms sold in the United States and in European countries are fresh, frozen, or canned.

Mushrooms also may be preserved by freezing or by freeze-drying, and in this condition they may be kept for approximately one year with little change in appearance. Large portions of the commercial mushroom harvest in the United States are canned in brine or other liquid such as tomato sauce. Pickling is another popular method used to preserve many kinds of mushrooms. Mushrooms that are broken or damaged during washing, grading, and preserving often are made into mushroom soup or into one of many mushroom creams or mushroom sauces.

Mushrooms are cultivated in some countries as a food product and in others as a profit crop. West Germany produces approximately 40,000 metric tons and the United States approximately 150,000 metric tons of *A. bisporus* per year and consume approximately 150,000 and 200,000 metric tons, respectively.

In contrast, China produces 35,000 metric tons and Taiwan 75,000 metric tons of the same mushroom but consume barely 1,000 and 2,000 metric tons per year, respectively. West Germany imports some 110,000 metric tons and the United States 50,000 metric tons per year, respectively. China exports approximately 34,000 metric tons and Taiwan 73,000 metric tons each year, respectively. France exports some 60,000 metric tons per year of the more rare mushrooms including several varieties of truffles. Because of its unique position as the world's major supplier of truffles, France considers its mushroom industry one of its commercial cornerstones in world markets.

The Mushroom Meal

The reason for finding, raising, or buying mushrooms lies in the joy of cooking and eating them. Although the expert can travel his or her own road in this regard, the beginner will do much better to stay to the third choice and simply go to the grocery store for a supply of raw mushrooms. Hundreds, and perhaps thousands, of recipes for mushroom dishes are available in cookbooks, advertisements, personal files, and in people's memories. A few from the fourth category are given below simply to elucidate the versatility of these microorganisms on the list of favorite foods of human beings.

Mushroom Soup (for six)

Place 4 oz of dried mushrooms in cold water in saucepan and soak until soft; slice mushrooms and return to water. Add 5 cups of beef stock and simmer 3 hours. Melt 2 tsp butter in frying pan and sauté 1 cup of chopped onions until brown. Add onions to soup, simmer for 30 min, add salt to taste. Serve with light white wine.

Coprinus Parmesan (for four)

Mix one egg with 1 Tbsp water. Slice one lb shaggy manes and dip in egg-water mixture. Cover shaggy manes with grated Parmesan and sauté in frying pan with 2 tsp butter for 10 min or until brown; turn once or twice for even cooking. Dust with a mixture of equal parts oregano, parsley, and dry garlic. Salt to taste and serve with full-bodied red Burgundy.

Mushrooms with Sausage (*Hongos con Chorizo*) (for four)

One pound hot, spicy sausage, preferably Mexican, but any hot sausage will do well. Remove sausage from casing, mix well, and heat in frying pan until all fat is melted; drain fat and discard. Place 2 tsp fresh olive oil in clean frying pan, add 1 clove garlic, one tsp parsley, dash cumin seed, and warm. Slice 1 lb common white mushrooms and sauté in oil 4 min; drain oil and discard. Add half of the mushrooms to the sausage, mix, and mold in serving dish. Layer remaining mushrooms over sausage, and garnish with white (ranchero) Mexican cheese. Serve with beer, hardy Burgundy, or, preferably, champagne.

ALGAE AS FOOD

Those who advocate algae as the ultimate source of food for human beings state that human beings, throughout their life on Earth, have been searching

Table 15.4   Chemical Constituents of Algae

Component	% of Dry Weight
Carbohydrate	15–40
Protein	50–60
Lipid	4–20
Vitamins	all
Minerals	all

for their food in the wrong place. This idea is based on the fact that only one-third of the surface of the Earth is dry land and two-thirds of it is covered with water. Furthermore, much of the Earth's dry land is not suited for the production of food. As a consequence, these theoreticians say, the water must contain more food than dry soil does. Although this theory holds much wisdom, many experts find fault with it.

Since algae and aquatic plants grow as well in water as land plants grow on the land, it is not unreasonable to assume that food can be grown in the water as well as it is grown on the land, and this on an acre-per-acre basis. Another important consideration is that while soil easily becomes depleted of nitrogen, iron, magnesium, phosphorous, and other minerals needed for plant growth, the water is homogeneous in chemical composition at all times and localized depletions are difficult to produce and then last only for a brief time.

Soil minerals must be replenished constantly at great cost; otherwise the land has to be abandoned as it will no longer yield food for man. The water, on the other hand, is in motion constantly and in this way represents an inexhaustable source of plant nutrients. Also, the oceans have no fallow seasons, no droughts, no floods, no locust plagues, and no possibilities of dust bowls.

Had human beings spent the same amount of time learning the management of **aquaculture** as they have spent in learning land management, food shortages possibly could have been avoided. Even without the attention of human beings, approximately 1×10^{10} metric tons of algae are produced in the Earth's oceans each year. This algal biomass represents an unexplored potential source of food for human beings.

A figure somewhat analogous to the total production of algae might be 1×10^{10} metric tons of vegetable food products that could be harvested on land each year at absolutely no cost in terms of money, labor, or even concern. This translates to approximately two metric tons of food for each person on Earth, and it would be free of cost. Algal cells are made of the same chemical substances that human beings require for their nutrition. Table 15.4 gives the approximate composition of a mixture of different kinds of algae.

As the population of human beings increases out of control—that is, without regard to the availability of food—it soon will become evident that in the 21st century one of the best promises for an adequate food supply must be the Earth's oceans.

Aquatic Animals versus Aquatic Plants

Early human beings assuredly used the food nearest at hand rather than that which was out of sight. As an extension of this, algae and aquatic plants may have been considered only infrequently as a usable food source and, up to recent times, only faint exploration of food from the sea likely has been attempted in isolated communities of human beings who live in coastal areas.

Human beings seem to have a curious apprehension about making a commitment to investigate the sea as a major source of food. This is an unreasonable apprehension in that human beings have searched and explored all the waters of Earth for fish, shellfish, marine mammals, and crustaceans. A significant amount of evidence suggests that many kinds of fish have been depleted and in the future, it may be algae that become the major food from the ocean. Some evidence suggests that human beings used water plants and algae as food as long ago as the 8th century B.C.

Use of Aquatic Plants

Some communities have left records of limited uses of algae and sea plants both as food and as fuel. References in the early literature indicate the use of **Irish moss** (an alga named *Chondrus crispus*) as a staple food in the British Isles and **sea lettuce** (*Ulva lactuca*) as a staple food in the western coast of Europe. The Japanese have consumed algae from earliest times, and by the 2nd century A.D. were so well-versed in the science of aquaculture that a compendium of edible algae was compiled. Many of the algae listed were gathered on a commercial scale by Japanese workers and sold on the open market by Japanese merchants, but the extent of experimentation with algae and aquatic plants as food was limited. In all of human experience at acquiring food from the various sources available, human beings have been timid to a fault in exploiting the value of algae and seaweeds as a natural resource.

Food and Fuel from the Sea

Japan is the prime producer and consumer of algae and aquatic plants as food. These constitute approximately 12% of the food supply of the Japanese, and the variety and quantity of algae used increases constantly. Japanese coastal villages began to seed and harvest algae in the 17th century, and their technology has improved steadily since that time. At present, approximately 40,000 people are employed in raising some 20 different species of sea algae used for food in Japan. So advanced is the Japanese industry that one of their major plants is located in Shannon, Ireland.

Other countries—notably, Brazil, Peru, Canada, China, and Korea—are investing large sums of money in establishing algal farms. Some countries seek

Figure 15.5

Open ocean kelp farm of the future.

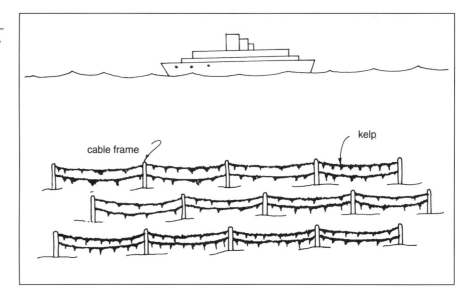

to produce mass quantities of algae not only for food but also for use as fuel for generating electric power. The latter is to be accomplished either by direct combustion of the algae or, alternatively, by converting the algal matter into methane gas first and then using the methane gas as boiler fuel.

Mass Culture of Algae

Marine Culture   Algae can be grown in the ocean with all the advantages of aquaculture noted above. A project has been designed in which giant sea kelps are to be grown in an ocean "farm" of more than 250,000 hectares. The project is to show the full potential of sea farming as a source of food and also as a source of fuel for generating electricity in the future. The proposed sea farm will consist of a cable-and-wire structure roughly in the shape of a circle, with a diameter of approximately 3 kilometers. The circular structure will surround an array of vertical poles and horizontal cables not unlike a wire fence (Figure 15.5). Kelps of the genus *Macrocystis* will be seeded on the cable network and allowed to reach full size before harvesting.

Such a system can be expected to produce approximately 3×10^9 metric tons of food per year, and roughly 10 times that much algal material which could be used for power generation. From this very small area of ocean, three-fifths of 1 metric ton of food could be produced for each human being on Earth. Ten such farms would produce 6 tons of food per human being or, roughly, the entire need for food of the world's population.

Other attractive aspects of this system include the fact that it can be expanded infinitely without conflict of property lines, highways, or land

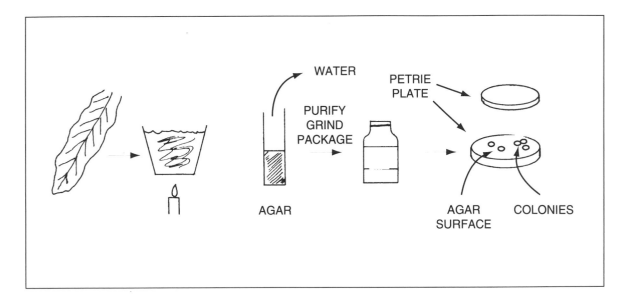

titles. As stated, fertilizers, herbicides, and insecticides will not be needed, and crops will grow at the same rate all year long. Projections made from the parameters of the proposed project indicate that the entire food and energy requirements of the United States in 1990 could be met by farming an area of some 6.5×10^7 hectares of suitable ocean surface—an area approximately covered by the waters of the Intercoastal Waterway from Galveston to Brownsville, Texas.

Although some of the crops harvested from the sea have been used for food, most have been used for other purposes. Since the early 1700s, people along the English coast have harvested both natural and cultivated seaweed in quantities greater than 1 million metric tons per year. This algal biomass is a source of many products including certain minerals such as iodine and potassium. Algal polysaccharides such as agar and carrageenan have been harvested for local use and also as export products for many years.

Japan, the Philippines, and Canada also cultivate sea algae as a source of agar and carrageenan, along with other products extractable from the algae. The price for these products on world markets ranges from $100 to $1,000 per kilogram depending on the quality of the product. Agar generally is both the cheapest and the most abundant product obtained from sea algae. It is employed universally in home canning, in preparing medicinals and cosmetics, and in all microbiology laboratories for cultivating bacteria (Figure 15.6). Both agar and carrageenan are used as thickening and suspending agents in medicines and cosmetics.

Freshwater algae are used as food in small amounts in most parts of the world. In some societies of the Orient, North Africa, Mexico, and Central

Figure 15.6

Preparation and use of agar in microbiology laboratory.

and South America, fresh water algae are cultivated, albeit on a small scale. They are used in larger quantities for decoration and as seasoning in many local foods.

Cultured Algae   Growing algae in effluent water from sewage plants fired the vision and imagination of human beings long ago because this practice would solve two of the major problems that modern human beings face:

1. The elimination of sewage effluent.
2. The production of vast quantities of food.

Sewage plant effluent contains all the minerals necessary for the growth of plants, including all the algae. The source of energy for plant growth is sunlight, and a large part of the Earth's surface receives the requisite amount of light daily. The technology associated with growing algae in sewage effluent is well understood, and few problems are predicted.

The processes envisioned require neither expensive equipment nor complex techniques, and the expectations for success are high. The organisms deemed most attractive for mass cultivation are the **microalgae**, microscopic algae which grow as single cells in much the same manner as bacterial cells.

Handling large quantities of single cell algae is much easier than manipulating equal quantities of kelp or kelplike algae. Tests to probe the feasibility of mass cultures of microalgae as a means of producing food were started in the 1930s and continue to the present. By the 1950s the United States, Russia, Germany, Israel, Britain, France, Italy, and many other countries were testing various approaches to the mass culture of microalgae. *Chlorella, Scenedesmus, Porphyrium, Carteria, Chlamydomonas, Rhodomonas, Nannochloris,* and more than two dozen other species have been cultivated in large amounts and used for animal feed supplement. These algae also have been used as sources of chemicals including glycerol, carotene, and linoleic acid.

One of the organisms that gave the most promising results in these tests was the halophilic alga *Dunaliella salina* and another was the cyanobacterium *Spirulina platensis.* Many of these algae now are being grown in mass cultures as sources of various products. Glycerol, a primary material for the production of medicinals, cosmetics, lubricants, and antibiotics is a major product of algal cultures. Many of the algae listed above have been tested for nutritive value and also for toxicity in long-term studies.

Mass Culture of *Dunaliella*   *Dunaliella* can be grown in sea water in shallow ponds. Some studies have indicated that 50 grams of dry cells of dunaliella can be harvested each day from an area of 1 square meter. The amount of algal cells that could be obtained from a salt water pond 1 kilometer on each side would be on the order of 5×10^4 kilograms dry weight of algal cells per day. If the algal cells yield 30% of the dry weight as edible protein, this would be a harvest of some 1,500 kg of edible protein per calendar day.

Sea water contains many kinds of protozoa that feed on algae, but *Dunaliella salina* is able to grow in water containing more than 20% salt. If water is allowed to evaporate to the point where the salt content is increased to 10%, protozoa, bacteria, fungi, and almost all other algae would fail to grow. This makes it possible to maintain an **axenic culture** of dunaliella even if exposed to the air.

One of the major problems associated with this vision is the extraction of protein. Proteins would have to be separated by chemical procedures requiring expensive chemicals and novel apparata. The protein would have to be combined with other edible parts of the alga to make it stable and presentable for cooking and eating. In contrast to this, the protein available from livestock is ready to consume as soon as the animal is sacrificed. The difference in cost, however, is large enough to make algal proteins a promising substitute for animal proteins in the near future.

Mass Culture of Chlorella *Chlorella pyrenoidosa* and several other species of fresh water algae can be grown in mass cultures in the laboratory or in controlled systems outside of the laboratory. One of the methods proposed for growing large crops of chlorella is based on using sewage plant effluent water as the nutrient medium. In this proposal water emerging from the sewage treatment plant would be pasteurized and then moved to a holding lagoon. Its chemical composition would be analyzed, and any existing chemical deficiencies would be amended by addition of the proper substances. The pond then would be inoculated with the desired organism and the temperature of the pond maintained at an optimal value by addition or subtraction of heat. Sunlight falling on the surface of the pond would be maximized by mirrors or other light-gathering devises.

The pond would be operated as a **continuous culture** by finding the rate at which algal cells could be removed from the pond and replaced with sewage effluent without altering the rate of growth of the algal population in the pond. Once equilibrium is established, cell yield is constant over extended periods. The growth rate constant can be changed by changing the volume of incoming nutrient solution.

Algae As Animal Food Several kinds of algae have been studied with regard to animal nutrition. The greatest success was attained when 5% to 10% of the animal's feed was substituted with raw algal cells. Cattle, pigs, chickens, and laboratory rats and mice showed no ill effects after months of such feeding.

On the other hand, rats given 10% to 20% algal cell substitution for normal rat feed lost weight over a period of 6 months. When fed solely on cells of *C. pyrenoidosa*, white laboratory rats suffered chronic diarrhea, lost weight, were irritable, turned green, and died prematurely. Chlorophyll in large amounts accumulates in the tissues of mammals because it is degraded slowly.

When algal cells were fractionated so that proteins, carbohydrates, lipids, and other constituents were isolated from the nonedible parts of the cell and the isolated substances prepared to yield a palatable, balanced diet, test animals thrived even after many months of unsupplemented feeding. In an extension of this study, undernourished animals in poor physiological condition, gained weight and improved in condition after a few weeks of eating feed made from algal cells.

Algae As Food for Human Beings   Few well-designed studies have been conducted on the use of algae in the nutrition of human beings. The relative amounts of cellulose, chlorophyll, and nucleic acids in single-cell algae are much greater, by far, than those found in plant or animal cells. A paradigm of biology is that the smaller the living organism, the greater is the relative amount of nucleic acid it contains. Some viral particles contain approximately 50% of their dry weight as nucleic acid, bacterial cells 5%, and animal and plant cells less than 0.1%.

Controlled studies demonstrated that human beings tolerate approximately a 10% substitution of whole algal cells in place of 10% of the normal diet. When the ratio was increased, diarrhea and weight loss were noticed quickly. A team of Venezuelan physicians showed that malnourished patients in a leper colony could be maintained successfully with a supplement of 35 g per day of algal cells, and that preexisting conditions of malnutrition were corrected after approximately 6 months of algal supplement.

Other studies showed clearly that subjects who consumed 100 g of algal cells daily suffered symptoms of ill health almost from the beginning of the feeding regimen. To the contrary, when the algal cells were rendered chemically and the isolated protein used as food, it was found to be as adequate for human nutrition as was any other protein.

Considerations regarding the use of algal cells for the nutrition of human beings begin with the premise that human beings cannot consume algal cells in large amounts but that cell proteins and other cell constituents are sufficient to satisfy all the nutritional requirements of human beings. Again, this implies that the edible parts of the algal cells will be used and the inedible parts will be discarded.

BACTERIA AS FOOD

Bacteria are part of all the food that human beings consume. In some foods, such as thoroughly washed and cooked fresh vegetables eaten immediately after cooking, the number of bacterial cells may be very small, while in others, such as buttermilk and sauerkraut, the number is quite large. The vast majority of bacteria are killed in the stomach and simply become part of the food ingested. Unlike algae, bacteria do not have undigestible cell walls and chlorophyll. Like algae, however, they contain a relatively large amount of nucleic acid per unit of protein.

Potential

Rapid growth and their ability to grow on a variety of substrates has made the bacteria favorite microorganisms for consideration as the food of the future. The advantage of bacteria over other microorganisms is seen in terms of the amount and cost of protein that can be grown in a given period of time. Mass cultures of bacterial cells for food production are envisioned as laboratory processes. Unlike algae, bacteria that can be used for food are not photosynthetic and must be provided with a source of energy. Some of the energy sources that have been considered include household garbage, sewage, oil refinery wastes, abattoir wastes, sawdust, and other such materials.

These energy sources would have to be processed to make them into suitable bacterial growth media. Part of the preparation would include removing substances that would inhibit or diminish cell growth. Mineral nutrients required for the medium could be provided by soil extract, sewage plant effluent, or plant leaf ash. Once completed, the medium would be placed in a culture vessel not unlike a swimming pool. Ideally, the culture tank would have automated means of controlling temperature, acid content, oxygen demand, and all the other parameters that must be maintained to obtain optimal bacterial growth.

Plant Design

Once the medium is prepared, it would have to be inoculated with the selected bacteria. Without doubt, the culture will be mixed and will contain a consortium of both aerobic and anaerobic organisms. Aeration and mixing would give the highest growth rates, and the system would be operated as a continuous culture for maximal yield. Upon reaching equilibrium, it could be maintained by harvesting cells and adding nutrient solutions in coordinated steps to maintain growth at maximal rate. This equilibrium would be monitored and maintained by computers, and the system would run continuously for long periods of time.

Cell Harvest

Cells harvested from such a culture vessel would first be treated with a small amount of acid to make precipitation easy and then separated from the liquid medium by centrifugation. From the centrifuge the cells would go to a holding vessel for digestion with several enzymes. The enzymes would release cell wall and other carbohydrates usable in the nutrition of human beings, and other enzymes would convert cell nucleic acids into chemicals of low

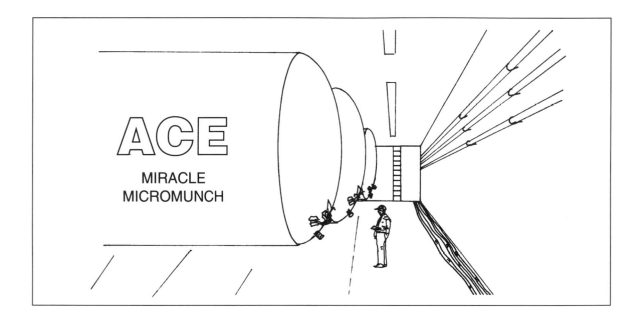

Figure 15.7

Production of bacterial cells, hypothetical facility.

molecular weight that are not toxic to human beings. Cell proteins, carbohydrates, lipids, and minerals would be isolated and purified by automated chemical processes.

Food Production   Cell chemicals isolated from the bacterial cells would be blended and mixed in measured portions to produce foods of predetermined composition. By this means, the nutritive value, calorie content, mineral content, flavor, aroma, consistency, texture, and all other characteristics of the food can be established—food produced to taste and need. In the future it likely will be possible to make sirloin steak from bacterial cells and to control precisely the amount and type of fat, cholesterol, protein, and carbohydrate in the steak. This will be no more impossible in the future than the making of butter from corn oil is today.

Operation   A plant designed for the production of food from bacterial cells could produce many tons of the basic food material every day (Figure 15.7).

A plant with a volume of 10,000 liters (a very small plant) operated on a continuous culture basis where the cells are harvested each time that the culture doubles could yield:

Mass of cells in culture: = 100 kg
Time of doubling of culture: = 1 hr
Rate of cell production: = 100 kg/hr
Production in 24 hours: = 2,400 kg per day

If this were a plant with a large volume of culture space and the cells were allowed to continue dividing for 24 hours before harvesting the entire crop, the cell yield would be:

$$N_t = ?$$
$$N_0 = 100 \text{ kg}$$
$$\text{G.T.} = 1 \text{ hr}$$
$$T = 24 \text{ hr}$$

then:

$$\text{G.T.} = \frac{0.301 \times T}{\log N_t - \log N_0} \qquad 1 = \frac{0.301 \times 24}{\log N_t - \log 100}$$

and:

$$\log N_t - 2 = 7.224$$
$$\log N_t = 7.224 + 2 = 9.224$$
$$N_t = 1{,}674{,}490{,}000 \text{ kg cells per day}$$

This calculation is more to show the potential for growth than to describe an actual situation. It represents a theoretical potential only; consider the fact that a culture with 1,674,490,000 kg of cells would have to be in 167,490,000,000 L of water; approximately 30 times more water than used by the entire city of New York per day (see p. 310).

If only 5% of the cell mass is edible, the rate of food production would be 8,300 metric tons per day. Such a system could operate at only a small fraction, e.g., $\frac{1}{1{,}000}$, of theoretical and still yield 8.3 metric tons of food per day or as in the previous example, 2.4 metric tons per day. The initial cost of such a system would be very high but productivity would offset the cost in a short time. The efficiency of conversion of raw materials to food is greater by this means than by any other means. Also, the rate at which substances without value, such as sewage, brewery, or cannery wastes, can be converted to valuable food products is much greater than any other known method.

Food Preparation As mentioned, the basic food substance produced from microbial cells would have to be textured, flavored, colored, aromatized, and molded to desired specifications. The food could be made to taste like the best beef steak but have no gristle, like ripe pears but have no bruises, and like the best cheese but have no cholesterol, fat or dry spots. All the food produced by this method would be a complete nutrient, balanced for the needs of specific, individual human beings. It would be formulated on a daily basis according to the computer's assessment of the individual's nutritional needs for that day. It would not have too much cholesterol but just the optimal amount. It would not be high in starch but would have just the right quantity, and it would not be rich in calories but have just the quantity needed. The

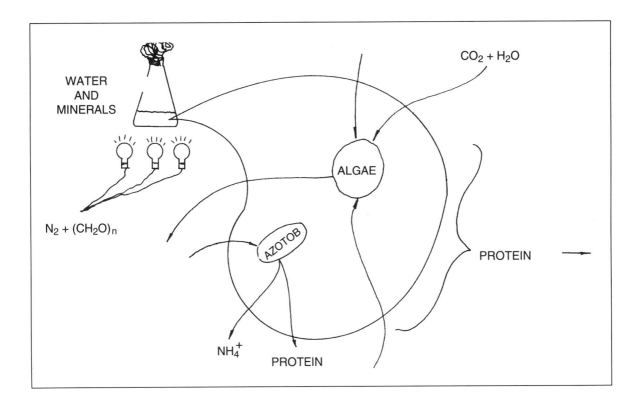

Figure 15.8

Culture for obtaining food from the air.

price of food produced in a plant like this would not fluctuate, and the supply would be constant the year round.

FOOD FROM THE AIR

Bacteria of the genus *Azotobacter*, as well as many other microorganisms, have the ability to **fix nitrogen**, and all algae and some bacteria have the ability to **fix carbon dioxide**. In this sense, "to fix" means to remove from the air and to use for nutrition and subsequently cell growth. Nitrogen fixers take nitrogen from the air and convert it into proteins, and carbon dioxide fixers convert carbon dioxide from the air into cell carbohydrates. Carbon dioxide fixation requires energy from light for the production of carbohydrates by the process of photosynthesis, and nitrogen fixation requires energy from the metabolism of carbohydrates for the production of proteins. The two types of organisms combined, therefore, should be able to produce both carbohydrates and proteins, and from this beginning, everything else necessary for cell growth.

One of the most intriguing challenges of modern science is to find the method whereby nitrogen-fixing bacteria and carbon-dioxide fixing algae can live in symbiosis that would allow them jointly to convert nitrogen and carbon dioxide from the air into microbial cells. The only other substance needed for growth would be a small quantity of minerals and the water necessary for growth. Such microbial cells would serve as the ideal food for human beings. The culture shown in Figure 15.8 was first established by a Japanese investigator and has been used as a demonstration in the author's laboratory many times.

QUESTIONS TO ANSWER

1. Hen of the woods (*Polyporus frondosus*) is a mushroom that may weigh up to 30 kg. How can this be a microorganism?
2. *Gyromitra esculenta* contains lethal toxins. How can it be an edible mushroom?
3. How do mushrooms grow on dead logs?
4. Why is manure used in the commercial production of mushrooms?
5. What is casing soil?
6. What is the major difference between algae and aquatic plants?
7. Describe agar and carrageenan and their uses.
8. Briefly describe an algal farm.
9. Briefly describe a kelp farm.
10. How is *Dunaliella salina* different from *Chlamydomonas*?
11. What is a halophilic organism?
12. What is sewage plant effluent?
13. Why does *Chlorella pyrenoidosa* grow in sewage plant effluent?
14. How can human beings eat bacterial cells?
15. What is nitrogen fixation?
16. What is carbon dioxide fixation?
17. Why does one get more cells out of a culture when it is harvested once per day instead of twice per day?
18. How can a culture yield 165,000 metric tons of cells in 24 hours?
19. Why will it be difficult to get a culture to produce 165,000 metric tons of cells in 24 hours?
20. When will human beings need food from microorganisms?

FURTHER READING

Beuchat, L. R. ed. 1987. FOOD AND BEVERAGE MYCOLOGY, 2nd ed. Van Nostrand Reinhold, New York.

Chang, S. T. and Hayes, W. A. 1978. THE BIOLOGY AND CULTIVATION OF EDIBLE MUSHROOMS. Academic Press, New York.

Borowitzka, M. A. and Borowitzka, L. A. eds. 1988. MICRO-ALGAL BIOTECH-NOLOGY. Cambridge University Press, New York.

Marteka, V. 1980. MUSHROOMS: WILD AND EDIBLE. W. W. Norton & Company, New York.

Richmond, A. ed. 1986. HANDBOOK OF MICROALGAL MASS CULTURE. CRC Press, Inc., Boca Raton, Florida.

Singer, R. 1961. MUSHROOMS AND TRUFFLES; BOTANY, CULTIVATION, AND UTILIZATION. Interscience Publishers, Inc., New York.

Adulteration of Foods and Beverages

16

*a dul' ter ate, v.t.   To corrupt, debase, or
make impure by an admixture of a
foreign or baser substance; to prepare,
esp. for sale, with an ingredient
included which is not part of the
professed substance, or, according to
certain statutes, with an essential
ingredient abstracted, . . .*

Webster

Adulteration is commonly interpreted as any act that changes the nature of food or beverage in such a manner that the consumer pays more and receives less and, most important, that this is done without notification. In its most usual form this means that foods or beverages are mixed with inferior materials for the sake of profit. The first example that comes to mind is the age-old, and probably revered by merchants, act of adding water to milk. The logic to this dishonest act is based on the fact that water is cheaper than milk. The definition of adulteration given above also carries with it the notion that the adulterant may be noxious or harmful to the consumer. The history that depicts adulterations of foods and beverages is replete with stories of loss of health and life as the result of the actions of unscrupulous operators.

In a stricter, more literal interpretation, the term *adulteration* simply means that extraneous matter is added to the food or beverage and no further qualifications need be given. And again, in the legal sense of the word, change of any kind is considered adulteration. For example, milk is considered to be adulterated if butterfat is removed and replaced by cottonseed oil so the chemical analyses will indicate the expected quantity of fat. Also, in legal terms, adulteration includes the removal of just a small part of the butterfat from the milk even if the largest part of it is left undisturbed. This is a common practice today, as low-fat milk is demanded by a health-conscious population.

The law makes no distinction as to whether the adulteration was the result of an intentional or unintentional act, and the penalty upon conviction is the same in either case. Milk is said to be adulterated if it bears filth, whether the filth was introduced by slothful or inattentive operators or by natural acts such as the introduction of animal hair or animal secretions into the milk.

LEGAL DEFINITION

The legal definition of adulteration in the United States is stated in the Food, Drug, and Cosmetic Acts of 1938 and 1968, as follows:

> A food shall be considered to be adulterated if it contains poisonous or deleterious substances which may render it injurious to health or if it consists in whole or in part of any filthy, putrid, or decomposed substance or it has been prepared, packaged, or held under unsanitary conditions whereby it may have become contaminated with filth or rendered injurious to health.

In this book, the definition of adulteration given above will be employed, in both the legal and the moral sense. All adulterations are considered undesirable, a threat to good nutrition, and dangerous to the consumers' health. It will also be assumed that all adulterations to food and drink are made solely for profit. Therefore, all adulterations are considered detrimental to the consumer.

The United States and almost all other countries have laws designed to prevent adulteration of foods and beverages. The laws that protect the public from unethical operators in the United States were first formulated in the Federal Food and Drug Act of 1906 and later enlarged and made all-inclusive by the Food, Drug, and Cosmetic Act of 1938.

NEED FOR LAWS

Historically, laws regarding the production, preparation, and selling of foods and beverages were enacted in response to the alarming state of conditions

Figure 16.1

Publications such as "The Jungle," focused public outcry for protective laws in the United States.

that resulted from universal disregard for the public interest and well-being of the consumer.

Food adulteration and misadvertisement reached the zenith in the period between the Civil War and the turn of the century. Commercial greed and immoral practices in the food industry became so exaggerated that sales to foreign merchants almost disappeared and many countries banned the import of food products from the United States on the basis that they were a threat to health and proper nutrition. Reform came as the result of investigative stories and books published by authors such as H. L. Mencken, Upton Sinclair, H. W. Wiley, and Samuel Hopkins Adams (Figure 16.1).

ENFORCEMENT

Although the laws that have evolved from the original Acts of 1906 and 1938 are totally adequate to protect the public well-being and to assure fair practices in the selling of food and drink, dishonest and unscrupulous operators still can manage to increase their profit by stealth and legal maneuver. Law enforcement in the United States was stringent during the first three quarters of the 20th century, resulting in the greatest amount of protection ever accorded by any government to its constituents. From 1980 on, however, a great deal of neglect and permissiveness on the part of the enforcement agencies at all levels of government has permitted generalized decay of the guarantees and protection possible under the law. Further, the number of legal charges for violations against food adulterators declined precipitously from previous years.

As a result of withdrawal of funds, the Food and Drug Administration Office in Dallas, Texas, was reduced from some 58 food inspectors in 1980 to only eight in 1987, and the Food Microbiology Laboratory lost approximately 30 specialists during the same period before being closed in 1989. Although the degree of compliance with the law is still high, this may well be attributed more to the fear of lawsuits than to fear of the law. On the other hand, many food and beverage providers perform their work at the highest level of integrity out of a spirit of pride in their enterprise and in the goods they present to the public.

HISTORICAL ORIGINS

Adulteration of foods is a deception practiced from the time that human beings first began to trade one item for another with the intent of receiving more value than that given, i.e., making a profit. All ancient literatures chronicle the various forms of adulteration of foods and beverages common at the time, and some describe the punishment given to those caught and found guilty of adulterating foods and other materials.

Marcus Gavius Apicius, who lived in Rome from 14 B.C. to 34 A.D., is credited with having compiled a large collection of recipes into a book which may have been named *Qui Dicuntur de Opsoniis et Condimentis Sive Arte Coquinaria Quae Extant,* now called Cookery and Dining in Imperial Rome. In his writings Apicius described his ability to make rosé wine without roses, how to fix spoiled honey so it could be sold as fresh honey, and how to turn ". . . birds with a goatish smell . . ." into a delicious and appetizing dish. He also gave instructions for the use of blackbirds, crows, and buzzards, in place of rare game birds. For this trick he recommended boiling the birds with several spices including pepper, mint, roasted almonds, honey, and vinegar, among others. He described the final product in the following words: "The birds will be more luscious and nutritious," but he did not say how his guests received those luscious animals. In the case of Apicius, adulteration meant to serve inferior meat while claiming to be selling a superior one and of selling spoiled honey for the price of good honey.

Adulteration, as defined, means adding or removing substances from foods and beverages without the consumer's knowledge. If an operator sells beer for $10.00 per barrel and makes a profit of $4.00 on each barrel, the profit can be increased to $ 4.40 per barrel if one-tenth of the barrel is filled with water and only nine-tenths with beer. If caught and charged with selling watered-down beer, the person may well be found guilty and punished accordingly. On the other hand, if another person does exactly the same but hires an advertising agency to tell everyone that the watered-down product is really "lighter beer" and describes it as containing fewer calories than regular beer, no charge of adulteration will be forthcoming. To the contrary, this person probably will be hailed as a marketing genius and given a prize for having an enterprising spirit. The second person's profit is $4.40 per barrel, the same as in the first example but his marketing approach saved him. If the cost of advertising was equivalent to 10¢ per barrel of beer sold, the additional profit of 30¢ per barrel was still a significant gain. In one view, this operator succeeded in selling water at $3 per barrel (one-tenth of 10 barrels).

ADULTERATIONS FOR PROFIT

Bad honey may be turned into a salable product by mixing one part of spoiled honey with two parts of good honey, according to Marcus Gavius Apicius. Obviously, the reason for selling spoiled honey as if it were fresh honey is simply to make money. Throughout the ages profit has been the driving force of all practices of food adulteration. In another point of view, adulteration may be necessary simply for the survival of the poor merchant. In the book *Coffee from Plantation to Cup*, written in 1884, it is stated "The high prices that have prevailed in the coffee market since 1871 . . . have greatly stimulated the adulteration of coffee in the United States." It further stated that a prominent manufacturer said the average basis for this business was the following proportion to every 1,000 pounds:

Roasted peas	400 pounds
Roasted rye	200 ″
Roasted chicory	100 ″
Other ingredients	50 ″
Roasted coffee (best quality)	250 ″

Whether adulteration is strictly for profit or simply by necessity, it is performed to augment profit.

Adulteration of Meat and Meat Products

In some common but illegal practices, producers or distributors of meat and meat products seek to increase profits by substituting cheaper materials for the ones advertised. The number and variety of adulterations is limited only by the adulterator's imagination. Too many forms of adulteration became commonplace long ago and now are considered normal practice and, in many cases, sanctioned by law.

Beef, pork, lamb, and other cuts, together with fish, shellfish, and poultry, are difficult to adulterate successfully because of the characteristic appearance of the meat. In addition, the consumer's familiarity with the characteristics and properties desired in the product offered for sale leaves little room for

Figure 16.2

Deception as a sales technique then and now.

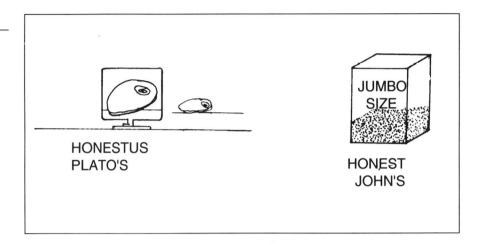

change. On the other hand, it is almost an article of faith that, in being prepared for sale, meat products are made to carry all the water they are capable of holding without affecting texture, color, and appearance adversely.

Another common practice is to bathe certain cuts of beef with a solution containing small amounts of nitrate and nitrite, as this treatment retards spoilage and preserves the red, fresh color of the meat. A form of misrepresentation practiced universally today is to display red meats in refrigerated cases illuminated with colored lights that make the meat appear redder in color and fresher than it actually is. As testimony that humans have changed little over the centuries, one may compare this practice to that found in Pompeii just prior to the eruption of Mount Vesuvius in 79 A.D., where there were ". . . steaks of sacrificial meat displayed behind enlarging glasses." Figure 16.2 depicts deception then and now.

Ground Meats

Ground meats can be adulterated easily because all identifying form and texture are obliterated as the meat is ground and blended. The normal practice of preparing ground beef includes the addition of substantial amounts of suet. The title "ground beef" need not be changed, as only beef suet is used. In previous times beef was mixed with cheaper meats such as horse, donkey, goat, dog, and other animals without fear of detection by the authorities. As long as the amount of adulterant was not so much as to affect taste and odor, it was a fairly easy matter to substitute 20% to 30% of the beef with other meats. Also in an earlier time flour or cereal often was added to ground meat, sausage meat, and other such products.

In the early part of the 20th century, it became possible to identify flour, starch, cereal, and other materials containing carbohydrates by selective digestion of the meat with chemicals or with enzymes. The residual starchy materials also could be identified positively by the iodine-starch reaction described below.

1. Mix ground meat well in a clean container free of any trace of cereal, starch, or flour.
2. In the same way, prepare a sample of meat known to not be mixed with adulterants.
3. Place a small amount of each meat in separate centrifuge tubes, label these Control and Test, add an equal volume of water, and mix thoroughly by vigorous shaking.
4. Centrifuge mixtures until clear, then place one drop of supernatant on a spot plate and add a drop of iodine solution.
5. If neither sample turns a deep purple color, the test is negative for starch.
6. If both samples turn a deep purple color, take 1 ml of supernatant from each tube and add to tubes with 9 ml of water. Place a drop of these dilutions on the spot plate, add iodine solution, and mix.
7. If both samples give a purple color, dilute and test with iodine again, until the Control gives a negative test. At this point, if the Test solution gives no color, the results are negative.
8. If the Test solution gives the purple color, dilute again, at least 1:1, and if this dilution gives a purple color, the results are positive.
9. Return to centrifuge tube. Place 5 ml of meat-water mixture in centrifuge tube and carefully add 5 ml of conc. sodium hydroxide. Mix well and let stand several hours with intermittent mixing.
10. When meat has dissolved, search sediment for cereal, flour, or starch. Test sediment for starch as above. Compare with Control.

In the 1940s it became possible to identify non-beef meats used as adulterants. The procedure depicted in Figure 16.3, the serological test, enabled food inspectors to identify mixtures of ground meats in almost any form. However, once meat is cooked, or even heated to obtain a partly cooked product, the technique has little value. When cooked, meat can be identified by microscopic and chemical analyses, but the results are neither exact nor as sensitive as the serological method. This test is extremely sensitive, and care must be taken that a meat is not declared positive simply because it was cut with utensils that had been used in cutting other meat or that the working surfaces contained even traces of other meats.

The implication that horse, goat, dog, donkey, or rabbit meat is not as good a food as is beef is not intended here. As a matter of fact, some of these animals may provide better nutrition than beef in some respects, but the point in question is legal, not biological. If the producer sells the meat as beef, to substitute the meat of another animal is a violation of the law.

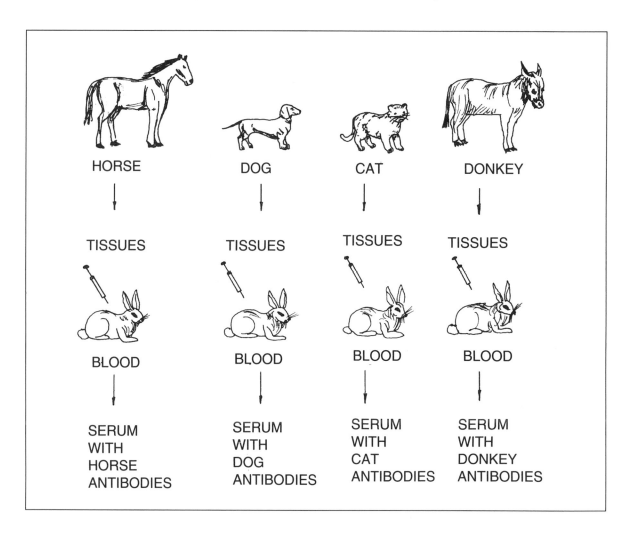

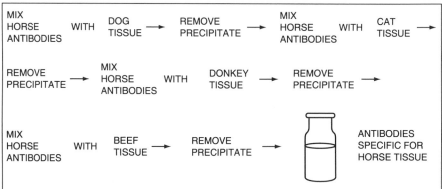

Figure 16.3

Procedure for detection of non-beef meats mixed with beef.

Step 1. Production of antisera specific for non-beef meats.

Step 2. Cross absorption to remove cross-reacting antibodies.

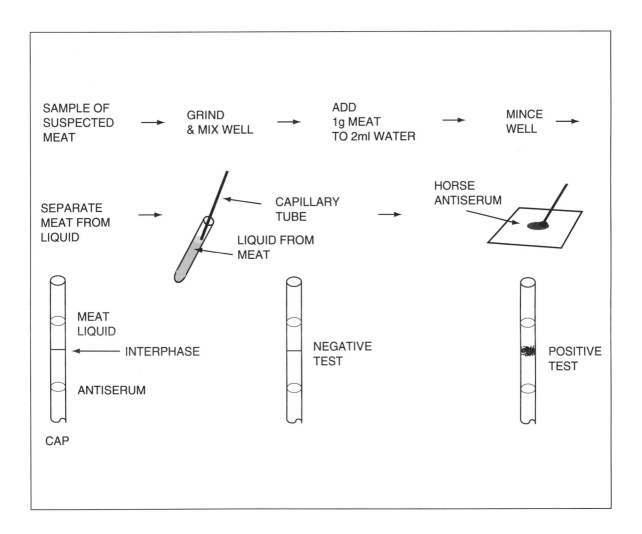

Processed Meats

Processed meats such as bologna, chopped ham, salami, and sausages are easy to adulterate because the processing plus the spices, flavoring agents, additives, and colorants disguise the tastes and odors of adulterants. On the other hand, the cuts or parts of the beef used to produce processed meats are so inexpensive that to substitute them for inferior meats is hardly worthwhile. Generally, processed meats are made from all the parts of the animal that are not otherwise eaten by the public. By law in the United States, specific cuts used in the preparation of processed meats must be given in the list of ingredients, but generally they are referred to euphemistically as "selected beef cuts."

Step 3. Use of specific antisera for identifying meats.

The laws that specify the ingredients that may be used in preparing processed meats permit certain amounts of flour or cereal to be incorporated. Although the laws are designed to be permissive in this regard, some manufacturers will add amounts well beyond allowable limits. The rationale is that if you can sell cereal for the price of meat, you will make a greater profit than you would by selling just the meat. Again, detecting this type of adulteration is easy, and this practice is generally limited to small, marginal operators.

One of the most dangerous practices of the unscrupulous adulterator is that of incorporating into meat for human consumption the flesh of animals that died of infectious disease or under other untoward circumstances. Human beings are naturally predatory animals and, as such, normally eat only meat they kill. Most societies have laws, customs, religious practices, or other means of prohibiting the consumption of the flesh of animals not killed specifically for food. Although many rationalizations can be offered to justify using the flesh of dead animals for food, in the final analysis, if not aesthetics, good health practices are sufficient reason to avoid this practice.

Unfortunately, the fresh meat from animals that died of disease cannot be distinguished from that of animals sacrificed for use as food. In the past, epidemics of anthrax, relapsing fever, tuberculosis, and many other diseases occurred frequently from the use of the flesh of animals that died of disease.

Adulteration of Bakery Products

Adulteration of bakery goods generally means that flour, shortening, sugar, eggs, or other ingredients of bread and pastries have been substituted for substances of lower quality. It also implies that the substitutions made are not advertised on the package or literature describing the product. Mixing approximately equal parts of flour and sugar can make fillings and decorations appear to be made from pure sugar. Adding flour or starch to thin jellies makes them look like fruit preserves. Canned fruits or damaged and overripe fruits of low value often can be passed for expensive confectioner's candied fruit. In the past, adding sawdust, flour mill sweepings, and other such matter to flour and using fat from deceased animals in place of high-quality lard or cooking oil in bread and many other products was not unusual.

Ironically, the more complex and expensive bakery items are easier to adulterate than the simpler ones such as plain white bread. Fancy items, sweet rolls, cakes, pies with whole fruit, and cookies with raisins, nuts, and candy can hide many adulterants. Using heavy and inexpensive molasses in place of refined sugar, cottonseed oil in place of shortening, egg whites in place of whole eggs, and nut tailings instead of nuts can make large differences in the cost of fancy breads. Again, however, the violation is not in the ingredients used but, rather, in the labeling. If bread is made with the more expensive ingredients, it will sell for more than if it is made with the inferior ones. The

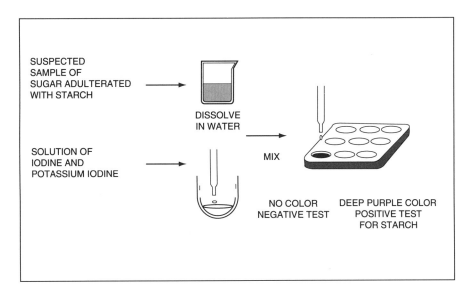

Figure 16.4

Use of iodine for detecting starch in sugar.

adulterator's gain, therefore, lies in using the cheaper ingredients and selling the product at the higher price. Figure 16.4 illustrates the use of iodine in detecting starch in sugar.

Adulteration of Dairy Products

All products made from adulterated milk are, by definition, considered adulterated, too. Sour milk, buttermilk, cheese, ice cream, custards, puddings, and all other products can carry adulterants from the milk used in their manufacture, or they may be adulterated during production. Examples of well-known practices are cheese made from good pure milk adulterated by the addition of flour or starch, ice cream made with skim milk instead of cream, and coffee cream made by thickening milk with starch.

Presence of Pathogens   Milk is said to be adulterated if it contains any foreign substance or if it contains more bacteria than indicated by the grade designation of the milk. If dairy products carry coliform bacteria, the bacteria that cause tuberculosis, or those that cause relapsing fever, they also are said to be adulterated. Although such milk is not acceptable because it contains pathogenic microorganisms, it also may be classified as adulterated because it contains substances deleterious to the health and well-being of human beings. Substances such as toxins, pyrogens, and pigments may affect the human adversely. The addition of bactericidal or bacteriostatic substances

Figure 16.5

Detection of old milk in fresh milk by chemical analyses.

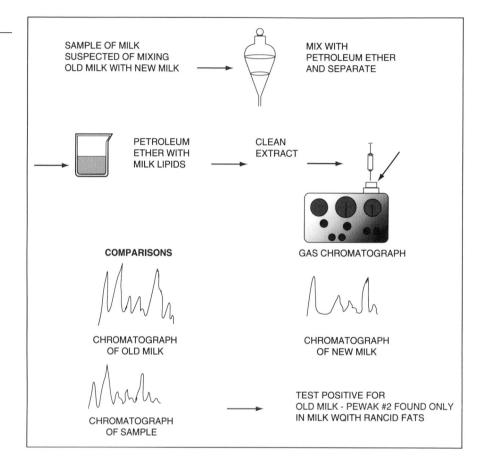

such as soap, detergents, and antibiotics to milk to suppress bacterial growth also constitutes adulteration.

Addition of Other Adulterants Addition of water to increase the volume, detergents and quaternary ammonium compounds to retard bacterial growth, and plant lipids to increase the butterfat content are all adulterations and therefore illegal in most communities. Large amounts of detergents, quaternary ammonium compounds, and other chemicals designed to deter microbial growth also may be dangerous to the health of the consumer and, therefore, are not permitted. All of these malpractices can be detected easily in even a modest analytical laboratory today. In the past most kinds of adulterants, including the age-old rite of adding water to the milk, could not be detected. Figure 16.5 shows the chemical analysis to detect old milk in fresh milk.

ADULTERATIONS BY ACCIDENT

Personal Items

Many food substances go to the public in adulterated form not by design of the producer but, rather, as the result of accidents and other unforeseen and inevitable occurrences. In canning plants, foreign items sometimes fall accidentally into food preparation utensils and even into the final container (see Figure 16.6). Even though many safety measures have been implemented and many detector devices have been put in place, foreign items still are found in canned foods. The food containers can pass many tests and eventually be sealed, sterilized, and shipped out without detecting the foreign item.

If a catalogue of such items were to be made, it would contain all the articles in the following list:

hair pins	lint balls	combs	safety pins
gum wrappers	glasses	contact lenses	pocket knives
false teeth	pencils	animal parts	pens
chewing gum	paper clips	coins	rosaries
beads	money bills	hearing aids	cigars
cigarette lighters	glass eyes	nail clippers	animal parts
tie clips	cuff links	severed thumbs,	medals
hair	dry cell batteries	fingers, human	insects
small animals	animal excreta	skin	animal hair
false nails	wrist watches	small toys	wrist watch links
straight pins	rubber toys	snack wrappers	

Figure 16.6

Inspection of the inside of a baby food capping machine for glass. One of several preventative measures taken to prevent accidental contamination and adulteration during the capping process. Photograph courtesy of FDA

In addition to the objects listed, almost everything else that human beings carry on or in the body has been found in foods preserved in cans, jars, and packages. Unfortunately for the producers of packaged foods, these items typically are not discovered until the consumer opens the can or jar and, in many instances, not until half of the contents have been consumed.

When a worker loses a finger at his or her work station, the routine practice is to stop the production line and search until the entire digit is recovered lest it end up in the food being packaged. Nevertheless, such relics do go out of the food processing plant and are found later by some unfortunate consumer. The repugnancy and disgust that must arise in a consumer who opens a can of food in his kitchen and pours the contents into a saucepan only to find a severed human finger usually is great enough to cause that person to bring legal action against the producer. In the past, such legal claims usually were settled in favor of the producer, but now they often are settled in favor of the consumer.

Insect Fragments

When wheat or any other grain is harvested from the field, excluding all the insects and other small animals that feed on the plants is nearly impossible. Even hand picking, selecting, and cleaning the grain crop fails to remove all the pests associated with the grain as it is harvested. This is a normal condition, and nothing can be done to reduce the number of insects that come with the harvest. Although many insects simply fly from plant to plant, many others are attached to the plant in such intimate contact that they cannot be successfully removed. When the harvested grain is ground into flour, it will bear all the insects and insect parts that were on it in the field.

In addition to this, the grain or flour also attracts rodents and insect pests including beetles, mites, spiders, grubs, and worms of various kinds. The finished flour or other cereal meal invariably will contain small insects and pieces of insects such as legs, antennae, and other body parts in a number proportionate to the insect populations in the field where the grain was harvested or the conditions under which it was stored. The presence of small insects and insect fragments, while displeasing to most individuals, provides little or no hazard to those who consume the grain. Therefore, all regulatory agencies make allowances by permitting a certain, small number of insects, insect fragments, and vermin hairs in all food products made from cereal grains.

Rodent Hairs and Excreta

Animal hairs and the excreta of small animals, usually rodents, is almost always found in cereal grains and the food products made from cereals. Although these are considered adulterants, raising, harvesting, and processing a crop of

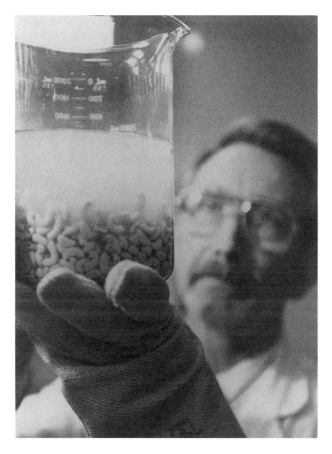

Figure 16.7

Richard Trauba, a food specialist with the Food and Drug Administration, holds a beaker of macaroni in an acid solution that when heated, under pressure, in an autoclave for 30 minutes, eliminates the starches, leaving soluble sugars and carbohydrates. Hot water, mineral oil and HCl are then added to the mixture in a glass vessel with a bottom drain. The mineral oil selectively coats the insect fragments and other animal material, such as hair and feathers. These float to the surface with the oil. The contents of the container are then washed out with alcohol into a beaker and filtered through paper.

The fragments on the filter paper are identified using a microscope. The viewer compares the fragments with slides or pictures of known insects and insect parts such as mandibles and antennae. Fragment analysis is important to determine if a food product is contaminated or falls within permissible levels.

cereal grains without attendant small animals is almost impossible. Vermin and their parts fall under the general category of filth and, as such, limits exist that may not be surpassed without making the product unacceptable. The number of animal hairs or pieces of animal excreta represent a history of the conditions under which the grain was grown, harvested, processed, and stored.

If 100 grams of flour are dissolved in water with a small amount of amylase to degrade starch, and the solution then is centrifuged to collect all solid materials, these can be identified and counted with a microscope.

Blending

Because the price of the flour, meal, and other cereal products often is affected by the number of insect fragments, bakery products sometimes are made from mixtures of flours. Removing animal hairs and similar objects from cereal products is impossible, so the only alternative often is to mix flours containing large numbers of insect fragments with flours containing few fragments in efforts to bring the mix to acceptable standards. As this practice reduces the quality of the superior product, this kind of blending obviously has to be done with great knowledge of prices and legal requirements, and also with great care, lest the quality of both batches of flour be lowered instead of raised.

METALS AND OTHER TOXINS

Mercury

During the 1960s and the 1970s, fish in deep waters off the coast of Japan were found to contain large amounts of mercury in their flesh. Many cases of mercury poisoning, including 48 fatalities, were traced to mercury wastes from Japanese industrial operations. Mercury enters the sea as metallic waste and then is converted chemically to mercuric sulfide, a form of the metal not harmful to fish life. However, the mercuric sulfide in the intestinal tract the tissues of fish can be converted to methylmercury by bacteria. Methylmercury is deposited in the flesh of fish and accumulates there. If the level of mercury in fish flesh is low, it is difficult to detect by chemical tests and the poisoned fish appear to be normal and healthy in every respect. They are approved for use as food on the criteria of appearance alone and usually not examined by chemical analyses. Fish from ocean areas surrounding industrial nations that discharge effluents containing mercury into the ocean should be tested for mercury before being approved for food use.

Arsenic

Until recent times, arsenic residues in food were quite common. Although arsenic poisonings from food sources were the exception rather than the rule, they were not uncommon. Many arsenic poisonings in the past resulted from the inadequate washing of plant crops that had been sprayed with pesticides containing arsenic. Food purveyors depended entirely on the fresh and healthy appearance of the food only to find later that they contained lethal amounts of arsenic. Producers now perform chemical analyses on all items they handle for which the slightest doubt of accidental arsenic adulteration exists.

Pentachlorophenol

Pentachlorophenol is used as a universal preservative in hundreds of items of everyday use. It also is used to cause wilting of plant crop foliage to facilitate machine harvesting. In normal use, food crops treated with pentachlorophenol are cleaned until no trace of this toxic material remains. In many communities it no longer is used for spraying food crops of any kind. Its use on non-food crops, however, still presents a hazard because the pentachlorophenol remains active in the soil for long periods and can be moved from one location to another by water. Many deaths have resulted from eating food contaminated inadvertently with pentachlorophenol.

An outbreak of pentachlorophenol poisoning in California was traced to contaminated well water used to wash table vegetables. In other cases of accidental adulteration with pentachlorophenol, the poison was in water runoff from lands where it was used on non-food vegetation. Because microbial degradation is not effective, the pentachlorophenol runs into the sea, where fish become exposed to it. It concentrates in fish tissues and then is passed on to human beings who consume the fish. Some 80 fatalities have been recorded in the United States from drinking contaminated water, and many more have been recorded in Japan from consuming contaminated fish.

ADDITIVES AND PRESERVATIVES

Chemical substances are added to foods and beverages:

1. To prevent microbial growth.
2. To extend shelf life.
3. To prevent chemical changes that affect taste and odor.
4. To enhance the appearance of food.
5. To improve taste, aroma, and texture.
6. To make handling and processing easier.

Before 1958, additives were added to foods and beverages very much at the producer's discretion and very much to satisfy needs associated with profit or market conditions. By legal definition, all additives are adulterants and, as such, fall under purview of the same laws that regulate all food and beverage production.

In 1958, the Food, Drug, and Cosmetic Act was amended and all food and beverage additives were brought under control of this regulatory agency. The new laws require that no additives be added to foods and beverages unless the additive can be shown to not be harmful. In addition, all additives now must be approved by the FDA and must be identified on the package.

The FDA does not approve additives for foods and beverages unless the additive is shown to enhance the quality of the food item. Additives are not approved if their sole purpose is to hide flaws in the food, increase profit, deceive the consumer, or make an inferior product appear better than what it actually is. Additives that are harmful to human beings or that are designed to circumvent good manufacturing practices are not permitted under any circumstance.

The provision in these laws that has caused major turmoil since its inception in 1958 is the Delaney Clause. This clause states that any substance that causes cancer in animals or in human beings may not be added to foods or beverages at any level.

Many additives that were in use prior to 1958 and that were shown to be valuable and without harm to human well-being were continued and are still in use without further action by the FDA. This "grandfathering" of additives was available only for substances that had been in use for many years without incident. They were placed in a separate category and described as "generally recognized as safe."

Major Additives

The major food additive in the United States at this time is sugar. Sugars of various kinds constitute 82% of all additives, and salt is a distant second (11%) in terms of total weight. Each individual in the United States consumes an estimated 58 kilograms of sugar and salt in the form of food additives each year. This does not include consumption of sugar and salt as foods themselves. Both are used to retard spoilage, improve taste, enhance appearance, and increase the nutritive quality of the product.

The remaining 7%, or 3.5 kilograms of food additives consumed per person per year, is made up of thousands of different chemical substances. These chemicals are used for many different purposes such as preservation, taste enhancement, increase in shelf life, and prevention of rancidity. Thousands of food additives are in use today, but only a few are described in very brief detail below.

Antibiotics

Many antibiotics may be present in meats because antibiotics are mixed with animal feeds to protect the animals from bacterial infection. Animals fed antibiotics probably yield meat that is more resistant to spoilage than meat from animals fed normal diets. Mixtures of antibiotics also are used as dips for poultry and fish meats because this practice has been shown to increase shelf life beyond that of nontreated meat. The amount of antibiotic present on the meat at the time it is consumed is small, but the effects it exerts on microbial populations are significant.

Chlorine

Flour is particularly vulnerable to insect pests and can easily be rendered useless. Adding chlorine gas to flour protects it from insect depredations and also affects the texture of the flour in a yet unknown manner. Flour treated with chlorine has been tested extensively and found to be safe for human use. It has been approved by the World Health Organization Committee on Food Additives for any use as a food product.

Propionic Acid

This acid is obtained from the fermentation of sugar with the bacterium *Propionibacterium* and is used in the food and beverage industry because it is a powerful fungistatic and bacteriostatic agent. It is used as a preservative in bakery products, canned vegetables, fruits, jams, jellies, and various beverages. Propionic acid has a strong, pungent, disagreeable odor, but its sodium salt, sodium propionate, has only a faint odor.

Sorbic Acid

Sorbic acid and its potassium salt, potassium sorbate, are used to inhibit the growth of bacteria and fungi in a wide variety of foods including bakery products, fats, oils, cheese, puddings, and sauces. Although it has a strong odor, it is used in small amounts, and its odor is not objectionable under these conditions.

Sodium Benzoate

Sodium benzoate is one of the most widely used preservatives in the food and beverage industry. It is used universally as a preservative in canned and bottled drinks, in many kinds of pastry products, and in preventing spoilage in fresh fish. In the latter case, sodium benzoate is used in combination with borax or with salt in solution for dipping fresh fish transported in ice over long distances. Those who employ this preservative for fish say that benzoate also keeps fish fillets from becoming rancid as a result of chemical changes that take place in certain fish oils.

Sulfur Dioxide

In the past, sulfurous acid and its vapor, sulfur dioxide, were used as fumigants for the safe storage of cereals and grains, but their high degree of toxicity in human beings resulted in their being abandoned. Sulfurous acid, however, still is used in the preparation of grapes for wine production. Grapes come to the winery with large numbers of different bacteria, fungi, and yeast acquired in the field. If these fungi are allowed to participate in fermentation of the grape juice, the finished wine would have an unpredictable taste and bouquet because of the fermentation products released by these "wild" or "natural" yeasts. To avoid this, grapes are treated with sulfur dioxide and then inoculated with the specific yeast whose fermentation characteristics are predictable and desirable.

Sodium Bisulfite

Sodium bisulfite is a salt of sulfurous acid and also is used in fermenting wine and beer. It is used in beer fermentation to inhibit the growth of bacteria because sodium bisulfite inhibits bacterial growth while not interfering with the growth of beer yeasts.

Thaumatin

Thaumatin is a sweetener used in place of sugar in many different foods. It can be used as a sweetener or as a flavor enhancer depending on the quantity used and the manner in which it is combined in food products.

Vanillic Acid

Vanillic acid is a chemical substance related closely to benzoic acid. It is used as a preservative in canned and bottled drinks, in bakery products, and in retarding spoilage in fresh fish. Its preservative properties are superior, and its toxicity for certain animals is lower than that of benzoic acid. Although vanillic acid has a pungent, sweet odor, it is used in the salt form, which is nonodorous.

Hormones

Until quite recently, meat producers in the United States used hormones to obtain larger animals that matured earlier. The hormones had been assumed to have no effect on human beings, but the protests of European consumers in 1988 and 1989 brought about a reconsideration of the entire situation. The hormones most used are **diethylstilbestrol** and **progesterone** because they are more effective than any other agent in causing rapid weight increase in beef and other animals.

It has been known for some 30 years that diethylstilbestrol (DES) causes cancer in human beings, but its use was permitted on the basis that the hormone disappeared from beef muscle within 48 hours of administration. On this basis it was assumed that if the hormone was not injected 48 hours prior to sacrifice, it would not be present in meat when it was consumed. Even if this were so, subsequent research has shown that diethylstilbestrol causes cancer in animals, including human beings, that ingest the flesh of those that were treated. The use of diethylstilbestrol in any form in meat intended for human consumption was banned in 1975 by application of the Delaney Clause.

PACKAGING

The packaging of food items is extremely important in marketing. In designing the proper package, one of the major objectives is to make the food as attractive as possible, although in reality it is no better or worse than that of the competitor. Another major importance of packaging resides in the information given to the consumer.

In the past, producers often took advantage of this and used the information on the package as a means of advertising. Exaggerations or fabrications often were used to make consumers think they were receiving something of value while in reality they were receiving very little. These claims have ranged from the turn-of-the-century advertising on foods items with claims of curing snake bite and the consumption to recent labels that proclaim long life and smooth skin. No exaggeration seems to have been too much for the advertising genius.

With the advent of the truth in advertising laws, false claims have decreased significantly. To say that eating Sky Giant Green Beans makes you jump higher is no longer possible unless the producer is willing to inform the public of what higher means and also provide the data that show that the claim is true. The data given on packages of food and beverage generally indicate the name and quantity of the additive included and the reason for its use, and, in addition, the name of the food included, all ingredients used, and a list of **inert** ingredients. The only justifiable *caveat* lies in the fact that the proprietary names often used give little information on the chemical nature of the items included in the food.

QUESTIONS TO ANSWER

1. If peanut oil is as good a nutrient as butterfat is, why is replacing butterfat with peanut oil in ice cream illegal?
2. How does the definition of adulteration of foods given by *Webster's Dictionary* differ from that given in the 1968 Act of the Food and Drug Administration?
3. What do Sinclair Lewis and Samuel Hopkins Adams have to do with foods and beverages?
4. Is lite beer an example of adulteration of beverages? Explain.
5. Is skim milk an example of adulteration of beverages? Explain.
6. What is cross-absorption of antibodies?
7. What is a species-specific antiserum?
8. If horse meat is as good as beef, why is it wrong to mix the two when making hamburger?
9. Why is it against the law to use a cow for food that was alive and well the day before but found dead in the barn the next morning?

10. How can it be shown that old, outdated milk was mixed with fresh milk?
11. Might wheat flour contain insect fragments? If you answer yes, explain.
12. Might bread contain animal hairs? If you answer yes, explain.
13. How does mercury get into fish?
14. How does arsenic get into cattle?
15. What is pentachlorophenol, and what does it have to do with the food and beverage industry?
16. Diethylstilbestrol is expensive; why was it used in cattle in the past?
17. What does the Delaney Clause have to do with foods and beverages?
18. How is chlorine, a toxic gas, used in the food industry?
19. What is the use of antibiotics in the food industry?
20. What is the truth in advertising law?

FURTHER READING

Banwart, G. J. 1979. BASIC FOOD MICROBIOLOGY. AVI Publishing Company, Inc., Westport, Connecticut.

Committee on Scientific and Regulatory Issues, Board on Agriculture, National Research Council. 1987. REGULATING PESTICIDES IN FOOD. National Academy Press, Washington, D.C.

Crompton, T. R. 1979. ADDITIVE MIGRATION FROM PLASTICS INTO FOOD. Perga-mon Press, New York.

Gilbert, J. ed. 1984. ANALYSIS OF FOOD CONTAMINANTS. Elsevier Applied Science Publishers, New York.

Joint FAO/WHO Expert Committee on Food Additives. 1987. TOXICOLOGICAL EVALUATION OF CERTAIN FOOD ADDITIVES AND CONTAMINANTS. Cambridge University Press, Cambridge.

Rules and Regulations

17

In all his actions:
The intelligent man is guided by reason,
The less fortunate by experience,
And the brute, by law.

Anonymous

All through history the quality of food has been determined more by price than by any other criterion. An establishment with a good reputation for high-quality products could charge more for its merchandise. The definition of quality, even as it does today, implied authentic ingredients free of all adulterants and sufficiently fresh to give a reasonable shelf life.

Cleanliness always has been a mark of high standards, although the criteria have changed with the advent of microbiology and food chemistry. Today, cleanliness in milk production is ascertained by total plate counts and coliform content, whereas in the past it was determined by personal knowledge of the integrity of the purveyors and a long history of use of their products.

In the time before food chemistry became established as a science, assessment of the possibility of adulteration of meat was based entirely on the producer's established honesty, as no objective criteria existed by which to determine adulteration. The producers' personal reputation was the major, and often the only, impetus for maintaining quality and product integrity. The benefit to the producers came from the fact that they could charge more for products than could producers who did not enjoy the public trust. In those times the food producer or food purveyor, as well as the profitability of the enterprise, rested on reputation.

RATIONALE

The major problem with the system described above is that quality and price went hand in hand. The better the quality, the higher the price. In such a system, the rich eat well and the poor fare less well. Although this may well be much of the reality of life, it is not consistent with modern sociological thought, which states that all human beings merit, and should have, the basic essentials of life including adequate and healthful food. In the sense of food quality this means safe, wholesome food free of pathogenic microorganisms, toxins, and poisons.

ORIGIN OF LAWS

The original notion of protecting the consumer from the effects of contaminated or poisoned food probably started when human beings began to buy food instead of producing it themselves. In the absence of knowledge of microorganisms, toxins, and poisons, food safety was based on freshness, appearance, and odor. Conventions that proscribed tainted meat, measly flour, and musty vegetables early became part of community convention. Evolution of such conventions eventually acquired the force of law and eventually the laws became universal.

By the 20th century, many societies created complex, elaborate, and expensive governmental agencies designed only for the purpose of safeguarding the quality of food. It goes without saying that people of more advanced and richer communities are better protected against misrepresentation, adulteration, and danger of intoxication and infection than are those of the poorer and less advanced communities. The protection and guarantees offered always have been in the form of agreements, regulations, and laws.

In many communities protection from the hazards and dangers of eating contaminated foods has become part of the religious system. The best example of the efficacy of this system may be the proscription of pork in the Jewish and Islamic religions as a means of protecting the population from trichinosis.

INFERIOR PRODUCTS

The laws that guarantee safety and well-being in the consumption of foods and beverages do not necessarily guarantee quality. That is, an item of food may not be sold if it contains pathogenic bacteria, toxins, or mineral poisons. On the other hand, it may be sold if it contains large amounts of cholesterol and fat, has much gristle, little nutritive value, and came from an old and sinewy animal, as long as there is no hazard in eating it.

The quality of certain foods and beverages, however, is maintained by the enactment of laws that specify traits and characteristics. Some of the oldest of these laws served for many years to guarantee the quality of beer. The discovery that hops gave beer a distinctive and characteristic flavor was considered so important an innovation in the 11th century that growing and harvesting hops was controlled in many parts of Europe by authorities ranging in rank from the king to the brewers' guild.

In efforts to compete with beers that contained hops, many producers resorted to adding various plant extracts that would imitate the flavor of hops. Some of the substances that were used were noxious and others toxic causing consumers to be wary of beer lest it was certified by credible authorities as safe. Generally, this meant that only beers with hops were certified and those flavored with other herbs were not. To protect the quality of beer, central European brewers' guilds caused laws to be enacted that specified the ingredients in beer.

By the 14th century, laws had been instituted in Germany that made unlawful the addition of anything in the production of beer other than water, malted barley, hops, and mother liquor. In the 20th century, protection of quality was extended even to the most subtle characteristics of the ingredients. For example, wines may not be called Champagne, Cognac, or Bordeaux unless they are made from grapes grown in those regions of France since it is assumed that only that soil will produce that particular quality in grapes and the wines obtained from them. The intent is to protect the consumer from an inferior product sold under an established name and label but can there be doubt that this practice also serves to protect the producer from competition.

MODERN LAWS

Although misrepresentation and adulteration are as old as humanity, modern methods of mass production and large-volume transportation took the problem from the few and made it the problem of the many. The products of an unscrupulous food producer in today's world can affect hundreds or perhaps thousands of consumers instead of just a few, as was the case in the last century. Without proper laws to protect consumers, the entire community, including producers, will suffer the consequences.

Unscrupulous practices in food production in the absence of adequate laws in many European countries and the USA during the middle of the 19th century led to widespread abuses and questionable practices that endangered the lives of entire communities. Selling diseased and spoiled beef to the army during the Civil War was not a rare occurrence in either North or South. By the end of the 19th century, reform measures had been instituted in many countries throughout the world, but the USA lagged far behind in this regard.

Table 17.1   Important Benchmark Laws for Consumer Protection

Year	Law	Intent
1890	Inspection of meats	Protection of meats for exportation
1895	Inspection of meats	Protection of American public
1906	Federal Food and Drug Act	Protection from deceptive practices
1906	New Meat Inspection Act	Provide wholesome meat
1910	New York City, Philadelphia, etc.	Pasteurization of milk
1938	Federal Food, Drug and Cosmetic Act	Increase protection and include safe cosmetics
1954	Miller Pesticide Law	Prevent incorporation of pesticides in food
1957	Poultry Inspection Law	Guarantee quality of poultry and poultry products
1958	Food Additives Amendment	Exclude harmful additives
1967	Wholesome Meat Act	Avoid impurities in prepared meats
1968	Poultry Inspection Law	Inspection of poultry slaughter
1972	Truth in Labeling Act	Ingredients and weights must be stated
1972	Allowable Limits of Filth	Number of insect fragments, vermin hairs, etc. permitted

Food adulteration, misrepresentation, and threat to the consumer had reached such levels that many European countries prohibited the entry of American food products into their domestic markets.

The many efforts by the federal government to improve the national image on foreign shores resulted in the Meat Inspection Act, a law enacted in 1890 whch required that all meat intended for export be inspected. In 1895, the law was extended to include meats previously exempted and to broaden the scope of the inspection. Two laws passed by Congress in 1906:

1. The Pure Food and Drug Act, and
2. The Meat Inspection Act.

formed the foundation on which future laws eventually would make the United States the world leader in food quality and safety. In effect, these were the same laws introduced in 1890 and 1895 both amended to apply to meat intended for domestic consumption. These laws were administered by the Department of Agriculture and extended to all facets of the food and beverage industry including production, sale, packaging, transportation, serving, and advertising. The laws (Table 17.1) now are administered by the Department of Health, Education and Welfare through its agency, the Food and Drug Administration (FDA).

The laws by which the production, transportation, storage, and sale of foods and beverages are regulated today in the United States guarantee the best protection possible to the consumer and are sensed as the ideal for all the world's communities. Many multinational agreements for the international transport and marketing of foods in the world are based on the laws developed in the United States from 1906 to the present.

A major revision of the laws of 1906 was introduced in 1938. This revision, presently called the Food, Drug, and Cosmetic Act of 1938, is the foundation for many of the modern laws that reg'late the food and beverage industry today. Subsequent changes include a l w regulating the use of pesticides in food crops, introduced in 1954 and ca ed the Miller Pesticide Amendment. The Food Additives Amendment of 19!,8 specifies the kinds and quantities of additives that can be added to foods without impeaching consumer safety. The 1959 Poultry Products Inspection Act and the 1972 Fair Packaging and Labeling Act round out the federal regulations by which the food processing industry is regulated in the United States.

INSPECTION

The heart of food inspection is the hazard analysis: critical control points (HACCP) concept. This method can be applied to any process for any food or beverage. HACCP is a formidable concept which begins with the idea that food and beverage are potentially dangerous. It is applied during the entire process of food preparation and includes many different forms. For example, all stages and manipulations that may diminish the quality of the food must be thoroughly understood by both the supervisory and line employees. Personnel must be able to recognize problems when they occur and must know how these may be eliminated before the food item leaves the production area.

In its best form, the HACCP concept and application are in effect from the time the food item enters the food preparation area until it exits as a finished product. Figure 17.1 shows the essentials of this process. The objective of the HACCP procedure is to identify all areas where problems may exist and to test for such problems during the production operation. This procedure preempts the old method of waiting for problems to develop before finding the solution. It is designed to prevent the old problem of finding faulty product and then searching for the cause.

Microorganisms are regarded as the major hazard in the food and beverage industry. These are classified according to the effect they may have on the consumer when ingested along with the food or beverage which carries them. The effects range from almost certain death from botulinum intoxication to subclinical discomfort from the ingestion of large numbers of *Bacillus cereus*. Table 17.3 is a list of the organisms which have deleterious effects when

Figure 17.1

HACCP protocol in a meat processing plant.

consumed with food or beverages by human beings. Only bacterial agents are listed in Table 17.3. To this list must be added hepatitis virus, poliovirus, Norwalk agents, *Entamoeba histolytica*, worm eggs such as *Trichura trichuris*, and worm larvae such as *Trichinella spiralis* and *Dibothriocephalus latus*.

ENFORCEMENT

Although the existing laws are quite adequate for the consumer's protection, they must be enforced before the public can enjoy the guarantee to which it is entitled. Enforcement has suffered severely from 1980 to the present and, in certain cases, the quality of food products has decreased in the United States.

   The laws by which food is presented to the consumer in the United States are enforceable in the nation's courts. These include the legal actions described in Table 17.2. Enforcement of federal laws is accomplished using federal marshals and the federal court system while enforcement of state, county, and municipal laws is accomplished by the appropriate legal system and police forces.

Table 17.2 Federal Actions Possible Under Food and Beverage Laws

Action	Meaning
Cease and desist	A mandate to stop a given practice
Recall	Removal from public display and sale any product deemed to be a threat to the public health
Seizure	Confiscation by the FDA of any product produced in violation of the law or any plant or property operating in violation of the law
Prosecution	Legal actions by the FDA against individuals violating any law pertaining to the production, sale, or transport of food

Table 17.3 Designation of Bacterial Diseases
According to Severity of the Illness

Severe hazard

Salmonella typhi

S. paratyphi

S. cholera-suis

S. sendai

Shigella dysenteriae

S. boydii

S. flexneri

S. sonnei

Vibrio cholerae

Brucella abortus

B. melitensis

B. suis

Clostridium botulinum (intoxication)

Mycobacterium tuberculosis var. *bovis*

Moderate hazard

Salmonella enteritides

Salmonella spp.

Escherichia coli 0157:H7

Streptococcus pyogenes

Staphylococcus aureus

Vibrio parahaemolyticus

Yersinia enterocolitica

Campylobacter jejuni

Clostridium perfringens

THE PROTECTIVE NET

Seventeen Food and Drug Administration districts in the United States are responsible for maintaining the public health by assuring that the requirements of the Federal Food and Drug Act of 1968 are carried out effectively by all persons engaged in the preparation and sale of food. In addition to this federal system of laws, each state has a body of laws that protect the consumer. The state laws generally are administered by the State Public Health Department through its Division of Food and Drink Inspection. Still another layer of interconnecting laws is the responsibility of each county in the state, and in addition to this, each municipality usually has a City Health Department. These agencies enforce a system of county and municipal laws that reflect local customs and demands.

Although these laws overlap greatly, by design, they are not in conflict and the consumer receives maximal protection. The objective of all these administrative units is to protect the public by assuring that food and drink are clean and safe. Ultimately, all the local laws depend on the federal statutes as their authority for legal action.

WORLD ORGANIZATIONS

Since the end of World War II, serious efforts have been made to ensure a safe and healthful food supply for the entire population of the planet. Much of this work has been devoted to bringing together the various national agencies that protect their respective constituencies from unscrupulous suppliers of foods and beverages and to work in enforcing food purity standards. Part of the work of the international agencies has been to establish standards of purity and safety in nations that have none and to encourage enforcement in nations where laws exist but are not enforced.

The international agency most active in these two areas of food and beverage safety is the World Health Organization (WHO). Its concern with the health and well-being of the world's population involves proper nutrition and a safe and wholesome food supply. Another international agency with the same concerns but with emphasis on the nutrition and health of the world's children is the United Nations International Children's Emergency Fund (UNICEF). The third major international agency involved in assuring pure food and beverages for the world's population is the United Nations Food and Agriculture Organization (UNFAO). This organization concerns itself with production and distribution of wholesome food to the world's peoples regardless of their ability to pay.

FEDERAL AGENCIES

In addition to the laws and regulations monitored by the FDA, other regulations are pertinent to producing and serving food. The U.S. Department of Health and Human Services, through its various agencies, guards the quality of the nation's food through several mechanisms:

1. *Public Health Service.*   This office is concerned with the health of the nation by controlling the dissemination of infectious diseases through foods and beverages.
2. *U.S. Department of Agriculture, Food Safety and Inspection Service.*   The efforts of this service are to produce clean, disease-free food animals. Research on and control of zoonotic diseases such as tuberculosis, anthrax, and typhoid fever are major responsibilities of the FSIS. In addition, this service enforces the:

   - Meat Inspection Act of 1906, with many amendments,
   - Food grading and standards,
   - Poultry Products Inspection Act of 1957, and
   - Egg Products Inspection Act of 1970.

Many other government agencies also focus on the quality of the nation's food and beverages. Often these independent agencies have jurisdictions that overlap, opening the possibility for conflict in enforcement, although such occurrences are rare. Among these agencies are the U.S. Department of Commerce (USDC) and the Environmental Protection Agency (EPA). The USDC, through its National Marine Fisheries Service, examines and enforces

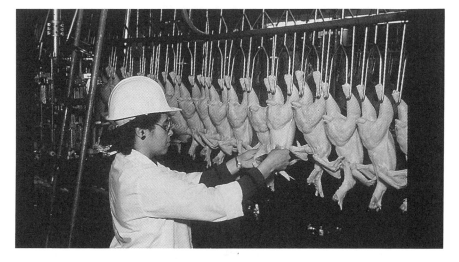

Figure 17.2

U.S. Department of Agriculture inspector checking quality and examining for possible contamination of chicken. Photograph courtesy of U.S. Dept. of Agriculture

standards for fishery products. The EPA establishes and enforces standards on use, application, and content of herbicides and pesticides in the production of grains, vegetables, and fruits.

STATE LAWS

Each state in the United States has the prerogative of enforcing federal laws or of formulating its own laws. The majority of states follow the latter course so they are able to represent local needs and preferences that may not be dealt with adequately in the federal system. As examples, consider the laws that prescribe the taking of lobsters in Maine and the obvious point that these laws are useless in New Mexico, while New Mexico needs laws that give the maximal amount of pesticides that may be applied to crops of hot, red peppers—which are not grown in Maine.

In essence, each state duplicates the system of laws, inspections, and court enforcement that exists at the national level. In almost all cases the state and the federal governments share jurisdiction so that violators face prosecution from one level or the other.

COUNTY AND MUNICIPAL LAWS

In much the same manner as federal and state laws, county and municipal governments apply and enforce laws made by their respective governing bodies. These laws too often reiterate state and federal laws, but they do have the value of expressing local needs.

NONCONFLICT

No law or requirement enacted by a lower body may conflict with that of a higher body. That is, no city may enact an ordinance, a regulation, or a law that is in conflict with that of any county. In turn, no county may enact laws in conflict with state laws, and so on to the level of national laws.

The objective of the many layers of rules, regulations, and laws is to provide maximum guarantees to all the citizens of the country while making all provisions necessary for special needs in local settings. The total effect of these laws, when they are enforced, is to make the American consumer the best-protected and best-treated of all consumers. The number of food poisonings, food-borne infections, and cases of financial loss resulting from fraud may be lower in the United States than in any other country in the world. The incidence of intestinal parasites in many communities and the constant "problem with the water" seen in so many countries emphasizes this contention.

APPLIED FOOD MICROBIOLOGY INDEX

NOTE: Page numbers followed by *f* refer to figures; page numbers followed by *t* refer to tables.

ALSO AVAILABLE FROM STAR PUBLISHING COMPANY

BASIC MICROBIOLOGY TECHNIQUES
by Susan G. Kelley and Frederick J. Post.

MICROBIOLOGY TECHNIQUES
by Susan G. Kelley and Frederick J. Post.

MICROBIOLOGY WITH HEALTH CARE APPLICATIONS
by Isaiah A. Benathen.

IMMUNOLOGY INVESTIGATIONS: A LABORATORY MANUAL
by Loreli A. Batina.

MEDICAL MICROBIOLOGY: A LABORATORY STUDY
by William G. Wu.

LABORATORY MANUAL FOR FOOD MICROBIOLOGY AND BIOTECHNOLOGY
by Frederick J. Post.

MEDICAL MYCOLOGY AND HUMAN MYCOSES
by Everett S. Beneke and Alvin L. Rogers.

IDENTIFYING FILAMENTOUS FUNGI: A CLINICAL LABORATORY HANDBOOK
by Guy St. Germain and Richard Summerbell.

WADSWORTH ANAEROBIC BACTERIOLOGY MANUAL
by Paula Summanen, Ellen Jo Baron, Diane Citron, Catherine Strong, Hannah Wexler, and Sydney Finegold.

A CLINICAL GUIDE TO ANAEROBIC INFECTIONS
by Sydney M. Finegold, Ellen Jo Baron, and Hannah Wexler.

PRINCIPLES AND PRACTICE OF CLINICAL ANAEROBIC BACTERIOLOGY by
Paul G. Engelkirk, Janet Duben-Engelkirk and V. R. Dowell, Jr.

MICROGAMES AND PUZZLES
by I. Edward Alcamo.